Prentice Hall Series in Process and Pollution Control Equipment

by Nicholas P. Cheremisinoff and Paul N. Cheremisinoff

Pumps and Pumping Operations

Compressors and Fans

Filtration Equipment for Wastewater Treatment

Carbon Adsorption for Pollution Control

Water Treatment and Waste Recovery: Advanced Technology and Applications

Materials and Components for Pollution and Process Equipment

Heat Transfer Equipment

Materials and Components for Pollution and Process Equipment

Nicholas P. Cheremisinoff
Paul N. Cheremisinoff

P T R Prentice Hall
Englewood Cliffs, New Jersey 07632

Library of Congress Cataloging-in-Publication Data

Cheremisinoff, Nicholas P.
   Materials and components for pollution and process equipment / by
Nicholas P. Cheremisinoff, Paul N. Cheremisinoff.
     p.   cm.
   Includes bibliographical references and index.
   ISBN 0-13-285776-6
   1. Pollution control equipment.   I. Cheremisinoff, Paul N.
II. Title.
TD192.C48 1993
660--dc20   92-40471
                                                                                  CIP

Editorial/production supervision: *Brendan M. Stewart*
Buyer: *Mary Elizabeth McCartney*
Acquisitions editor: *Michael Hays*

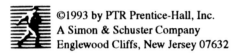

©1993 by PTR Prentice-Hall, Inc.
A Simon & Schuster Company
Englewood Cliffs, New Jersey 07632

The publisher offers discounts on this book when ordered
in bulk quantities. For more information, contact:

       Corporate Sales Department
       PTR Prentice Hall
       113 Sylvan Avenue
       Englewood Cliffs, New Jersey 07632

       Phone: 201-592-2863
       Fax: 201-592-2249

All rights reserved. No part of this book may be
reproduced, in any form or by any means,
without permission in writing from the publisher.

Printed in the United States of America
10 9 8 7 6 5 4 3 2 1

ISBN 0-13-285776-6

Prentice-Hall International (UK) Limited, *London*
Prentice-Hall of Australia Pty. Limited, *Sydney*
Prentice-Hall Canada Inc., *Toronto*
Prentice-Hall Hispanoamericana, S.A., *Mexico*
Prentice Hall of India Private Limited, *New Delhi*
Prentice-Hall of Japan, Inc., *Tokyo*
Simon & Schuster Asia Pte. Ltd., *Singapore*
Editora Prentice-Hall do Brasil, Ltda., *Rio de Janeiro*

Contents

Preface   xi

1   *System Design and Equipment Selection*   1

    Introduction, 1
    Planning Projects and Equipment Design, 2
    Equipment and Instrumentation Codes, 5
      Vessel Codes and Flange Ratings, 9
    General Process and Equipment Design Factors, 9
      Performance and Specifications, 24
      Capital and Operating Costs, 26

2   *Corrosion*   27

    Introduction, 27
    Types of Corrosion, 27
    Materials Evaluation and Selection, 32
    Design Guidelines, 52
    Glossary of Corrosion Terms, 60

3 Material Properties—Metals   65

Properties of Cast Irons, 67
  Gray Cast Iron, 68
  White Cast Iron, 69
  Malleable Cast Irons, 69
  Nodular Cast Iron, 70
  Austenitic Cast Iron, 70
Application Requirements of Cast Irons, 70
  Abrasion Resistance, 70
  Corrosion Resistance, 70
  Temperature Resistance, 71
  Welding Cast Iron, 72
Properties of Steels, 72
  Low-Carbon Steels (Mild Steel), 72
  Corrosion Resistance, 73
  Heat Resistance, 73
  Low Temperatures, 74
  High-Carbon Steels, 74
  Low-Carbon, Low-Alloy Steels, 74
  Properties, 74
  Oxidation Resistance and Creep Strength, 75
  Low-Temperature Ductility, 75
  High-Carbon, Low-Alloy Steels, 75
High-Alloy Steels, 76
  Chromium Steels (400 Series), Low-Carbon Ferritic
    (Type 405), 76
  Medium Carbon Martensitic (Types 403, 410, 414,
    416, 420, 431, 440), 76
  Medium Carbon Ferritic (Types 430 and 446), 76
  Chromium/Nickel Austentitic Steels (300 Series), 76
  Precipitation Hardening Stainless Steel, 77
  Chromium/Nickel/Ferrite/Austenite Steels, 78
  Maraging Steels, 78
Applications of High-Alloy Steels, 78
  Oxidation Resistance, 79
  Mechanical Properties at Elevated Temperatures, 79
  Mechanical Properties at Low Temperatures, 79
Corrosion-Resistant Nickel and Nickel Alloys, 79
  Nickel/Copper (Alloy 400), 80
  Nickel/Molybdenum, 80
  Nickel/Molybdenum/Chromium, 80
  Nickel/Chromium/Molybdenum/Iron, 80

Nickel/Chromium/Molybdenum/Copper,   81
Nickel/Silicon,   81
Heat-Resistant Nickel Alloys,   81
Nickel/Chromium,   81
Nickel/Chromium/Iron,   81
Copper and Copper Alloys,   82
Brasses,   84
Tin Bronzes,   84
Aluminum and Manganese Bronzes,   84
Silicon Bronzes,   87
Cupro-Nickels,   87
Corrosion Resistance,   87
Mechanical Properties of Lead and Lead Alloys,   87
Corrosion Resistance,   89
Aluminum and Aluminum Alloys,   90
Heat-Treatable Alloys,   91
Casting Alloys,   92
Corrosion Resistance,   92
Organic Acids,   93
Miscellaneous Precious Metals,   94
Titanium,   95
Tantalum,   96
Zirconium,   98
Precious Metals,   98
Silver,   98
Gold,   99
Platinum,   99
Metallic Coatings,   99
Electrodeposition,   99
Dip Coating,   100
Sprayed Coatings,   100
Diffusion Coatings,   100

4 Carbon, Glass, and Plastics   102

Carbon and Graphite,   102
Glass,   103
Cements, Bricks, and Tiles,   104
Cements,   104
Bricks and Tiles,   104
Plastic and Thermoplastic Materials,   105
Polyolefins,   111
Polyvinyl Chloride (PVC),   114

Rigid PVC (UPVC), 114
High-Impact PVC, 115
Chlorinated PVC (CPVC), 115
Chlorinated PVC (CPVC), 115
Plastic PVC, 115
Acrylonitrile-Butadiene-Styrene (ABS), 115
Fluorinated Plastics, 115
Polyvinyl Fluoride, 116
Acrylics, 116
Chlorinated Polyether, 116
Nylon (Polyamide), 116
Miscellaneous Engineering Plastics, 117
Acetal Resin, 118
Polycarbonate, 118
Polyphenylene Oxide, 118
Polysulfone, 118
Thermosetting Plastics, 118
  Phenolic Resins, 119
  Polyester Resins, 119
  Epoxy Resins, 121
  Furane Resins, 121
  Rubber Linings, 121
Organic Coatings and Paints, 122
Glossary of Fabrication and Plastic Terms, 124
Nomenclature, 135
Chemical Resistance of Thermoplastics, 135

5 Fabrication Methods   146

Fabrication and Design Principles, 146
Design of Cast Equipment, 149
Design of Welded Equipment, 164
  Influence of Machining on Half-Finished Parts, 180
  Sheet Materials, 188
  Assembly/Disassembly Considerations, 190
Nomenclature, 195

6 Design and Materials Properties   197

Design from High-Alloy Steels, 197
Design of Equipment Using Copper, 206

Design of Equipment Using Aluminum, 213
Design of Ceramic-Lined Equipment, 215
Design Using Molding Materials, 219
Nomenclature, 225

7 Pipes, Compensators, and Valves — 226

Pipes, 226
  Design of Pipes Operating Under External Pressure, 230
  Reinforced Plastic Piping, 230
  Service Considerations for Plastic Pipe, 249
  Pressure and Vacuum Service Considerations, 253
  Buried Pipe Installations, 258
  Line Testing, 265
Compensators, 267
  Lens Compensators, 268
  Stuffing-Box Compensators, 271
Valves and Their Selection, 272
  Gate Valves, 274
  Globe Valves, 277
  Check Valves, 284
  Ball Valves, 285
  Butterfly Valves, 285
  Valve Selection, 286
  Valve Ratings, 287
  Valve Installation, 288
Nomenclature, 289

8 Seals — 291

Introduction, 291
Sealing with Gaskets, 295
Sealing without Gaskets, 301
Special Obturation, 305
Special Seals, 306
Nomenclature, 315

9 Flanges and Threaded Connections — 316

Requirements for Flange Connections, 316
Flange Selection, 317

Welding-Neck Flanges, 318
Slip-On Flanges, 318
Lap-Joint Flanges, 321
Screwed Flanges, 321
Blind Flanges, 322
Nonstandard Flanges, 322
Flanges of Special Types, 325
Welded Around Flange Connections, 325
Detachable Flanges, 325
Flanges for Pipes and Shells from Fragile Materials, 326
Flange Facings, 327
Gasket Selection, 330
Optimum Bolt Selection, 337
Principles of Flange Design, 337
Additional Comments and Notes on Flange Design, 342
Threaded Connections, 342
Screwed Fittings, 345

10 Equipment Supports   349

Design of Supports for Vertical Vessels, 353
Skirt Supports, 353
Lug Supports, 357
Supports for Horizontal Vessels, 361

Index   363

Preface

Pollution control operations as well as process industries use numerous unit operations and processes involving a vast array of equipment and materials. Both chemical and physical methods are employed in such operations ranging from bulk storage to complex chemical reactions involving heat and/or mass transfer. To function properly process equipment performing under actual conditions must meet intended requirements. Such equipment must be mechanically reliable, which includes strength, rigidity, durability, and tightness. It must also be fabricated of proper materials of construction to ensure continuing integrity during the life of the operation. The choice of proper technical options results ultimately in economies of material selection and operation. These technical options are discussed for pollution equipment and industrial systems.

Technology continues to be the important driving force in pollution control solutions as well as rejuvenation and profitability for the chemical industry. This book discusses the basics of the technology in use, selection and systems design, and helps evaluate the parts that comprise equipment, namely materials and configuration.

As in previous volumes in this series, the emphasis is on practical engineering solutions. The book should form an overview and comprehensive reference source for plant managers, process engineers, those people specifying, purchasing, selecting, and using equipment. It is intended as a general reference source for such materials and equipment that are typically used in the pollution control, process chemicals, and allied industries.

—Nicholas P. Cheremisinoff
—Paul N. Cheremisinoff

1

System Design and Equipment Selection

INTRODUCTION

Pollution control, chemical processing, petroleum, and allied industries apply physical as well as chemical methods in handling chemicals and process streams. Because of the wide range and in many cases unique process conditions and requirements, equipment design is often case specific. Requirements of any piece of equipment are that it performs the function for which it was designed under the intended process operating conditions and does so in a continuous and reliable manner. Equipment must have mechanical reliability, which is characterized by strength, rigidness, steadiness, durability, and tightness. One or combinations of these characteristics may be needed for a particular application and piece of equipment.

Cost of equipment determines the capital investment for a process operation. However, there is no direct relationship to ultimate costs. More expensive equipment may mean more durability and, hence, longer service and less maintenance. These characteristics can produce higher operating efficiencies, fewer consumption coefficients and operational expenses and, thus, fewer net costs.

The desirable operating characteristics of equipment include simplicity, convenience, and low-cost maintenance; simplicity, convenience, and low-cost assembly and disassembly; convenience in replacing worn or damaged components; ability to control during operation a test before permanent installation;

continuous operation and steady-state processing of materials without excessive noise, vibration, or upset conditions; a minimum of personnel for its operation; and, finally, safe operation. Low maintenance may often be associated with more complex designs as well as cost. Automation is the most complete solution to problems associated with maintaining steady operation, easy maintenance, and a minimum of operating personnel. The addition of control devices must be considered as part of the overall design and a factor that adds to the capital investment of the project. Increased automation through the use of controls increases the degree of sophistication in equipment design but lowers operational expenditures while increasing production quality. The use of automatic devices influences the form and dimensional proportions of the equipment as well as imposing additional constraints on the design. It is justified by increased production efficiencies and added security during normal and emergency operations.

The first chapter provides an overview of process design strategies, fundamental definitions, and a brief review of preparing process flow plans.

PLANNING PROJECTS AND EQUIPMENT DESIGN

There are numerous stages of activities that must be conducted before an actual process, plant, or even small-scale pilot system reaches its operational stage. Figure 1-1 is a simplified flow diagram illustrating some of the major activities and their sequence. From the initial idea the engineer is directed to prepare a preliminary design basis. This includes a rough flow plan, a review of the potential hazards of the process, and an assimilation of available technical, economic, and socioeconomic information and data. At this stage of a project often the engineer or engineers are not the final equipment designers, but merely play the devil's advocate, by establishing the equipment requirements. Dialogue established between conceptual design and the process designer results in an initial process flow plan. From the flow plan, a preliminary cost estimate is prepared, many times by a different engineer whose expertise is cost estimation. Once management approval is received, the design engineer's work begins. In the initial stages the design engineer will help prepare a preliminary engineering flow plan, select the site, and establish operating and safety requirements.

This initial project stage is often considered a predesign period, which constitutes the basis of the conceptual design. Usually a collection of individuals is involved in discussions and planning. The cast of characters includes the project engineer, who oversees the entire project; the design engineer (with whom we are most concerned); the safety engineer; the environmental engineer; and, perhaps, a representative from management and additional support personnel.

Once the overall process has been designed conceptually, a more detailed engineering flow plan is prepared. This flow plan serves two purposes: (1) to

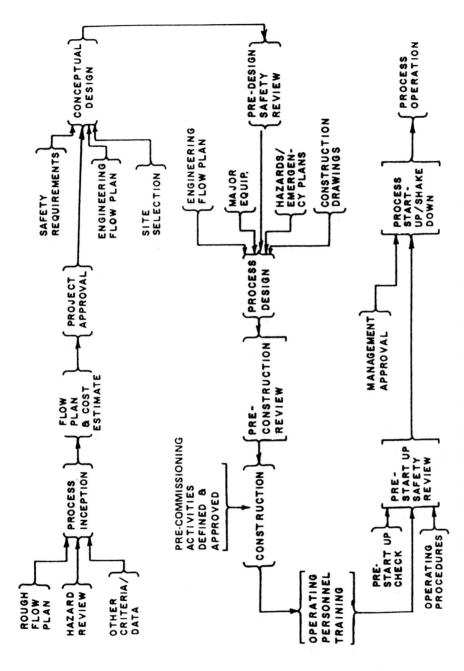

Figure 1-1 Simplified flow diagram of activities in planning and implementing process and plant design projects

TABLE 1-1 MAJOR ITEMS IN OPERATING GUIDELINES PLANNING

Purpose of process or operation

- General Discussion of Process
    - What will be done (brief summary)
    - Chemistry involved
    - Major unit operations
- Personnel Protection
    - Hazards involved—severity
    - Protective equipment—what, where, when
    - Area restrictions—what, where, when
    - Ventilation
- Startup
    - Preparation and handling
        - Feedstocks
        - Catalysts
        - Equipment
- Step-by-Step Description
    - Flow plans
    - Sketches
    - Labeled parts of units
    - Position of valves, control settings, and so on
- Sampling and Final Product Form
    - Description of equipment
    - Actions required
- Shutdown Procedure
    - Step-by-step description
        - Flow plans
        - Sketches
        - Labeled parts of unit
        - Position of valves, control settings, and so on
- Emergency Shutdown Procedure
    - Action required
    - Followup required
    - Emergency personnel/outside organizations
- Unit Cleaning Procedures
    - Description
    - Hazards or precautions
- Product or Waste Disposal

---

document the logic behind the process operation, and (2) to identify in detail major process equipment, including all control devices. A complete flow plan also will identify potential hazards and their consequences, in addition to how they are handled. After the environmental and safety engineers have reviewed all potential hazards related to handling toxic materials, noise, radiation, and so on, recommendations are outlined for safe and standard handling and disposal practices. These recommendations often affect the overall system design, resulting in revised plans.

The next stage is the actual construction of the unit according to the revised plans. By now, the design engineer is totally involved and has selected, sized, and designed most of the equipment and process piping, based (hopefully) on the standard practices outlined in this book. During the actual construction phase, the design engineer will list and review the plans with the project engineer.

At the completion of the unit or system construction, a prestart-up review is conducted by the designer and support personnel. This should include a review of all operating procedures, as well as emergency and shutdown procedures. The prestart-up review normally involves the following personnel in addition to the designer: project engineer, trained operating personnel, operations supervisor, the company environmental engineer, the division and company safety engineers, and representatives from management. At this point, any additional changes or recommendations to the process design are made. Major process revisions may be requested by the operations supervisor, project engineers, design engineer, safety and environmental coordinators, and/or plant operating personnel. Table 1-1 summarizes major items that are considered in the operating procedures planning.

The project planning activities may be much more complex than illustrated by the simple flow diagram of Figure 1-1. This depends, of course, on the magnitude of the project. Often, large complex system planning has numerous checkpoints at various stages where a continuous review of technical and revised economic forecasts is performed. Also not shown in this flow diagram is the legal framework for obtaining construction and operating permits as well as preparing the environmental impact statement and meeting local, state, and federal regulations.

EQUIPMENT AND INSTRUMENTATION CODES

Process and instrumentation flow diagrams (P & I diagrams) essentially define the control and operating logic behind a process as well as provide a visual record to management and potential users. In addition, P & I diagrams are useful at various stages of a project's development by providing:

- the opportunity for safety analysis before construction begins
- a tabulation of equipment and instrumentation for cost-estimating purposes
- guidelines for mechanics and construction personnel during the plant assembly stage
- guidance in analyzing start-up problems
- assistance in training operating personnel
- assistance in solving daily operating and sometimes emergency problems

P & I diagrams contain four important pieces of information, namely, all vessels, valves, and piping, along with a brief description and identifying specifications of all sensors, instruments, and control devices. Included also are the control logic used in the process and additional references where more detailed information can be obtained.

Information normally excluded from P & I diagrams includes electrical wiring (normally separate electrical diagrams must be consulted), nonprocess equipment (for example, hoist, support structures, foundations, and so on), and scale drawings of individual components.

There are basically two parts to the diagram. The first provides a schematic of equipment and the second details the instrumentation and control devices. The P & I diagram provides a clear picture of what each piece of equipment is, including identifying specifications, the sizes of various pieces of equipment, materials of construction, pressure vessel numbers and ratings, and drawing numbers. Equipment and instrumentation are defined in terms of a code consisting of symbols, letters, and a numbering system. That is, each piece of equipment is assigned its own symbol; a letter is used to identify each type of equipment and to assist in clarifying symbols, and numbers are used to identify individually each piece of equipment within a given equipment type. Table 1-2 illustrates common equipment symbols and corresponding letter codes.

When denoting instrumentation, it is important that definitions are understood clearly. Terms for instruments and controls most often included on P & I diagrams follow.

Instrument loop—A combination of one or more interconnected instruments arranged to measure or control a process variable.
Final control element—A device that directly changes the value of the variable used to control a process condition.
Transducer (converter)—A device that receives a signal from one power source and outputs a proportional signal in another power system. A transducer can act as a primary element, transmitter, or other device.
Fail closed (usually normally closed)—An instrument that will go to the closed position on loss of power (pneumatic, electric, and so on).
Fail open (usually normally open)—An instrument that will go to the open position on loss of power (pneumatic, electric, and so on).
Fail safe—An instrument that on loss of power (pneumatic, electric, and so on) will go to a position that cannot create a safety hazard.
Process variable—A physical property or condition in a fluid or system.
Instrument—A device that measures or controls a variable.
Local—An instrument located on the equipment.
Remote—An instrument located away from the equipment (normally a control cabinet).
Primary element—A device that measures a process variable.
Indicator—A device that measures a process variable and displays that variable at the point of measurement.

Equipment and Instrumentation Codes

TABLE 1-2 COMMON EQUIPMENT SYMBOLS AND TYPICAL LETTER CODES

Equipment	Symbol	Code	Information needs
Control valve		CV	Size, maximum flow rate, pressure drop
Piping			Material, size, wall thickness
Valves			Type: ball (B), globe (G), needle (N), and so on
Centrifugal Pump		P	Inlet/outlet pressure, flow rate
Rotameter		R	Tube, float, body, maximum flow rate
Reactor		R	Pressure vessel number, drawing number, size
Filter		FIL	Pore size
Backpressure Regulator			Range of gauge and loading source
Electric Heater			Shown on vessel with power pack and control signal
Tracing			Type: steam (S)/electric (E)
Spring-Loaded Relief Valve		PR	Relief pressure, orifice size

Transmitter—A device that senses a process variable through a primary element and puts out a signal proportional to that variable to a remotely located instrument.

TABLE 1-3 TYPICAL INSTRUMENT CODES AND EXAMPLES

General symbols

Symbol	Description	Symbol	Description
————	Instrument process piping		Electrically operated valve (solenoid or motor)
—#—#—#—	Instrument air lines		Piston-operated valve (hydraulic or pneumatic)
- - - - - -	Electrical leads		
—×—×—×—	Capillary tubing		3-way body for any valve
○	Locally mounted instrument (single service)		
⊗	Locally mounted transmitter		Safety (relief) valve
⊗	Board-mounted transmitter		
	Diaphragm motor valve		Manually operated control valve

Temperature symbols

Symbol	Description	Symbol	Description
(TW)	Temperature well	(TRC)	Temperature recording controller
(TI)	Temperature indicator		

Pressure symbols

Symbol	Description	Symbol	Description
(PI)	Pressure indicator (locally mounted)	(PA)	Pressure alarm
(PR)	Pressure recorder (board mounted)	(PC)	Pressure controller (blind type)

Flow symbols

Symbol	Description	Symbol	Description
(FI)	Displacement-type flow meter	(FR)	Flow recorder
(FI)	Flow indicator, differential type		

Controller—A device that varies its output automatically in response to changes in a measured process variable to maintain that variable at a desired value (setpoint).

General Process and Equipment Design Factors

Instrumentation normally is denoted by a circle in which the variable being measured or controlled is denoted by an appropriate letter symbol inside the circle. When the control device is to be located remotely, the circle is divided in half with a horizontal line. Table 1-3 gives various instrumentation symbols and corresponding letter codes. The specific operating details and selection criteria for various process instrumentation are not discussed in this book.

Piping normally is denoted by solid lines. Piping lines on the P & I diagram should be accompanied by the following identifying information:

1. line number
2. nominal pipe size and wall thickness
3. origin and termination
4. design temperature and pressure
5. specified corrosion allowance
6. winterizing or process protection requirements (that is, heat tracing via steam or electric)
7. insulation type and thickness
8. test pressure (indicate hydrostatic or pneumatic)
9. piping flexibility range (for example, the maximum or minimum operating temperature)

Vessel Codes and Flange Ratings

In the United States, the primary standard for pressure vessel design is that of the American Society of Mechanical Engineers (ASME). The ASME code is essentially a legal requirement. It provides the minimum construction requirements for the design, fabrication, inspection, and certification of pressure vessels. The ASME code does not cover: (1) vessels subject to federal control; (2) certain water and hot-water tanks, (3) vessels with an internal operating pressure not exceeding 15 psig with no limitation on size; and (4) vessels having an inside diameter not exceeding 6 inches with no limitation on pressure.

GENERAL PROCESS AND EQUIPMENT DESIGN FACTORS

Variations in Flow

Waste streams vary in their volume and chemical composition because of changes in the processes or in the events which generate these streams. These variations may exhibit clearly defined cycles. Industrial wastes may be influenced by the working hours of the plants, shift changes, weekend shutdowns, summer holidays, or fluctuations in production rates caused by seasonal marketing patterns. It is important to recognize that these fluctuations occur in the

TABLE 1-4 CONDITIONS AFFECTING INSTALLED COST OF CONTROL DEVICES

Cost category	Low cost	High cost
Equipment transportation	Minimum distance; simple loading and unloading procedures.	Long distance; complex procedure for loading and unloading.
Plant age	Hardware designed as an integral part of new plant.	Hardware installed into confines of old plant requiring structural or process modification or alteration.
Available space	Vacant area for location of control system.	Little vacant space requires extensive steel support construction and site preparation.
Corrosiveness	Noncorrosive	Acidic emissions requiring high-alloy accessory equipment using special handling and construction techniques.
Complexity of start-up	Simple start-up, no extensive adjustment required.	Requires extensive adjustments; testing; considerable downtime.
Instrumentation	Little required	Complex instrumentation required to assure reliability of control or constant monitoring of gas stream.
Guarantee on performance	None needed	Required to assure designed control efficiency.
Degree of assembly	Control hardware shipped completely assembled.	Control hardware to be assembled and erected in the field.
Degree of engineering design	Autonomous "package" control system.	Control system requiring extensive integration into process, insulation to correct temperature problem, noise abatement.
Utilities	Electricity, water, waste disposal facilities readily available	Electrical and waste treatment facilities must be expanded, water supply must be developed or expanded.
Collected waste-material handling	No special treatment facilities or handling required.	Special treatment facilities and/or handling required.
Labor	Low wages in geographical area.	Overtime and/or high wages in geographical area.

chemical character of the waste stream as well as in its volume. Other fluctuations are related to changes in the weather, such as those due to ground-water infiltration, illegal connections, and extraneous streams such as water entering the collection system. Some provisions must therefore be made for dealing with these variations when a pollution control plant is being designed. The major elements in conventional plants are usually designed on the basis of the average expected flow during the design period of the plant. However, smaller elements such as the internal piping are designed on the basis of peak flows. Peak flows are frequently assumed to be two to three times the average flow.

In cases of above-average flow wastewater treatment, wastewater is permitted to pass through the treatment plant as usual, but the increased flow

TABLE 1-5 UNIT OPERATIONS AND PROCESSES TYPICALLY EMPLOYED IN VARIOUS INDUSTRIES

Process industries / Plants that make	Fluid flow	Heat transfer	Evaporation	Humidification and dehumidification	Gas absorption	Solvent extraction	Adsorption	Distillation and sublimation	Drying	Mixing	Classification	Sedimentation and decantation	Filtration	Screening	Crystallization	Centrifugation	Disintegration	Materials handling
Chemicals	X	X	X	X	X	X	X	X	X	X	X	X	X	X	X	X	X	X
Coke and gas	X	X		X	X		X	X	X	X	X			X	X	X	X	X
Drugs, medicines, and cosmetics	X	X	X	X		X	X	X	X	X			X	X	X	X	X	X
Explosives	X	X		X	X	X		X		X	X			X	X	X	X	X
Fertilizers	X	X			X		X		X	X	X	X	X	X	X	X	X	X
Glass and ceramics	X	X							X	X	X	X	X	X		X	X	X
Leather	X	X							X	X								X
Lime and cement	X	X							X	X	X	X	X	X		X	X	X
Oils and fats	X	X	X			X	X	X	X	X		X	X		X	X		X
Paints, pigments, and varnish	X	X	X		X				X	X	X	X	X	X	X	X	X	X
Paper and pulp	X	X	X	X	X		X		X	X		X	X			X	X	X
Petroleum products	X	X	X		X	X	X	X	X	X	X	X	X		X	X	X	X
Rayon	X	X	X	X					X	X			X			X		X
Rubber and synthetic rubber goods	X	X			X			X	X	X				X	X	X	X	X
Soap	X	X	X						X	X		X			X	X	X	X

(continued)

TABLE 1-5 *Continued*

Process industries / Plants that make	Combustion or oxidation	Neutralization or silicate formation	Causticization	Electrolysis	Double decomposition	Calcination or dehydration	Nitration	Esterification	Reduction	Ammonolysis	Halogenation	Sulphonation	Hydrolysis	Hydrogenation	Alkylation	Friedel and Crafts	Condensation or polymerization	Diazotization and coupling	Fermentation	Pyrolysis
Chemicals	X	X	X	X	X	X	X	X	X	X	X	X	X	X	X	X	X	X	X	X
Coke and gas	X													X						X
Drugs, medicines, and cosmetics	X	X			X			X			X	X	X	X	X	X	X	X	X	
Explosives		X					X	X			X	X	X		X			X		
Fertilizers		X				X								X						
Glass and ceramics		X				X														
Leather					X															
Lime and cement		X				X														
Oils and fats			X			X		X				X	X	X						
Paints, pigments, and varnish	X				X							X				X	X	X		
Paper and pulp	X		X									X								
Petroleum products	X						X				X	X	X	X	X	X	X	X		X
Rayon								X					X							
Rubber and synthetic rubber goods											X				X	X	X			
Soap		X	X																	

General Process and Equipment Design Factors

TABLE 1-6 UNIT OPERATIONS AND EQUIPMENT FOR PROCESS ENGINEERING AND POLLUTION CONTROL

Unit operation	Equipment
Materials handling	
Gathering	Cranes, drag lines, industrial cars, pneumatic conveyors, car dumpers, power and hand trucks, stackers, skids
Storing	Bins, hoppers, sheds, barrels, drums, open storage, skids, stackers
Transporting	Automobile trucks, industrial cars, dry-flow cars, barrels, drums, kegs, power and hand trucks, skids, box and gondola cars
Conveying	Belts, screws, elevators, bucket and pallet conveyors, Redlers, solids pumps, pneumatic, flight and vibrating conveyors
Fluids handling and transport	
Gravity	Piping, flumes, conduits, valves, fittings
Pressure pumping	Plunger, piston and rotary pumps, pulsometers, blow cases, air lifts, screw pumps, piston compressors, pipe lines
Kinetic pumping	Centrifugal and turbine pumps and compressors, fans, blowers, jets, pipe lines
Size reduction	
Crushing	Jaw crushers, gyratories, disk crushers, Hadsel mills, rolls
Grinding	Ball and hammer mills, rolls, ring-roll mills, cage mills, chasers
Pulverizing	Hammer, ball and tube mills, stamps, colloid mills, ring-roll and ball-and-race mills, rod mills, roller, attrition and buhrstone mills
Shredding	Hammer and attrition mills, grindstones
Chipping	Hogs, chippers
Masticating	Masticators
Spraying	Spray nozzles, turbine-type dispersers, pumps, piping, valves, spray chambers
Mixing	
Blending	Chasers, flight and screw mixers, tumbling barrels, buhrstone mills, roller mills, turbine, paddle and propellor mixers
Agitating	Turbine, propellor and paddle mixers, air-lift and air-bubble agitators, pumps, compressors
Kneading	Rolls, double-arm mixers, chasers, epicyclic chasers (Lancaster)
Emulsifying	Turbine, paddle and propellor mixers, colloid mills, homogenizers, pumps
Dissolving	Tanks, vats, agitators, pumps
Suspending	Tanks, vats, agitators, pumps
Dispersing	Turbine, paddle and propellor mixers, colloid mills, homogenizers, pumps
Diffusing	Tanks, vats, diffusion cells, pumps
Heat transfer	
Direct heating	Electric arc, resistance and induction furnaces, fuel-fired furnaces, kilns, muffles, and so on (heating by radiation and furnace gases)
Indirect heating	Heat exchangers, furnaces, and so on used to supply heating medium such as hot air, water, or oil, vapors such as steam, mercury, diphenyl
Water cooling	Condensers, diesels, refrigerating equipment, process coolers

(continued)

TABLE 1-6 *Continued*

Unit operation	Equipment
Ice cooling	Process vessels
Spray cooling	Spray ponds, cooling towers
Refrigerating	Steam jet units, centrifugal vapor units, compression and absorption units
Separation	
Leaching	Tanks, vats, Pachuca and Shanks tanks, classifiers, ball, tube, and stamp mills, pumps
Extracting	Tanks, vats, classifiers, diffusion cells, agitators, gas equipment, pumps
Percolating	Percolators (solvent circulated), pumps
Washing	Tanks, agitators, classifiers, thickeners, filters, centrifugals
Gas absorbing	Packed towers, spray chambers, bubble and plate towers, Woulff bottles, tourills, blowers, exhausters, pumps
Crystallizing	Ponds, vats, agitators, agitated troughs, rotary and rocking crystallizers, salting evaporators, vacuum crystallizers, grainers
Condensing	Jet and surface condensers, spray chambers, heat exchangers, fractional condensors, fractionating columns, pumps
Distilling	Pot and pipe stills, packed, bubble and plate columns, pumps, condensers, separators
Evaporating	Tanks and kettles, atmospheric and vacuum evaporators (single and multiple effect), film and forced-circulation evaporators, water stills
Drying	Atmospheric and vacuum dryers, tunnel, truck, pan and tray dryers, conveyor dryers rotary, drum and festoon dryers
Subliming	Retorts, condensing chambers
Melting	Furnaces, kettles, crucibles, reaction equipment
Freezing	Molds, solid CO_2 presses, refrigeration equipment
Adsorbing	Char and contact filters, false-bottom and agitated tanks, packed towers, activated carbon, silica gel, and other adsorbents
Magnetic separating	Magnets (permanent and electro), magnetic pulleys and chutes, high-intensity separators
Electrostatic separating	Huff separator, Cottrell precipitator
Electrophoresis	Clay-extruding machinery (only presently known commercial application)
Gravity filtering	Slow sand filters, false-bottom tanks, Harding clarifier, Laughlin thickener, nutsches
Pressure filtering	Rapid sand filters, clarifiers, filter presses, leaf filters, edge filters, continuous filter presses (Thompson)
Vacuum filtering	Rapid sand filters, disk filters, leaf filters, drum-type filters, Dorrco filter, Genter filter, filter-type thickeners, savealls, nutsches
Centrifugal filtering	Continuous, semicontinuous, and batch centrifugals with perforated baskets
Screening	Grizzlies, trommels, gyrating, shaking, and vibrating screens
Sieving and bolting	Sieves, sifters, bolters (rotary, shaking, vibrating), turbine sifters
Pressing	Hydraulic and mechanical presses, curb presses, expellers, continuous filter presses (Thompson)
Draining	Draining decks, screens, classifiers, tables, dewaterers, filters, centrifugal filters, nutsches
Flotation	Air- and mechanically-agitated flotation cells, skimmers
Classifying	Reciprocating and spiral-rake classifiers, bowl and drum classifiers, hydraulic cones, multiple-spigot classifiers
Chamber (basin) settling	Settling basins, dust chambers (void, chain-filled, shelved), Spitzkastens

TABLE 1-6 Continued

Unit operation	Equipment
Air separating	Equipment employing change of air direction, velocity, or both, i.e., cyclones, multiclones, vacuum, fan, and dynamic-dype separators
Jigging	Hydraulic jigs
Tabling	Concentrating tables
Vanner separating	Vanners
Dialyzing	Animal membranes, vegetable parchment, cellulose film; used as bags or as diaphragms
Gas diffusing	Porous plates (unavoidable diffusion of certain gases, through metals at high temperatures)
Impinging	Air and gas washers, entrainment separators, spray chambers, air and gas filters (wetted or oiled surface), P & A tar extractor
Sedimentation	Settling basins, tanks, vats, thickeners
Liquids settling	Settling tanks with multiple or swing draw-off pipes
Centrifugal settling	Continuous, semicontinuous (rotojector), and batch centrifugals and centrifuges with solid baskets, cyclones

TABLE 1-7 PROCESS GROUPING FOR WASTEWATER COLLECTION AND DISPOSAL

Collection
    Force mains, transmission
    Lift stations, raw wastewater
    Sewers, gravity
    Sewers, pressure
    Sewers, vacuum

Disposal
    Aquaculture—water hyacinth
    Aquaculture—wetlands
    Rapid infiltration, underdrained
    Rapid infiltration, not underdrained
    Land treatment, slow rate, sprinkler, underdrained, not underdrained
    Overland flow, gravity, not underdrained

BIOLOGICAL TREATMENT PROCESSES

Suspended-growth Processes
    Activated sludge, conventional, diffused aeration
    Activated sludge, conventional, mechanical aeration
    Activated sludge, high rate, diffused aeration
    Activated sludge, pure oxygen, covered
    Activated sludge, pure oxygen, uncovered
    Activated sludge with nitrification
    Biofilter, activated (with aerator)
    Contact stabilization, diffused aeration
    Denitrification, separate stage, with clarifier
    Extended aeration, mechanical and diffused aeration
    Lagoons, aerated
    Lagoons, anaerobi
    Lagoons, facultative
    Nitrification, separate stage, with clarifier
    Oxidation ditch
    Phostrip

(continued)

TABLE 1-7 *Continued*

Fixed-film Processes
  Biological contactors, rotating (RBC)
  Denitrification filter, coarse media, fine media
  Intermittent sand filtration, lagoon upgrading
  Polishing filter for lagoon, rock media
  Trickling filter, plastic media, high rate, rock media
  Trickling filter, high rate, rock media

PHYSICAL PROCESSES
  Clarifier, primary, circular or rectangular with pump
  Clarifier, secondary, circular
  Clarifier, secondary, rectangular
  Clarifier, secondary, high-rate trickling filter
  Dissolved-air flotation
  Filtration, dual media
  Flow equalization
  Mixing/chlorine contact, high intensity
  Postaeration
  Preliminary treatment
  Pump stations, in plant
  Screen, horizontal shaft rotary

PHYSICAL-CHEMICAL PROCESSES
Nitrogen Removal
  Ammonia stripping
  Ammonia removal and recovery process
  Breakpoint chlorination
  Ion exchange for ammonia removal

Phosphorus Removal
  Lime recalcination
  Two-stage tertiary lime treatment without recalcination

Physical-Chemical System
  Independent physical-chemical treatment

Tertiary
  Tertiary granular activated carbon adsorption
  Activated carbon thermal regeneration
  Ozone oxidation (air and oxygen)

Miscellaneous
  Chlorination disinfection
  Dechlorination (sulfur dioxide)
  Ozone disinfection (air and oxygen)

CHEMICAL ADDITION
  Alum addition
  Ferric chloride addition
  Lime clarification of raw wastewater
  Polymer addition
  Powdered carbon addition

SLUDGE TREATMENT AND DISPOSAL
Transport and Disposal
  Dewatered sludge transport (rail/truck)
  Land application of sludge
  Sludge land filling—area fill
  Liquid sludge transport (pipeline)
  Liquid sludge transport (rail/truck)

General Process and Equipment Design Factors

TABLE 1-7 *Continued*

Sludge pumping/storage
Sludge land filling—sludge trenching
Sludge lagoons
Conversion Processes
  Coincineration of sludge, sludge incinerator, solid-waste incinerator
  Composting sludge, static pile
  Incineration of sludge, fluidized-bed furnace, multihearth furnace
  Codisposal of starved-air combustion
  Starved-air combustion of sludge
  Sludge drying
Thickening and Dewatering
  Centrifugal dewatering
  Drying beds, sludge
  Filter, belt
  Filter press, diaphragm, conventional filter press
  Thickening, dissolved-air flotation, gravity
  Centrifugal thickening
  Vacuum filtration, sludge
Treatment Processes
  Digestion, aerobic
  Digestion, autothermal thermophilic aerobic
  Digestion, autothermal thermophilic oxygen
  Digestion, two-stage anaerobic, thermophilic anaerobic
  Disinfection (heat)
  Heat treatment of sludge
  Lime stabilization
ON-SITE TECHNOLOGY
Complete Systems
  Aerobic treatment and absorption bed
  Aerobic treatment and surface discharge, on-site surface discharge
  Evaporation lagoons
  Evapotranspiration systems
  Septic tank, ion bed, mound system, polishing, surface discharge
  Septage treatment and disposal
System Components
  In-home treatment and recycle
  Nonwater carriage toilets

results in relatively poorer treatment and an unsatisfactory effluent is discharged for the duration of the surge.

One method of avoiding this impaired effluent quality is to construct a flow-equalization basin. Excess flows or highly concentrated wastes could then be accumulated during surges and later be allowed to enter the plant gradually without impairing treatment efficiency. Equalization basins are probably advantageous in most situations, although they may be more necessary in complicated process sequences than in simpler ones. Some form of flow equalization may be advisable in any situation where the processes themselves cannot accommodate the variations in flow.

TABLE 1-8 CHECK LIST OF WASTEWATER PROCESSES FOR PLANT TREATMENT

Pretreatment
   Screening and grit removal
   Equalization and storage
   Oil and grease separation
PRIMARY TREATMENT
Chemical Processes
   Neutralization
   Chemical addition
   Coagulation
Physical Processes
   Flotation
   Sedimentation
SECONDARY TREATMENT
Dissolved Organics (Biological Processes)
   Activated sludge
   Aerobic lagoons
   Trickling filters
   Anaerobic lagoons
   Stabilization basins
Suspended-solids Removal
   Sedimentation
Tertiary Treatment
   Coagulation and sedimentation
   Filtration
   Carbon adsorption
   Ion exchange
Liquid Disposal
   Receiving waters
   Controlled transport or discharge
   Surface application
   Deep-well injection
   Evaporation or incineration
Solids Treatment or Disposal
   Thickening
   Vacuum filtration
   Centrifuging
   Lagooning or drying beds
   Incineration
   Land filling
   Deep-well disposal
   Incineration

The requirements of flow equalization in some treatment systems are different from those in conventional plants or for other physical-chemical processes in two principal respects. First, since the process is usually preceded by one or more removal processes, these preliminary treatment stages act not only as pretreatment steps, but also to dampen variations in flow.

TABLE 1-9 INDUSTRIAL PROCESS AND CONTROL SUMMARY

Industry or process	Source of emissions	Particulate matter	Method of control
Iron and steel mills	Blast furnaces, steel making furnaces, sintering machines.	Iron oxide, dust, smoke.	Cyclones, baghouses, electrostatic precipitators, wet collectors.
Gray iron foundries	Cupolas, shake out systems, core making.	Iron oxide, dust, smoke, oil, grease, metal fumes.	Scrubbers, dry centrifugal collectors.
Metallurgical (non-ferrous)	Smelters and furnaces.	Smoke, metal fumes, oil, grease.	Electrostatic precipitators, fabric filters.
Petroleum refineries	Catalyst regenerators, sludge incinerators.	Catalyst dust, ash from sludge.	High-efficiency cyclones, electrostatic precipitators, scrubbing towers, baghouses.
Portland cement	Kilns, dryers, material handling systems.	Alkali and process dusts.	Fabric filters, electrostatic precipitator, mechanical collectors.
Kraft paper mills	Chemical recovery furnaces, smelt tanks, lime kilns.	Chemical dusts.	Electrostatic precipitators, venturi scrubbers.
Acid manufacture-phosphoric, sulfuric	Thermal processes, phosphate rock acidulating, grinding and handling systems.	Acid mist, dust.	Electrostatic precipitators, mesh mist eliminators.
Coke manufacturing	Charging and discharging oven cells, quenching, materials handling.	Coal and coke dusts, coal tars.	Meticulous design, operation, and maintenance.
Glass and glass fiber	Raw materials handling, glass furnaces, fiberglass forming and curing.	Sulfuric acid mist, raw materials dusts, alkaline oxides, resin aerosols.	Glass fabric filters, afterburners.
Coffee processing	Roasters, spray dryers, waste heat boilers, coolers, conveying equipment.	Chaff, oil aerosols, ash from chaff burning, dehydrated coffee dusts.	Cyclones, afterburners, fabric filters.

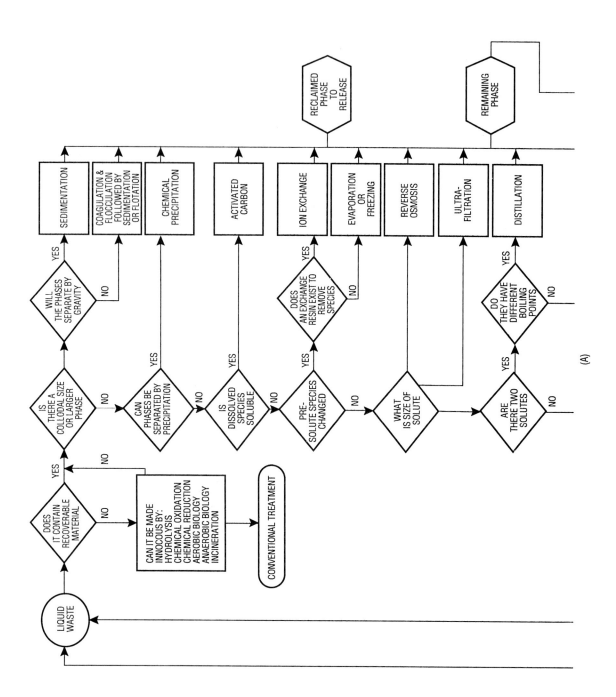

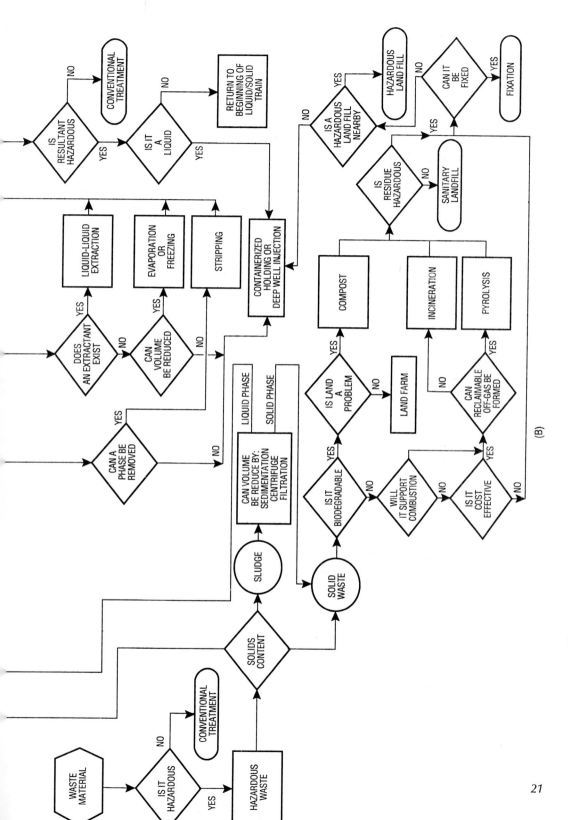

Figure 1-2 Decision tree for application of hazardous waste treatment and disposal technologies

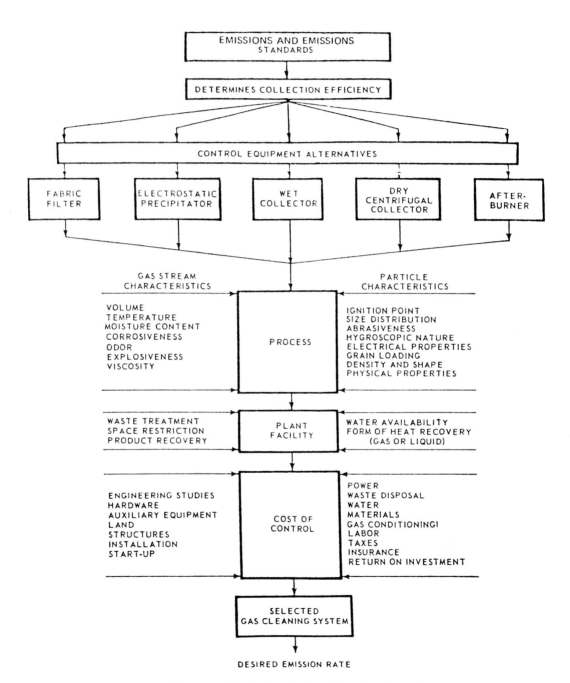

Figure 1-3 Criteria for selection of gas cleaning equipment

Second, it should be recognized that the process itself can accommodate significant variations in flow or loading without any substantial immediate disadvantage. At any given time, only a portion of the equipment or process is actively engaged in the bulk of the work. Any sort of excessive loading merely throws a greater burden on the portion of the equipment which is less active in treatment, that is, the downstream end. Shock loading may cause more rapid exhaustion even though effluent quality may not be seriously affected. If a process becomes overloaded, effluent quality will be seriously affected.

It is thus desirable to build excess capacity into the system for use during such flow surges in lieu of flow equalization. Surges in concentration which are not dampened by the pretreatment systems are accommodated by increased capacity.

The handling of sudden increases in the flow rate through equipment presents other problems for the designer. A pumped system will require additional capacity for flow and head in the feed pumps. In a gravity flow system, the requirements for increased flow can only be obtained by increasing the available head in equipment. It can be seen that the plant can be designed to handle considerable fluctuations in capacity and concentration through the relatively inexpensive technique of installing excess capacity. The accommodation of large flow surges may be more expensive because of the problems associated with allowing for increased head. A good economic comparison of the relative costs of flow-equalization basins versus increased plant capability for plants has not yet been developed. At the present time, it is necessary for designers of plants to evaluate several alternatives for a particular situation by specific studies for that situation. However, as long as waste streams show variations in quantity and concentration, the question of flow equalization cannot be ignored in wastewater treatment.

One of the more important design decisions which the engineer must make is the selection of the design flow for the treatment or process plant. This question is connected with those of handling variations in flow and of making an accurate forecast of the volume and character of waste which will be produced during the course of the design period. Selection of the proper design flow requires an analysis of the following factors:

- Useful operating lifetime of the plant as influenced by wear-and-tear and by technological obsolescence.
- Interest rates and the rate of currency inflation, during the life of the plant.
- Population and waste-load changes in the service area of the plant during its useful life.
- Changes in levels and types of industrial activity in the plant's service area during its useful life.
- Changes in requirements for treated effluent or emission quality during the plant's lifetime.

Anticipated useful lifetime for a plant is on the order of 20 years; at least from the standpoint of wear-and-tear, it is evident that accurate forecasts of the preceding parameters are difficult.

A better approach is to design for a relatively short term, taking into account only those future events and trends which can be foreseen with some precision. At the same time, the plant should be built so that its capacity can be increased as needed without scrapping or greatly changing the original equipment. The best way to accomplish this is by what might be called the modular approach to plant design. This means that plant expansion can be obtained by building a new plant beside the old one in parallel with it, with the flow proportioned between them. Physical-chemical treatment plants readily lend themselves to the modular design approach. In the original design, it is often relatively inexpensive to oversize the regeneration equipment and operate it initially on a part-time schedule. As the treatment loads grow, the utilization can be increased until running steadily at full capacity. Another piece of equipment can then be installed if needed. Of course, more will add to total replacement cost, but the cost per unit of treated material will not change greatly. Meanwhile, the capital cost is kept down.

The plant can also be expanded to accommodate a greater future flow by initially designing the equipment to have slightly more than the minimum necessary capacity. The initial capital cost for vessels is of course higher for this option. Resultant unused volume in the vessels may not interfere in any way with operations. As the flow to be treated increases in the course of time or as its character changes, the equipment in the plant can simply be increased to keep pace with the flow. It is also possible to increase the treatment plant's capacity by adding additional vessels as needed. The major requirement is that the initial design should provide space for the later addition of these vessels.

In summary, it seems best to design the major components of a process system for flow and quality specifications which are based upon the present situation and only those short-term changes which can be accurately estimated. It is the nature of the process itself that later expansion can generally be handled economically.

Performance and Specifications

It is important to prepare suitable performance specifications as part of any plant design. This is a consideration in the design of any waste treatment plant. Experience can be cited from existing plants. It is therefore important for the designer to conduct appropriate preliminary testing programs to cover the proposed situation. Municipalities or industries may be protected against deficient plant performance only by periodic reevaluations of influent waste characteristics and plant operations.

Operating costs may be divided into a number of categories, including power, fuel, labor, backwash water, maintenance, amortization, bond interest, and so on. Typical operating cost factors are:

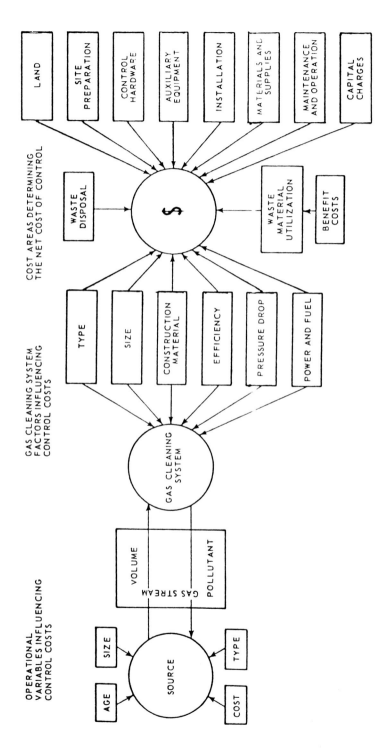

Figure 1-4 Diagram of cost evaluation of a gas-cleaning system

- cost component
- equipment replacement
- operating labor
- electric power
- fuel
- raw materials and chemicals
- maintenance
- amortization

The design flow should be stated exactly and the plant should be equipped with suitable flow controls. Design influent and effluent quality should also be clearly defined. Analytical and monitoring procedures to be used should be described in the specifications. The characteristics should be specified.

Capital and Operating Costs

In designing a process or treatment plant, many alternative schemes are available to the engineer. Optimizing the design becomes largely the problem of choosing among the alternatives. Capital and operating costs play important parts in this choice, but these considerations are tempered by considering ease of operation, availability of personnel, reliability, and so on. In making these decisions, it is helpful for the engineer to know the relative costs of various plant components. Cost-reduction efforts can then be directed to the more expensive items before considering the less expensive ones. Conditions affecting installed costs of control devices are shown in Table 1-4.

Table 1-5 lists unit operations and processes typically employed in various industries. Table 1-6 shows the unit operations for process engineering and pollution control. Shown is the function under unit operation and the equipment that might be used.

Table 1-7 lists a process grouping that can be used to categorize technologies for process design. Wastewater collection and disposal can be broken down into the categories shown. Table 1-8 is a check list of wastewater processes that might be used in plant treatment selection and design or waste control.

Figure 1-2 shows a decision tree for process selection for the application of hazardous waste treatment and disposal technologies.

Figure 1-3 lists criteria for the selection of gas-cleaning equipment for air pollution control and Figure 1-4 for costs involved in the selection of gas-cleaning systems. Table 1-9 lists industrial processes and typical particulate control methods available.

2

Corrosion

INTRODUCTION

Corrosion occurs in various forms and is promoted by a variety of causes related to process operating conditions. It is an ongoing problem that can lead to contaminated process streams in manufacturing which leads to poor product quality and unscheduled equipment shutdowns. This in turn leads to reduced production and high maintenance and equipment replacement costs. Minimizing corrosion is a key consideration for the design engineer and can be accomplished in two ways: (1) proper material selection for equipment, and (2) preventive maintenance practices. Both these approaches must be examined by the engineer. This chapter reviews principles of corrosion causes and control. It is important to recognize conditions that promote rapid material degradation to compensate for corrosion in design or prevent it altogether.

TYPES OF CORROSION

Corrosion is characterized by the controlling chemical-physical reaction that promotes each type. Descriptions of each of the major types of corrosion follow.

Uniform corrosion is the deterioration of a metal surface that occurs uniformly across the material. It occurs primarily when the surface is in contact with an aqueous environment, which results in a chemical reaction between the metal and the service environment. Since this form of corrosion results in a relatively uniform degradation of apparatus material, it can be accounted for most readily at the time the equipment is designed, either by proper material selection, special coatings or linings, or increased wall thicknesses.

Galvanic corrosion results when two dissimilar metals are in contact, thus forming a path for the transfer of electrons. The contact may be in the form of a direct connection (for example, a steel union joining two lengths of copper piping), or the dissimilar metals may be immersed in an electrically conducting medium (for example, an electrolytic solution). One metal acts as an anode and, consequently, suffers more corrosion than the other metal, which acts as the cathode. The driving force for this type of corrosion is the electrochemical potential existing between two metals. This potential difference represents an approximate indication of the rate at which corrosion will take place. That is, corrosion rates will be faster in service environments where electrochemical potential differences between dissimilar metals are high.

Thermogalvanic corrosion is promoted by an electrical potential caused by temperature gradients and can occur on the same material. The region of the metal higher in temperature acts as an anode and thus undergoes a high rate of corrosion. The cooler region of the metal serves as the cathode. Hence, large temperature gradients on process equipment surfaces exposed to service environments will undergo rapid deterioration.

Erosion corrosion occurs in an environment where there is flow of the corrosive medium over the apparatus surface. This type of corrosion is greatly accelerated when the flowing medium contains solid particles. The corrosion rate increases with velocity. Erosion corrosion generally manifests as a localized problem due to maldistributions of flow in the apparatus. Corroded regions are often clean, due to the abrasive action of moving particulates, and occur in patterns or waves in the direction of flow.

Concentration cell corrosion occurs in an environment in which an electrochemical cell is affected by a difference in concentrations in the aqueous medium. The most common form is crevice corrosion. If an oxygen concentration gradient exists (usually at gaskets and lap joints), crevice corrosion often occurs. Larger concentration gradients cause increased corrosion (due to the larger electrical potentials present).

Cavitation corrosion occurs when a surface is exposed to pressure changes and high-velocity flows. Under pressure conditions, bubbles form on the surface. Implosion of the bubbles causes local pressure changes sufficiently large to flake off microscopic portions of metal from the surface. The resultant surface roughness acts to promote further bubble formation, thus increasing the rate of corrosion.

Fretting corrosion occurs where there is friction between two metal surfaces. Generally, this friction is caused by vibrations. The debris formed by fretting corrosion accelerates the damage. Initial damage is done by contact welding. Vibrations cause contact welds to break, with subsequent surface deterioration. Debris acts to accelerate this form of corrosion by serving as an abrasive. Fretting corrosion is especially prevalent in areas where motion between surfaces is not foreseen. If allowances for vibration are not made during design, fretting corrosion may be a strong factor.

Pitting corrosion is a type of localized corrosion in which large pits are formed in the surface of a metal usually in contact with an aqueous solution. The pits can penetrate the metal completely. The overall appearance of the surface involved does not change considerably; hence, the actual damage is not readily apparent. Once a pit forms, it acts as a local anode. Conditions such as debris and concentration gradients in the pit further accelerate degradation. There are several possible mechanisms for the onset of pitting corrosion. Slight damage or imperfections in the metal surface, such as a scratch or local molecular dislocation, may provide the environment necessary for the beginning of a pit.

Exfoliation corrosion is especially prevalent in aluminum alloys. The grain structure of the metal determines whether exfoliation corrosion will occur. In this form of corrosion, degradation propagates below the surface of the metal. Corrosion products in layers below the metal surface cause flaking of the metal.

Selective leaching occurs when a particular constituent of an alloy is removed. Selective leaching occurs in aqueous environments, particularly acidic solutions. Graphitization and dezincification are two common forms of selective leaching. Dezincification is the selective removal of zinc from alloys containing zinc, particularly brass. The mechanism of dezincification of brass involves dissolving the brass with subsequent plating back of copper while zinc remains in solution. Graphitization is the selective leaching of iron or steel from gray cast irons.

Intergranular corrosion occurs selectively along the grain boundaries of a metal. This is an electrochemical corrosion in which potential differences between grain boundaries and the grain become the driving force. Even with relatively pure metals of only one phase, sufficient impurities can exist along grain boundaries to allow for intergranular corrosion. Intergranular corrosion is generally not visible until the metal is in the advanced stages of deterioration. These advanced stages appear as rough surfaces with loose debris (dislodged grains). Welding can cause local crystal graphic changes, which favor intergranular corrosion. It is especially prevalent near welds.

Stress-corrosion cracking is an especially dangerous form of corrosion. It occurs when a metal under a constant stress (external, residual, or internal) is exposed to a particular corrosive environment. The effects of a particular corrosive environment vary for different metals. For example, Inconel-600 exhibits stress-corrosion cracking in high-purity water with only a few parts per million of contaminants at about 300°C. The stress necessary for this type of corrosion to occur is generally of the residual or internal type. Most external stresses are not sufficient to induce stress-corrosion cracking. Extensive cold working or the presence of a rivet are common stress providers. Corrosion products also can build up to provide stress sufficient to cause stress-corrosion cracking. The damage done by stress-corrosion cracking is not obvious until the metal fails. This aspect of stress-corrosion cracking makes it especially dangerous.

Corrosion fatigue is caused by the joint action of cyclically applied stresses and a corrosive medium (generally aqueous). Metals will fail due to cyclic application of stress (fatigue). The presence of an aqueous corrosive environment causes such failure more rapidly. The frequency of the applied stress affects the rate of degradation in corrosion fatigue. Ordinary fatigue is generally not frequently dependent. Low-frequency applied stresses cause more rapid corrosion rates. Intuitively, low frequencies cause extended contact time between cracks and the corrosive medium. Generally, the cracks formed are transgranular.

Hydrogen blistering is caused by bubbling of a metal surface due to absorbed hydrogen. Monatomic hydrogen can diffuse through metals, whereas diatomic hydrogen cannot. Ionic hydrogen generated by chemical processes (such as electrolysis or corrosion) can form monatomic hydrogen at a metal surface. This hydrogen can diffuse through the metal and combine on the far side of the metal, forming diatomic hydrogen. The diffusion hydrogen also can combine in voids in the metal. Pressure within the void increases until the void actually grows (visibly apparent as a blister) and ultimately ruptures, leading to mechanical failure.

Hydrogen embrittlement is due to the reaction of diffused hydrogen with a metal. Different metals undergo specific reactions, but the result is the same. Reaction with hydrogen produces a metal that is lower in strength and more brittle.

Decarburization results from hydrogen absorption from gas streams at elevated temperatures. In addition to hydrogen blistering, hydrogen can remove carbon from alloys. The particular mechanism depends to a large extent on the properties of other gases present. Removal of carbon causes the metal to lose strength and fail.

Grooving is a type of corrosion particular to environmental conditions where metals are exposed to acid-condensed phases. For example, high concentrations of carbonates in the feed to a boiler can produce steam in the condenser to form acidic condensates. This type of corrosion manifests as grooves along the surface following the general flow of the condensate.

Biological corrosion involves all corrosion mechanisms in which some living organism is involved. Any organism, from bacteria and fungi to mussels, which can attach itself to a metal surface can cause corrosion. Biological processes may cause corrosion by producing corrosive agents, such as acids. Concentration gradients also can be caused by localized colonies of organisms. Some organisms remove protective films from metals, either directly or indirectly, leaving the actual metal surface vulnerable to corrosion. By selective removal of products of corrosion, biological organisms also can cause accelerated corrosion reactions. There are also some bacteria that directly digest certain metals such as iron, copper, or aluminum. Microorganisms also may promote galvanic corrosion by removing hydrogen from the surface of a metal, causing a potential difference to be created between different parts of the metal.

Stray current corrosion is an electrolytic degradation of a metal caused by unintentional electrical currents. Bad grounds are the most prevalent causes. The corrosion is actually a typical electrolysis reaction.

Gaseous corrosion is a general form of corrosion whereby a metal is exposed to a gas (usually at elevated temperatures). Direct oxidation of a metal in air is the most common cause. Cast-iron growth is a specific form of gaseous corrosion in which corrosion products accumulate onto the metal surface (and particularly at grain boundaries) to the extent that they cause visible thickening of the metal. The entire metal thickness may succumb to this before loss of strength causes failure.

Tuberculation occurs in aqueous solutions. Mounds form over metal surfaces providing for concentration differences, favorable environments for biological growth, and an increase in acidity leading to hydrogen formation.

Deposit attack occurs when there is nonuniform deposition of a film on a metal surface. The most common form appears as unequal scale deposits in an aqueous environment. Unequal film provides for concentration cells, which degrade the metal by galvanic means.

Impingement is corrosion caused by aerated water streams constricting metal surfaces. It is similar to erosion corrosion in which air bubbles take the place of particles. The pits formed by impingement attack have a characteristic teardrop shape.

Liquid metal corrosion occurs when a metal is in contact with a liquid metal. The main type of corrosion with highly pure liquid metals is simple solution. The solubility of the solid metal in the liquid metal controls the rate of damage. If a temperature gradient exists, a much more damaging form of corrosion takes place. Metal dissolves from the higher-temperature zone and crystallizes out in the colder zone. Transfer of solids to liquid metal is greatly accelerated by thermal gradients. If two dissimilar metals are in contact with the same liquid metal, the more soluble metal exhibits serious corrosion. The more soluble metal dissolves along with alloys from the less soluble metal. Metal in solutions may move by gross movement of the liquid metal or by diffusion. Depending on the system, small amounts of impurities may cause corrosive chemical reactions.

High-temperature corrosion is induced by accelerated reaction rates inherent in any temperature reaction. One phenomenon that occurs frequently in heavy oil-firing boilers is layers of different types of corrosion on one metal surface.

Causes of corrosion are the subject of extensive investigation by industry. Almost any type of corrosion can manifest itself under widely differing operating conditions. Also, different types of corrosion can occur simultaneously. It is not uncommon to see crack growth from stress corrosion to be accelerated by crevice corrosion, for example.

MATERIALS EVALUATION AND SELECTION

Materials evaluation and selection are fundamental considerations in engineering design. If done properly in a systematic manner, considerable time and cost can be saved in design work, and design errors can be avoided.

The design of any apparatus must be unified and result in a safe, functional system. Materials used for each apparatus should form a well-coordinated and integrated entity, which should not only meet the requirements of the equipment's functional utility, but also those of safety and product purity.

Materials evaluation should be based only on actual data obtained at conditions as close as possible to intended operating environments. Prediction of a material's performance is most accurate when standard corrosion testing is done in the actual service environment. Often it is extremely difficult in laboratory testing to expose a material to all of the impurities that the apparatus actually will contact. In addition, not all operating characteristics are readily simulated in laboratory testing. Nevertheless, there are standard laboratory practices that enable engineering estimates of the corrosion resistance of materials to be evaluated.

Environmental service is one of the most critical factors to consider. It is necessary to simulate as closely as possible all constituents of the service environment in their proper concentrations. Sufficient amounts of corrosive media, as well as contact time, must be provided for test samples to obtain information representative of material properties degradation. If an insufficient volume of corrosive media is exposed to the construction material, corrosion will subside prematurely.

The American Society for Testing Materials (ASTM) recommends 250 ml of solution for every square inch of area of test metal. Exposure time is also critical. Often it is desirable to extrapolate results from short time tests to long service periods. Typically, corrosion is more intense in its early stages (before protective coatings of corrosion products build up). Results obtained from short-term tests tend to overestimate corrosion rates which often results in an overly conservative design.

Immersion into the corrosive medium is important. Corrosion can proceed at different rates, depending on whether the metal is completely immersed in the corrosive medium, partially immersed, or alternately immersed and withdrawn. Immersion should be reproduced as closely as possible, since there are no general guidelines on how this affects corrosion rates.

Oxygen concentration is an especially important parameter to metals exposed to aqueous environments. Temperature and temperature gradients should also be reproduced as closely as possible. Concentration gradients in solutions also should be reproduced closely. Careful attention should be given to any movement of the corrosive medium. Mixing conditions should be reproduced as closely as possible.

The condition of the test metal is important. Clean metal samples with uniform finishes are preferred. The accelerating effects of surface defects lead to deceptive results in samples. The ratio of the area of a defect to the total surface area of the metal is much higher in a sample than in any metal in service. This is an indication of the inaccuracy of tests made on metals with improper finishes. The sample metal should have the same type of heat treatment as the metal to be used in service. Different heat treatments have different effects on corrosion. Heat treatment may improve or reduce the corrosion resistance of a metal in an unpredictable manner. For the purpose of selectivity, a metal stress corrosion test may be performed. General trends of the performance of a material can be obtained from such tests; however, it is difficult to reproduce the stress that actually will occur during service.

For galvanic corrosion tests, it is important to maintain the same ratio of anode to cathode in the test sample as in the service environment.

Evaluation of the extent of corrosion is no trivial matter. The first step in evaluating degradation is the cleaning of the metal. Any cleaning process involves removal of some of the substrate. In cases in which corrosion products are strongly bound to the metal surface, removal causes an inaccurate assessment of the degradation due to surface loss from the cleaning process. Unfortunately, corrosion assessments involving weight-gain measurements are of little value. It is rare for all of the corrosion products to adhere to a metal. Corrosion products that flake off cause large errors in weight-gain assessment schemes.

The most common method of assessing corrosion extent involves determining the weight loss after careful cleaning. Weight loss is generally considered a linear loss by conversion. Sometimes direct measurement of the sample thickness is made. Typical destructive testing methods are used to evaluate the loss of mechanical strength. Aside from the inherent loss of strength due to loss of cross section, changes brought about by corrosion may cause loss of mechanical strength. Standard tests for tensile strength, fatigue, and impact resistance should be run on test materials.

There are several schemes for nondestructive evaluation. Changes in electrical resistance can be used to follow corrosion. Radiographic techniques involving X-rays and gamma rays have been applied. Transmitted radiation as well as backscatter radiation have been used.

Radiation transmission methods, in which thickness is determined by (measured as) the shadow cast from a radioactive source, are limited to pieces of equipment small enough to be illuminated by small radioactive sources. There are several schemes for highlighting cracks. If the metal is appropriate, magnetic particles can be used to accentuate cracks. Magnetic particles will congregate along cracks too small to be seen normally. An alternate method involves a dye which can be used to soak into cracks preferentially.

General criteria for parallel evaluation of various materials that can assist in proper material selection are listed in general guidelines that can assist in material selection:

1. Select materials based on their functional suitability to the service environment. Materials selected must be capable of maintaining their function safely and for the expected life of the equipment at reasonable cost.
2. When designing apparatus with several materials, consider all materials as an integrated entity. More highly resistant materials should be selected for the critical components and for cases in which relatively high fabrication costs are anticipated. Often, a compromise must be made between mechanically advantageous properties and corrosion resistance.
3. Thorough assessment of the service environment and a review of options for corrosion control must be made. In severe, humid environments it is sometimes more economical to use a relatively cheap structural material and apply additional protection, rather than use costly corrosion-resistant ones. In relatively dry environments many materials can be used without special protection, even when pollutants are present.
4. The use of fully corrosion-resistant materials is not always the best choice. One must optimize the relation between capital investment and cost of subsequent maintenance over the entire estimated life of the equipment.
5. Consideration should be given to special treatments that can improve corrosion resistance (for example, special welding methods, blast peening, stress relieving, metallizing, sealing of welds). Also, consideration should be given to fabrication methods that minimize corrosion.
6. Alloys or tempers chosen should be free of susceptibility to corrosion and should meet strength and fabrication requirements. Often a weaker alloy must be selected than one that cannot be reliably heat treated and whose resistance to a particular corrosion is low.
7. If, after fabrication, heat treatment is not possible, materials and fabrication methods must have optimum corrosion resistance in their as-fabricated form. Materials that are susceptible to stress-corrosion cracking should not be employed in environments conducive to failure. Stress relief alone does not always provide a reliable solution.
8. Materials with short life expectancies should not be combined with those of long life in nonseparable assemblies.
9. For equipment in which heat transfer is important, materials prone to scaling or fouling should not be used.
10. For service environments in which erosion is anticipated, the wall thickness of the equipment should be increased. This thickness allowance should insure that various types of corrosion or erosion do not reduce the apparatus wall thickness below that required for mechanical stability of the operation. Where thickness allowance cannot be provided, a proportionally more resistant material should be selected.
11. Nonmetallic materials should have the following desirable characteristics: low moisture absorption, resistance to microorganisms, stability through

Materials Evaluation and Selection 35

temperature range, resistance to flame and arc, freedom from outgassing, resistance to weathering, and compatibility with other materials.

12. Fragile or brittle materials whose design does not provide any special protection should not be used under corrosion-prone conditions.

Thorough knowledge of both engineering requirements and corrosion control technology is required in the proper design of equipment. Only after a systematic comparison of the various properties, characteristics, and fabrication methods of different materials can a logical selection be made for a particular design. Tables 2-1 through 2-5 can assist in this analysis. Table 2-1 lists general

TABLE 2-1 PARAMETERS TO ANALYZE IN MATERIALS SELECTION

Metals	Nonmetals
General Physical Characteristics	
1) Chemical composition (%)	1) Anisotropy characteristics (main and cross-direction)
2) Contamination of contents by corrosion products	2) Area factor (in.^{2}/lb/mil)
3) Corrosion characteristics in:	3) Burn rate (in./min)
Atmosphere	4) Bursting strength (Mullen points)
Water	5) Change in linear dimensions @ 100°C for 30 min. (%)
Soil	6) Clarity
Chemicals	7) Color
Gases	8) Creep characteristics @ temperature range—creep apparent modulus (10^6 lb$_f$/in.^{2}).
Molten metals	
4) Creep characteristics @ temperature range	
5) Crystal structure	9) Crystal structure
6) Damping coefficient	10) Crystalline melting point
7) Density (g/cm^3)	11) Damping coefficient
8) Effect of cold working	12) Decay characteristics in:
9) Effect of high temperature on corrosion resistance	Atmosphere
	Alcohols
10) Effect on strength after exposure to:	Chemicals
Hydrogen	Gases
High temperatures	High relative humidity
11) Electrical conductivity (mho/cm)	Hydraulic oils
12) Electrical resistivity (Ω/cm)	Hydrocarbons
13) Fire resistance	Solvents
14) Hardenability	Sunlight
15) Maximum temperature not affecting strength (°C)	Water

(continued)

TABLE 2-1 *Continued*

Metals	Nonmetals

General Physical Characteristics

Metals	Nonmetals

16) Melting point (°C)
17) Corrosion factor (rapidity of corrosion)
18) Susceptibility to corrosion:
   General
   Hydrogen damage
   Pitting
   Galvanic
   Corrosion fatigue
   Fretting
   Stress corrosion cracking
   Corrosion/erosion
   Cavitation damage
   Intergranular
   Selective attack
   High temperature
19) Thermal coefficient of expansion ($°C^{-1}$)
20) Thermal conductivity (W/m °C)
21) Wearing quality:
   Inherent
   Via heat treatment
   Via plating
	13) Deflection temperature (°C)
	   264 ($lb_f/in.^2$) fiber stress
	   66 ($lb_f/in.^2$) fiber stress
	14) Density (g/cm^3)
	15) Dielectric constant
	16) Dielectric strength: short time/step-by-step
	17) Dissipation factor (1 MΩ)
	18) Effect on decay from: high temperature/low temperature/exposure to heat
	19) Electrical loss factor (1 MΩ)
	20) Electrical resistivity
	   arc/sec
	   insulation (96 hρ 90% RH and 35°C)MΩ
	21) Combustion properties/fire resistance
	22) Flammability
	23) Fillers
	24) Gas permeability (cm^3/100 $in.^2$/mil thick/24 hr/atm at 25°C): CO_2, H_2, N_2, O_2
	25) Heat distortion temperature at 264 $lb_f/in.^2$ (°F)
	26) Thermal coefficient of expansion ($in.^{-1}$ °F)
	27) Thermal conductivity (Btu/ft^2 h °F $in.^{-1}$)
	28) Light transmission, total white (%)
	29) Maximum service temperature (°C)
	30) Melt index (dg/min.)
	31) Minimum and maximum temperatures not affecting strength (°C)
	32) Softening temperature (°C)
	33) Stiffness—Young's modulus
	34) Susceptibility to various forms of deterioration:
	   General
	   Cavitation/erosion
	   Erosion
	   Fatigue
	   Fouling

Materials Evaluation and Selection 37

TABLE 2-1 *Continued*

Metals	Nonmetals

General Physical Characteristics

	Galvanic (metal-filled plastics)
	Impingement
	Stress cracking and crazing
	35) Thermal conductivity (W/m°C)
	36) Wearing quality:
	Inherent
	Via treatment

Strength and Mechanical Characteristics

Metals	Nonmetals
1) Bearing ultimate (N/mm^2)	1) Abrasion resistance
2) Complete stress-strain curve for tension and compression	2) Average yield (lb$_f$/in.^{2})
3) Compression modulus of elasticity (kg/mm^2)	3) Bonding strength (lb/thickness)
4) Fatigue properties	4) Brittleness
5) Hardness (Vickers)	5) Bursting pressure (lb$_f$/in.^{2})
6) Impact properties (Charpy kg/cm^2 @ 20°C):	6) Compressive strength:
Notch sensitivity	Flatwise (lb$_f$/in.^{2})
Effect of low temperature	Axial (lb$_f$/in.^{2})
Maximum transition temperature (°C)	at 10% deflection (lb$_f$/in.^{2})
7) Poisson's ratio	7) Deformation under load
8) Response to stress-relieving methods	8) Elongation (%)
9) Shear modulus of elasticity (kg/mm^2)	9) Elongation at break (%)-75°F (24°C)
10) Shear ultimate (Pa)	10) Fatigue properties
11) Tension modulus of elasticity (Pa)	11) Flexibility and flex life
12) Tension-notch sensitivity	12) Flexural strength (N/mm^2)
13) Tension yield	13) Hardness (Rockwell)
	14) Impact strength, Izod (ft lb^{-1} in.$^{-1}$ notch)
	15) Inherent rigidity
	16) Modulus of elasticity (lb$_f$/in.^{2} or kg/mm^2)
	In compression
	In flexure
	In tension
	In shear
	17) Resistance to fatigue
	18) Safe operating temperature (°C)
	19) Shear ultimate (Pa)
	20) Tear strength:
	Propagating (g/mil)
	Initial (lb/in.)
	21) Tensile strength (lb$_f$/in.^{2} or kg/mm^2)
	22) Vacuum collapse temperature

TABLE 2-2 FABRICATION PARAMETERS TO ANALYZE IN MATERIALS SELECTION

General Subject	Parameter
Metals	
Brazing and soldering	Compatibility
	Corrosion effect
	Flux and rod
Formability at elevated and room temperature	Aging characteristics
	Annealing procedure
	Corrosion effect of forming
	Heat treating characteristics
	Quenching procedures
	Sensitivity to variation
	Tempering procedure
	Effect of heat on prefabrication treatment
Formability in annealed and tempered states	Apparatus stress × local stream curve
	Characteristics in:
	Bending
	Dimpling
	Drawing
	Joggling
	Shrinking
	Stretching
	Corrosion effect of forming
	Elongation × gauge length
	Standard hydropress specimen test
	True stress-strain curve
	Uniformity of characteristics
Machinability	Best cutting speed
	Corrosion effect of:
	Drilling
	Milling
	Routing
	Sawing
	Shearing
	Turning
	Fire hazard
	Lubricant or coolant
	Material and shape of cutting tool
	Quality suitability for:
	Drilling
	Routing
	Milling

Materials Evaluation and Selection

TABLE 2-2 *Continued*

General Subject	Parameter
	Metals
	Sawing
	Shearing
	Turning
Protective coating	Anodizing
	Cladding
	Ecology
	Galvanizing
	Hard surfacing
	Metallizing
	Need of application for:
	Storage
	Processing
	Service
	Paint adhesion and compatibility
	Plating
	Prefabrication treatment
	Sensitivity to contaminants
	Suitability
	Type surface preparation
Quality of finish	Appearance
	Cleanliness
	Grade
	Honing
	Polishing
	Surface effect
Weldability	Arc welding
	Atomic hydrogen welding
	Corrosion effect of welding
	Cracking tendency
	Prefabrication treatment effects
	Electric flash welding
	Flux
	Friction welding
	Heat zone effect
	Heli-arc welding
	Pressure welding
	Spot welding

(continued)

TABLE 2-2 *Continued*

General Subject	Parameter
Metals	
	Torch welding
	Welding rod
Torch cutting	Cutting speed
Nonmetals	
Molding and injection	Compression ratio
	Compression molding pressure ($lb_f/in.^2$)
	Compression molding temperature (°C)
	Injection molding pressure ($lb_f/in.^2$)
	Injection molding temperature (°C)
	Molding qualities
	Mold (linear) shrinkage (in./in.)
	Specific volume (lb^3)
Lamination	Lamination pressure ($lb_f/in.^2$)
	Lamination temperature (°C)
Formation at elevated temperatures	
Machinability	Adverse effects of:
	Drilling
	Milling
	Sawing
	Shearing
	Turning
	Best cutting speed
	Fire hazard
	Machining qualities
	Material and shape of cutting tool
Protective coating	Cladding
	Painting
	Plating
	Sensitivity to contaminants
	Suitability
	Type surface preparation
Quality of finish	Appearance
	Cleanliness
	Grade

TABLE 2-2 Continued

General Subject	Parameter
Nonmetals	
	Polishing
	Surface and effect
Joining	Adhesive joining
	Bonding
	Cracking tendency
	Heat zone effect
	Welding

physical and material characteristics, as well as characteristics of strength, that should be considered when comparing different metals and/or nonmetals for a design. Table 2-2 is a listing of fabrication parameters that should be examined in the materials comparison process. In addition to the characteristics listed in Tables 2-1 and 2-2, an examination of design limitations and economic factors must be made before optimum material selection is accomplished. Design limitations or restrictions of materials might include:

- size and thickness
- velocity
- temperature
- composition of constituents
- bimetallic attachment
- geometric form
- static and cyclic loading
- surface configuration and texture
- special protection methods and techniques
- maintainability
- compatibility with adjacent materials

Economic factors that should be examined may be divided into three categories: (1) availability, (2) cost of different forms, and (3) size limitations and tolerances. More specifically, these include:

1. Availability
   - in required quantities (single, multiple, limited, unlimited)
   - in different forms (bar, casting such as sand, centrifugal, die, permanent mold, and so on, extrusion, forging, impact extrusion, pressing, sintered, powder pressing)

TABLE 2-3 GENERAL PROPERTIES OF THE CORROSION RESISTANCE OF METALS TO VARIOUS CHEMICALS [11]

R = Recommended
M = Moderate Service
L = Limited Service
U = Unsatisfactory
Blank = No information

	Carbon Steel	Cast Iron and Ductile Iron	304 Stainless Steel	316 Stainless Steel	347 Stainless Steel	Nickel-Resist Iron	Carpenter 20; Durimet 20	Worthite	Durriron-Durichlor	Monel	Inconel	Hastelloy B	Hastelloy C	Hastelloy D	Chlorimet 3	Aluminum and Alloys	Copper and Cu Alloys	Brass	Lead	Nickel
Aluminum Chloride	U	U	U	U	U	L	R	U	R	U	L	R	M	R	R	U	L	L	U	L
Aluminum Hydroxide	U		R	R	R		R	R	M	M	R	M	R	R	R	R	R	R	R	R
Aluminum Sulfate	U	U	L	L	L	L	R	R	R	L	L	R	R	R	R	R	L	L		L
Alums, Dilute	U		R	R	R	R	R	R	R	R	R	R	R	R	R	L	L	L		R
Amines (various)	R	R	R	R	R	R	R	R	R	R	R	R	R	R	R					R
Ammonia Gas	R	M	R	R	R	R	R		R	M	M	R	R	M	R	U	U	L	L	R
Ammonium Carbonate	M	U	M	M	R	R	R	R	R	U	M	M	M	M	R	L	L	U	M	R
Ammonium Chloride	U	R	U	L	L	R	R	R	M	R	R	L	L	U	R	L	L	U	L	M
Ammonium Hydroxide	U	R	R	R	R	L	R	R	R	U	L	U	R	R	R	U	U	U	M	U
Ammonium Nitrate	U		R	R	R	R	R	R	R	R	M	M	M	R	R					
Ammonium Sulfate	R		L	L	M	R	R	R	R	M	R	R	M	R	R	L	L	L	L	M
Benzene	M	M	R	R	R	R	R	R	R	M	R	M	M	M	R	R	R	R	M	M
Calcium Carbonate	R		R	R		R	R		R	R	R	R	R	R	R	U	R	R	R	R
Calcium Chlorate			L	L	U		M	L	M	M	M	L	R	M	M	L	L	L		R
Calcium Chloride	R	R	R	R	L	L	M	R	R	M	M	M	R	M	R	L	M	M	U	R

Chemical	1	2	3	4	5	6	7	8	9	10	11	12	13	14	15	16	17
Calcium Hydroxide	R	L	L	R	R	R	R	R	R	R	M	R	L	M	L	L	R
Calcium Hypochlorite	L	U	U	U	U	U	L	R	R	U	R	U	L	U	U	U	U
Calcium Sulfate	L	R	R	R	R	R	R	R	M	M	M	R	M	R	R	R	R
Carbon Dioxide (dry)	R	L	R	R	R	R	R	R	R	R	R	R	L	L	R	L	R
Carbon Dioxide (wet)	L	L	R	R	R	R	R	R	R	R	L	L	L	L	L	L	R
Chlorine (wet)	U	U	U	U	U	U	U	L	U	U	L	U	U	U	L	L	L
Chromic Acid Solution	L	M	U	L	L	L	M	R	R	D	M	L	M	U	U	M	U
Copper Chloride	U	U	U	U	U	U	U	R	U	D	L	M	U	U	U	U	U
Copper Sulfate	U	U	U	R	R	R	R	R	R	L	R	R	R	R	U	L	L
Fatty Acids	U	R	R	R	R	R	R	R	L	M	M	L	R	L	R	R	R
Ferrous Chloride	U	U	U	U	U	U	R	R	R	U	R	M	U	U	D	D	U
Ferrous Sulfate	U	M	R	M	R	R	R	M	M	D	M	M	U	U	R	R	R
Hydrochloric Acid (conc.)	U	U	L	U	U	U	U	R	L	L	L	L	U	U	U	U	U
Hydrochloric Acid (dilute)	U	U	L	U	U	L	L	R	L	L	L	R	D	D	D	D	D
Hydrogen Chloride (dry gas)	R	R	R	R	R	R	M	R	L	M	M	L	D	D	D	D	D
Hydrofluoric Acid	L	L	M	L	M	R	M	U	M	M	M	R	U	U	D	D	D
Hydrocarbons (aliphatic)	R	R	R	R	R	R	M	R	L	L	L	R	R	R	D	R	R
Hydrogen Peroxide (conc.)	U	M	L	M	L	L	L	L	L	L	R	L	U	U	U	L	L
Hydrogen Sulfide (dry)	M	L	M	L	L	L	R	L	L	L	M	L	D	D	D	D	D
Hydrogen Sulfide (wet)	L	L	L	L	L	L	M	R	L	L	L	M	U	D	D	R	R
Nitrating Acid (>15% H₂SO₄)	R	U	R	R	R	R	R	R	R	R	R	R	U	D	D	U	U
Nitrating Acid (<15% H₂SO₄)	U	R	R	R	R	R	R	R	R	R	R	R	D	D	D	R	D
Nitrating Acid (<15% HNO₃)	U	U	L	U	U	U	L	R	U	L	M	L	U	U	D	D	D
Nitrating Acid (<1% acid)	U	U	M	L	U	U	U	R	U	U	R	L	U	U	D	U	D
Nitric Acid (conc.)	U	U	R	R	R	R	R	R	R	R	R	R	U	R	U	U	U
Nitric Acid (dilute)	U	U	R	R	R	R	R	R	R	R	R	R	R	R	R	D	D
Nitrous Acid	M	M	L	L	L	L	R	M	R	L	M	R	L	U	L	L	M
Phenol (conc.)	U	U	U	U	L	L	U	R	U	U	L	L	U	U	D	M	M
Phosphoric Acid (100%)	D	U	U	U	U	U	U	R	U	U	L	U	U	U	U	M	M
Phosphoric Acid (hot >45%)	U	U	U	U	U	U	U	R	U	U	R	U	U	U	U	D	U

(continued)

TABLE 2-3 Continued

R = Recommended
M = Moderate Service
L = Limited Service
U = Unsatisfactory
Blank = No information

	Carbon Steel	Cast Iron and Ductile Iron	304 Stainless Steel	316 Stainless Steel	347 Stainless Steel	Nickel-Resist Iron	Carpenter 20; Durinet 20	Worthite	Durrion-Durichlor	Monel	Inconel	Hastelloy B	Hastelloy C	Hastelloy D	Chlorimet 3	Aluminum and Alloys	Copper and Cu Alloys	Brass	Lead	Nickel
Phosphoric Acid (cold >45%)	U	U	L	R	R	L	R	L	R	L	L	M	R	M	R	U	U	U	M	L
Phosphoric Acid (<45%)	U	U	L	R	R	L	R	L	R	L	L	M	R	M	R	L	U	U	M	L
Potassium Carbonate	L		R	R		L	R	R	R	R	M	R	R	R	R	U	L	L	R	R
Potassium Chlorate	R		R	L		R	R	L	R	M	M	U	R	U	R	M	L	L		M
Potassium Chloride	R	M	L	L	M		R	R	R	M	M	R	R	R	R	L	L	L		M
Potassium Permanganate	M	M	M	M	R		R	R	R	L	M	U	R	U	R	R	R	R		M
Sodium Bicarbonate	L	U	R	R	R	R	R	R	R	R	R	R	R	R	R	R	L	L	R	R
Sodium Bisulfate	U		L	R			R	R	L	M	L	R	R	R	R	U	L	L	R	
Sodium Bisulfite			R	R	R	R	R	R	L	M	L	M	M	U	R	R	L	L	L	M
Sodium Carbonate	M	M	R	R	R	R	R	R	R	M	M	M	R	L	R	U	M	M	R	M
Sodium Chlorate	L	M	R	L	L	R	L	L	R	M	M	U	R	U	R	M	M	M		M
Sodium Chloride	M	L	L	R	R	L	R	R	M	R	L	M	M	L	R	L	L	L	M	M
Sodium Hydroxide (conc.)	L	M	L	R	M	R	R	R	M	R	R	R	R	R	R	U	R	L	M	L
Sodium Hydroxide (dilute)	R		L	M		R	R	R	L	L	L	L	R	L	R	R	L	R	U	R
Sodium Hydrosulfite	R	R	M	M			R	R	R	U	U	L	R	L	R	R	L	L	U	U

44

Chemical																				
Sodium Hypochlorite	U		L	L	L	L	L	L		U	U	D	U	D	U	L	U	U	U	
Sodium Nitrate	R	U	R	R	R	R	R	R	R	R	M	R	M	R	M	R	U	M	M	
Sodium Phosphate	R	M	R	R	M	R	R	R	R	R	M	R	U	U	L	L	R	L	R	
Sodium Silicate	R		R	L	R	M	R	R	R	R	M	R	L	L	L	L	R	L	R	
Sodium Sulfate	R	R	L	L	L	L	R	R	R	R	M	R	L	L	L	R	R	R	R	
Sodium Sulfide	M	R	L	L	R	R	R	R	R	U	L	R	L	L	L	L	R	L	M	
Sodium Sulfite	M	R	L	L	L	L	R	R	R	R	R	R	L	L	L	L	R	L	L	
Stearic Acid	L	R	R	R	R	R	L	R	R	L	R	R	R	L	R	R	R	R	R	
Sulfur Dioxide (dry)	R	R	R	R	R	R	R	R	R	U	U	U	U	U	U	U	R	U	U	
Sulfur Dioxide (wet)	U		L	R	L		R	R	U	D							R			
Sulfur Trioxide	R		R	R	R	R	R	L	R	R	U	R	R	L	L	R	R	L	U	
Sulfuric Acid (fuming 98%)	L	L	U	D	L	R	L	U	L	U	L	U	U	L	U	U	L	R	U	
Sulfuric Acid (hot, conc.)	U		U	U	L	M	U	M	R	R	M	U	D	D	U	L	M	L	U	
Sulfuric Acid (cold, conc.)	M	M	R	R	L	R	R	R	R	R	R	R	U	U	D	R	R	R	U	
Sulfuric Acid (75-95%)	M	L	U	U	L	R	R	R	M	R	L	R	U	U	U	U	R	R	L	
Sulfuric Acid (10-75%)	U	U	D	D	L	R	L	L	L	L	L	R	L	D	D	D	R	R	R	
Sulfuric Acid (<10%)	U	U	D	L	L	L	L	L	L	R	L	R	D	D	D	D	R	L	R	
Sulfurous Acid	U	U	L	L	L	U	R	R	R	R	R	R	R	R	R	R	M	R	R	
Toluene	R	R	R	R	R	R	R	R	R	U	U	R	U	R	R	R	R	R	R	
Water (fresh)	R	R	R	R	R	R	R	R	R	R	R	R	R	R	R	R	M	L	R	
Water (distilled)	U	U	R	U	U	L	R	U	R	R	U	U	U	U	U	U	M	U	L	
Zinc Chloride	U	U	U	L	U	R	L	L	M	M	M	L	M	D	M	L	M	L	M	
Zinc Sulfate	U	U	M	M	M	R	R	R	R	M	M	M	M	M	M	M	R	M	M	

TABLE 2-4 GENERAL PROPERTIES OF THE CORROSION RESISTANCE OF NONMETALS TO VARIOUS CHEMICALS [11]

R = Recommended
M = Moderate Service
L = Limited Service
U = Unsatisfactory
Blank = No information

	Carbon Ceramics			Rubbers			Plastics															
	Carbon and Graphite	Glass (Pyrex®)	Chemical Porcelain	Natural Rubber	Neoprene	Butadiene	Asphaltic Bitumastic	Acrylic (Lucite, Plexiglas)	Polyethylene	Polyvinylchloride	Saran	Kel-F	Teflon®a	Penton	Polystyrene (Styron)	Haveg 41	Heresite	Molded Phenol Formaldehyde (Durez)	Epoxy Resins	Nylon	Durcon 6	Woods—Maple, Oak, Pine
Aluminum Chloride	R	R	R	R	R	R	R		R	R	R	R	R	R	R	R	R	R	R	R	R	
Aluminum Hydroxide	R	R	R	R	R	R	R	R	R	R	R	R	R	R	R	R	R	R	R	R	R	R
Aluminum Sulfate	R	R	R	R	R	R	R	R	R	R	R	R	R	R	R	R	R	R	R	R	R	R
Alums, Dilute	R	R	R	L	R	R		R	R	U	U	R	R	R	U	R		R		R		
Amines (various)	R	R	R																			
Ammonia Gas	R	R	R	L	R	R	R	R	R	L	U	R	R	R	U	R	R	R	R	R	R	
Ammonium Carbonate	R	R	R	R	R		L	R	R	R	R	R	R	R	R	R	R	R	R	R	R	
Ammonium Chloride	R	L	R	R	L	R	R	R	R	R	R	R	R	R	R	R	R	U	R	U	M	
Ammonium Hydroxide	R	R	R	L	L		L	R	R	R	U	R	R	R	R	L	R	U	R	R	R	
Ammonium Nitrate	R	R	R	L	L	R	R	R	R	R	R	R	R	R	R	R	R	R	R	R	R	
Ammonium Sulfate	R	R	R	R	R		R	R	L	R	L	R	R	M	M	L	R	R	R	U	U	
Benzene	R	R	R	U	U		U	U	L	U	L	R	R	R	U	L	R	R	R	R	R	

Chemical	C1	C2	C3	C4	C5	C6	C7	C8	C9	C10	C11	C12	C13	C14	C15
Calcium Carbonate	R	R	R	R	R	R	R	R	R	R	R	R	R	R	R
Calcium Chlorate	R	R	R	R	R	R	R	R	R	R	R	R	R	R	R
Calcium Chloride	R	R	R	R	R	R	R	R	R	R	R	R	R	U	R
Calcium Hydroxide	R	R	R	R	R	R	R	R	R	R	R	R	R	R	R
Calcium Hypochlorite	R	R	R	U	R	R	R	R	L	R	R	R	L	U	R
Calcium Sulfate	R	R	R	R	R	R	R	R	R	R	R	R	R	R	R
Carbon Dioxide (dry)	R	R	R	R	R	R	R	R	R	R	R	R	R	R	R
Carbon Dioxide (wet)	R	R	R	R	R	R	R	R	R	R	R	R	U	R	R
Chlorine (wet)	R	R	R	U	L	R	R	R	L	L	L	R	R	R	R
Chromic Acid Solution	U	R	R	U	R	U	L	R	L	L	U	L	U	U	U
Copper Chloride	R	R	L	R	R	R	R	R	R	R	R	R	R	R	R
Copper Sulfate	R	R	R	R	R	R	R	R	R	R	R	R	R	R	R
Fatty Acids	R	R	U	R	U	L	R	R	R	R	R	L	R	R	R
Ferrous Chloride	R	R	R	R	R	R	R	R	R	R	R	R	R	R	R
Ferrous Sulfate	R	R	R	R	R	R	R	R	R	R	R	R	R	M	R
Hydrochloric Acid (conc.)	R	R	L	L	R	R	R	L	L	R	R	R	R	R	R
Hydrochloric Acid (dilute)	R	R	L	R	R	R	R	L	R	R	R	L	R	R	U
Hydrogen Chloride (dry gas)	R	R	R	U	R	R	R	R	R	R	R	R	R	R	R
Hydrofluoric Acid	U	U	U	D	U	U	R	L	D	U	U	R	R	R	R
Hydrocarbons (aliphatic)	R	U	R	R	U	R	R	R	R	R	L	M			
Hydrogen Peroxide (conc.)	R	R	L	R	R	R	R	R	U	R	R	R	R	R	R
Hydrogen Sulfide (dry)	R	R	R	R	R	R	R	R	R	R		R	R		
Hydrogen Sulfide (wet)	R	R	R	R	R	R	R	R	R	R		R			
Nitrating Acid (>15% H₂SO₄)	U	R	U	U	U		R	R			U	U	U	R	R
Nitrating Acid (<15% H₂SO₄)	U	R	U	U	U		R	R			L	L	R	L	R
Nitrating Acid (<15% HNO₃)	U	R	U	U	U		R	R			L	L	R	L	R
Nitrating Acid (<1% acid)	U	R	L	L	L		R	R			L	U	R	U	M
Nitric Acid (conc.)	U	R	R	U	U	U	R	R		L	U	R	R	R	R

(continued)

TABLE 2-4 Continued

R = Recommended
M = Moderate Service
L = Limited Service
U = Unsatisfactory
Blank = No information

	Carbon Ceramics			Rubbers				Plastics														
	Carbon and Graphite	Glass (Pyrex®)	Chemical Porcelain	Natural Rubber	Neoprene	Butadiene	Asphaltic Bitumastic	Acrylic (Lucite, Plexiglas)	Polyethylene	Polyvinylchloride	Saran	Kel-F	Teflon®	Penton	Polystyrene (Styron)	Haveg 41	Heresite	Molded Phenol Formaldehyde (Durez)	Epoxy Resins	Nylon	Durcon 6	Woods—Maple, Oak, Pine
Nitric Acid (dilute)	L	R	R	U	U	R	U	R	L	R	R	R	R	M	R	L	R	L	U		R	
Nitrous Acid		R	R	L	L	R	R			L	L	R	R	R							R	
Phenol (conc.)	M	L	R	L	L	R	U		R	R	L	R	R	M	U	L	R	L	R	L	R	
Phosphoric Acid (100%)	R	L	U	L	R	R	R	R	L	R	R	R	R	R		R	R	U		L	R	
Phosphoric Acid (hot >45%)	R	R	R	L	L		R		R	R	R	R	R	R	R	R	R		R		R	
Phosphoric Acid (cold >45%)	R	R	L	R	L				R	L	R	R	R	R		R	R			U	R	
Phosphoric Acid (<45%)	R	R	R	R	R				R	R	R	R	R	R	R	R	R		R	R	R	
Potassium Carbonate	R	R	R	R	R		R		R	R	R		R	R	R			R	R	R	R	
Potassium Chlorate	R	R	R	R	R		U			R	L		R	R	R			R	R	U	R	
Potassium Chloride	R	R	R	R	R		R		R	R	R	R	R	R	R	R	R	R	R	R	R	
Potassium Permanganate	R	R	R	R	R		R		R	R	R		R	R	R	R		R	R	R	R	R
Sodium Bicarbonate	R	R	R	R	R				R	R	R		R	R	R			R	R	R	R	
Sodium Bisulfate	R	R	R	R	L					R	R		R	R	R			R	R	R	R	
Sodium Bisulfite	R	R	R	R	R				R	R	R	R	R	R	R		R	R	R	R	R	
Sodium Carbonate	R	L	R	R	R			U	R	R	R	R	R	R	R			R	R	R	R	M

48

Reagent	1	2	3	4	5	6	7	8	9	10	11	12	13	14
Sodium Chlorate	R	R	R	R	R	R	R	R	R	R	R	R	U	R
Sodium Chloride	R	R	R	R	R	R	R	R	R	U	R	R	R	R
Sodium Hydroxide (conc.)	R	L	U	R	R	R	U	R	R	L	U	R	R	U
Sodium Hydroxide (dilute)	L	L	L	R	R	R	L	R	R	R	R	R	R	
Sodium Hydrosulfite	R	R	R	R	R	R	R	R	R	R	R			
Sodium Hypochlorite	U	R	R	R	R	R	R	R	R	L	U	R	U	
Sodium Nitrate	R	R	R	R	R	L	R	R	R	R	U	R	R	
Sodium Phosphate	L	L	R	R	R	R	R	R	R	R	R	R		
Sodium Silicate	L	R	R	R	R	R	R	R	R	R	R	R		
Sodium Sulfate	R	R	R	R	R	R	R	R	R	R	R	R		
Sodium Sulfite	R	L	R	R	R	R	R	R	R	L	R	M		
Sodium Sulfide	R	R	R	U	R	R	R	R	R	R	R	R		
Stearic Acid	R	L	U	L	R	R	R	L	L	R	R	R		
Sulfur Dioxide (dry)	R	L	U	R	R	R	R	L	R	R	U	R		
Sulfur Dioxide (wet)	R	L	U	R	R	R	R	R						
Sulfur Trioxide	R	R	R	R	R	R	R	U	U	R				
Sulfuric Acid (fuming 98%)	U	R	U	R	R	R	U	U	L	U	D	U		
Sulfuric Acid (hot, conc.)	R	U					R	L	L	U	U			
Sulfuric Acid (cold, conc.)	U	R	R	L	R	R	L	M	L	L	L	L		
Sulfuric Acid (75-95%)	R	R	R	L	R	R	R	M	R	R	L	R		
Sulfuric Acid (10-75%)	R	R	R	R	R	R	R	R	R	R	L	L	L	R
Sulfuric Acid (<10%)	R	R	R	U	R	R	R	L	R	R	R	L	U	R
Sulfurous Acid	R	R	R	R	L	R	L	U	L	L	U	U	R	M
Toluene	R	R	R	R	L	R	R	R	R	R	R	R	R	R
Water (fresh)	R	R	R	R	R	R	R	R	R	R	R	R	R	R
Water (distilled)	M	R	R	R	R	R	R	R	R	R	R	R	R	R
Zinc Chloride	R	R	R	R	R	R	R	R	R	R	R	R	L	R
Zinc Sulfate	R	R	R	R	R	R	R	R	R	R	R	R	R	R

[a]Registered trademark of E. I. du Pont de Nemours and Company, Inc., Wilmington, Delaware.

TABLE 2-5 CORROSION RATES OF STEEL AND ZINC PANELS EXPOSED FOR TWO YEARS

		mils/yr		
No.	Location	Steel	Zinc	Environment[a]
1.	Norman Wells, NWT, Canada	0.06	0.006	R
2.	Phoenix, AZ	0.18	0.011	R
3.	Saskatoon, Sask., Canada	0.23	0.011	R
4.	Vancouver Island, BC, Canada	0.53	0.019	RM
5.	Detroit, MI	0.57	0.053	I
6.	Fort Amidor, Panama C.Z.	0.58	0.025	M
7.	Morenci, MI	0.77	0.047	R
8.	Ottawa, Ont., Canada	0.78	0.044	U
9.	Potter County, PA	0.81	0.049	R
10.	Waterbury, CT	0.89	0.100	I
11.	State College, PA	0.90	0.045	R
12.	Montreal, Que., Canada	0.94	0.094	U
13.	Melbourne, Australia	1.03	0.030	I
14.	Halifax, NS, Canada	1.06	0.062	U
15.	Durham, NH	1.08	0.061	R
16.	Middletown, OH	1.14	0.048	SI
17.	Pittsburgh, PA	1.21	0.102	I
18.	Columbus, OH	1.30	0.085	U
19.	South Bend, PA	1.32	0.069	SR
20.	Trail, BC, Canada	1.38	0.062	I
21.	Bethlehem, PA	1.48	0.051	I
22.	Cleveland, OH	1.54	0.106	I
23.	Miraflores, Panama C.Z.	1.70	0.045	M
24.	London (Battersea), England	1.87	0.095	I
25.	Monroeville, PA	1.93	0.075	SI
26.	Newark, NJ	2.01	0.145	I
27.	Manila, Philippine Islands	2.13	0.059	U
28.	Limon Bay, Panama C.Z.	2.47	0.104	M
29.	Bayonne, NJ	3.07	0.188	I
30.	East Chicago, IN	3.34	0.071	I
31.	Cape Kennedy, FL 1/2 mile	3.42	0.045	M
32.	Brazos River, TX	3.67	0.072	M
33.	Pilsey Island, England	4.06	0.022	IM
34.	London (Stratford), England	4.40	0.270	I
35.	Halifax, NS, Canada	4.50	0.290	I
36.	Cape Kennedy, FL 180 ft.	5.20	0.170	M
37.	Kure Beach, NC 800 ft.	5.76	0.079	M
38.	Cape Kennedy, FL 180 ft.	6.52	0.160	M
39.	Daytona Beach, FL	11.7	0.078	M
40.	Widness, England	14.2	0.400	I

Materials Evaluation and Selection

TABLE 2-5 *Continued*

		mils/yr		
No.	Location	Steel	Zinc	Environment[a]
41.	Cape Kennedy, FL 180 ft.	17.5	0.160	M
42.	Dungeness, England	19.3	0.140	IM
43.	Point Reyes, CA	19.8	0.060	M
44.	Kure Beach, NC 80 ft.	21.2	0.250	M
45.	Galetea Point, Panama C.Z.	27.3	0.600	M

[a] R rural   SI semi-industrial   M marine   IM industrial-marine
   RM rural-marine   SR semi-rural   I industrial   U   urban

- in metallized and pretreatment forms (galvanized, plastic coated, plated, prefabrication treated)
- in cladded forms
- uniformity of material
- freedom from defects
- delivery time

2. Cost in different forms—bar, shape, plate, sheet
   - casting (sand, centrifugal, die, permanent mold, and so on)
   - extrusion
   - Forging
   - impact extrusion
   - pressing
   - sintered
   - powder pressing

3. Size limitations and tolerances in different forms
   - gauge
   - length
   - weight
   - width

Tables 2-3 through 2-5 give general corrosion-resistance ratings of different materials. Table 2-3 lists various metals and Table 2-4 gives ratings for various nonmetals.

DESIGN GUIDELINES

Often complex equipment and systems, process piping arrangements, and even support structures utilize different metals, alloys, or other materials. These are often employed in corrosive or conductive environments and, in practice, the contact of dissimilar materials cannot be avoided totally. It is important that the designer minimize the damaging effects of corrosion by optimizing the compatibility of materials either by selection or arrangement in the overall design. Compatible materials are those that will not cause an uneconomic breakdown within the system, even though they are utilized together in a particular medium in appropriate relative sizes and compositions. In addition to material influences on each other by virtue of inherent or induced differences of electric potentiality, adverse chemical reactions can occur as a result of changes in materials caused by environmental variations. All these possibilities must be examined thoroughly by the designer.

The following general considerations should be followed in designing all types of process equipment:

1. Dissimilar metals should be in contact (either directly or by means of a conductive path such as water, condensation, and so on) only when the functional design so dictates.
2. Scales of galvanic potentials are useful indicators of galvanic corrosion; however, information is needed on the amount of current flowing between dissimilar metals.
3. To ensure compatibility, detailed engineering descriptions of all materials and their metallurgical properties are needed. General information (for example, mild steel) does not provide sufficient data to establish compatibility in conductive or corrosive media.
4. Galvanic corrosion of dissimilar metals can be minimized by controlling humidity near such bimetallic connections. In general, continuously dry bimetallic joints do not corrode.
5. Avoid faying surfaces of dissimilar metals by separating them completely. Examples of poor and proper connections are given in Figure 2-1. Note that dielectric separation can be provided in several manners, for example, insulating gaskets (synthetic rubber, PTFE, and so on), spreadable sealants, and coatings.
6. The formation of crevices between dissimilar metals should be avoided. Corrosion at such connections is generally more severe than either galvanic or crevice corrosion alone. Also, crevices between metals and certain types of plastics or elastomers may induce accelerated rates of combined crevice and chemical attack. Testing is recommended prior to establishing final design specifications.
7. Noble metals should be specified for major structural units or components, particularly if the design requires that these are smaller than adjoining

units. There is an unfavorable area effect of small anode and large cathode. Corrosion of a relatively small anodic area can be 100 to 1,000 times more severe than the corrosion of bimetallic components, which have the same area submerged in a conductive medium. Hence, less noble (anodic) components should be made larger or thicker to allow for corrosion. In addition, provision should be made for easy replacement of the less noble components.

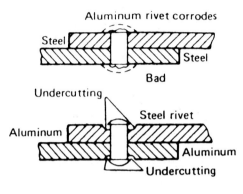

POOR DESIGN

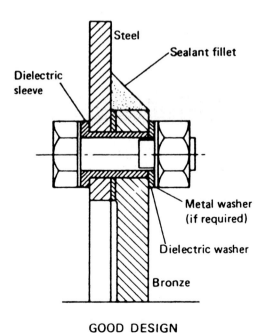

GOOD DESIGN

Figure 2-1 Examples of poor and proper connections of dissimilar metals

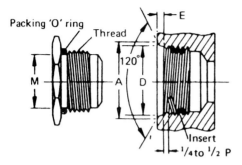

Figure 2-2 Example of a corrosion-resistant steel insert used in an aluminum casting

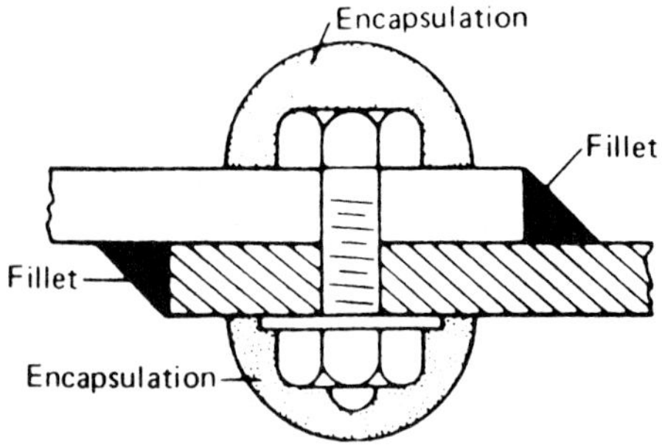

Figure 2-3 Encapsulation of exposed metal connections

8. Brazing or welding alloys should be more noble (that is, cathodic) than at least one of the joined metals. Also, these alloys should be compatible to both the other metals.
9. Fasteners made of dissimilar metal should be insulated completely from both metals of the joint (or at least the one that is least compatible with the metal of the fastener).
10. Clad metals are candidates for galvanic corrosion along exposed edges. An example is copper/aluminum clad to aluminum.
11. Proper system and sequences of welding attachment of bimetallic pads for structures and equipment should be specified to avoid distortion and input stresses.
12. In aluminum castings, integral corrosion-resistant steel inserts may be used. An example is shown in Figure 2-2.
13. Sources of mercury (for example, mercury thermometers) should be avoided in the vicinity of aluminum and copper alloy equipment.
14. Avoid coupling carbon or graphite components with other metals in conductive environments.
15. Designs that establish large temperature gradients in equipment resulting in adverse polarization of metals should be avoided.
16. If dielectric separation of fasteners in noncompatible joints cannot be implemented readily, fasteners should be coated with a zinc chromate primer and exposed ends encapsulated. This is illustrated in Figure 2-3.
17. Use sealing (encapsulating or enveloping type with shrinkable plastic) on bimetallic joints if geometrical arrangements prohibit access to such joints for replacement.

The following general guidelines are most applicable to piping system designs.

1. Ensure effective separation between piping sections of dissimilar metals. Examples of this are illustrated in Figure 2-4. As shown, dielectric nonab-

Design Guidelines

sorbent gaskets of adequate thickness can be inserted between dissimilar pipe connections. Note that graphite packings and gaskets should not be used for dielectric separation except for steam service or similar applications at elevated temperatures, as with nonconductive media.

2. Piping should not be directly attached to dissimilar metal structures via conductive materials.
3. Graphite and carbon packing should not be used in pipe systems containing conductive media upstream of heat exchangers and other critical equipment. Graphite particles can deposit onto tube bundles in heat exchangers and promote galvanic corrosion. Where possible, use insert seals and packing.
4. Avoid fitting copper alloy pipes upstream of carbon-steel equipment. Salts of carbon from copper-base pipes can dissolve in solution and pose problems to carbon-steel components and vessels downstream. If the use of copper alloy pipes is unavoidable, sacrificial sections of mild steel pipe can be inserted between such connections. These sacrificial sections should be easily accessible to enable replacement and thus should be provided with

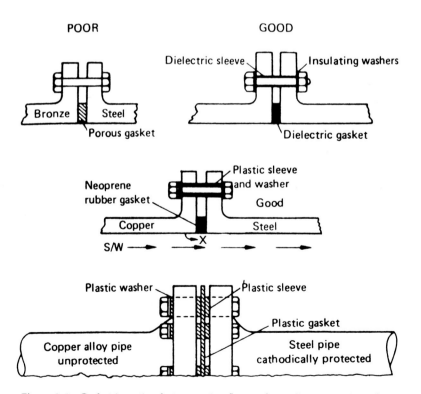

Figure 2-4 Gasket insertion between pipe flanges for sealing purposes and to minimize galvanic corrosion between dissimilar piping metals

adequate wall thickness to meet a well-planned maintenance program frequency.

5. Pickling and passivation of Monel and stainless steel pressure vessels should be specified to prevent deep pitting.
6. In situations in which piping protrudes partitions or bulkheads of dissimilar metals, proper precautions should be taken against galvanic corrosion. Possible solutions include the use of dielectric gaskets or sleeves and the use of plastic adhesive tapes. Examples are illustrated in Figure 2-5.
7. In buried pipeline installations, avoid contact of piping with structures of dissimilar metals. Also, where possible, specify uniform quality, grade,

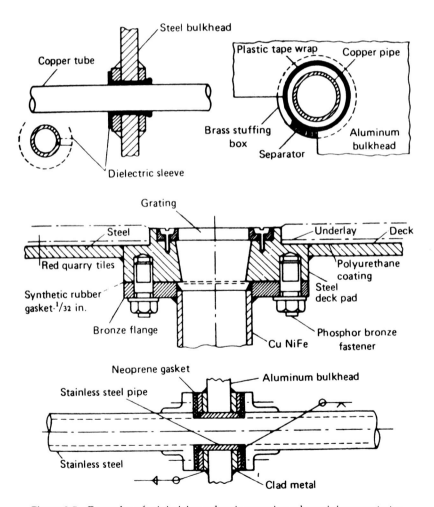

Figure 2-5 Examples of minimizing galvanic corrosion when piping penetrates partitions and bulkheads

Design Guidelines

and surface conditions. Various quality sections should not be welded together in buried installations.

8. Tinning of copper piping or components is a good approach toward minimizing galvanic action between dissimilar metals.
9. Heat exchangers that utilize copper coils are potential candidates for galvanic corrosion due to dissolved copper salts interacting with the galvanized steel shell. This problem can be avoided by nickel plating the coils. The coils then can be separated from direct contact with the vessel via insulation. Also, it is preferable to conduct the water on the tube side of heat exchangers.

The preceding factors represent considerations that the design engineer must account for to ensure compatibility between components and equipment materials. In addition to these, there are geometric considerations that can minimize corrosion problems if accounted for in design. The following are general guidelines pertaining to geometry in a design aimed at minimizing corrosion. The overall design approach involves the selection of the optimum geometry for a piece of equipment that is less likely to undergo certain types of corrosion, either directly or indirectly. Such shapes, forms, combinations of forms, and their method of attachment along with their fabrication technique and treatment should not aggravate corrosion.

1. For structures and equipment, the utility should be located where it cannot be affected by natural and climatic conditions. This includes corrosive pollution that may be airborne, prevalent winds, and surface water currents from near or remote sources.
2. Undrainable traps that accumulate liquids and absorbent solid wastes should be avoided. Structures should be designed to be self-draining.
3. Provisions should be made for the removal of moisture or other corrosive media from critical areas.
4. Laps and crevices should be avoided if possible. If they cannot, then effective seals should be used (particularly in areas of heat transfer) between metal and a porous material or where aqueous environments contain inorganic chemicals or dissolved oxygen.
5. Laps should be faced downward on exposed surfaces.
6. Effort should be made to design shapes that will reduce the effects of high fluid velocity, turbulence, and the formation of gas bubbles.
7. Asymmetrical shapes of unequal thickness should be avoided for galvanizing. Extremes in weight and cross sections of design members also should be avoided.
8. Impellers should have shapes that minimize high turbulence formation and reduce low-pressure buildup at their tips, which can lead to cavitation.

For piping arrangements and vessels, the following geometric considerations are recommended.

9. Piping systems should be designed for an economic flow velocity. For relatively clean fluids, a recommended velocity range where minimum corrosion can be expected is 2 to 10 fps. If piping bores exist, maximum fluid velocities may have a mean velocity of 3 fps for a 3/8-in bore to 10 fps for an 8-in-diameter bore. Higher flow velocities are not uncommon in situations that require uniform, constant oxygen supply to form protective films on active/passive metals.
10. Condensate filters, deaerators, traps, drains, and other means should be provided for the removal of dust, debris, and any other contaminants from the system that may promote corrosion.
11. The interior of piping systems should be streamlined for easy drainage. Stubs and dead ends should be avoided, and pipelines should be sloped continuously downstream to their outlets. Elbows should be sloped for drainage purposes.
12. Turbulence, rapid surging, excessive agitation, and impingement of fluids onto piping walls should be avoided. Throttle valves, orifices, and similar flow-regulating devices should be employed only where necessary. Control devices should be selected partly on the basis of minimum resistance to the flow. For example, a venturi tube is preferable to an orifice plate. In general, one should avoid using flow-controlling devices in close proximity to bends or changes of flow direction downstream.
13. When using soft metals such as lead, copper, and their alloys, avoid sudden changes in the flow direction, such as sharp bends.
14. To minimize nucleation, piping systems should be designed to maintain absolute pressure as high as possible.
15. The bend radii of pipes should be designed to be as large as possible. A minimum of three times the pipe diameter is recommended to maintain safe, economic flow velocities.
16. For heat exchangers, coolers, heaters, condensers, and related equipment, welding of tubes in tube sheets is recommended over the rolling-in system. Tubes should extend beyond the tube sheets. Also, the design should be such that cooling-water starvation at the periphery of the tube bundle is avoided (as illustrated in Figure 2-6). Heat-exchanger tubes should be slanted to provide proper drainage.
17. Condensers should be designed to provide a realistic amount of excess auxiliary exhaust steam with reasonable velocity steam inlets and exhausts. Also, steam baffles should be slanted away from condenser bracing and other critical areas.
18. For tanks and vessels, welded units are preferred over riveted or bolted designs. Fastener joints provide sites for crevice corrosion. Undrainable

Design Guidelines

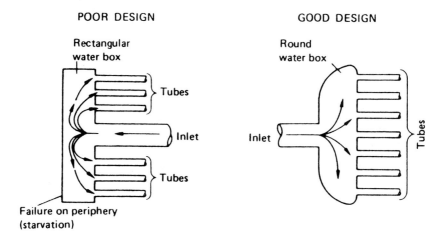

Figure 2-6 Poor and good designs for heat-exchanger inlets

horizontal flat tops of tanks should be avoided unless proper drainage schemes are included in the design. Tank bottoms should be sloped toward drain holes to eliminate the collection of liquids and sludge after emptying. Examples of good and poor designs for this latter case are shown in Figure 2-7.

19. Inlet pipes to vessels should be directed toward the center. Also, the inlet pipe should protrude into the tank and as close as possible to the normal level. This minimizes splashing during filling. Splashing causes precipita-

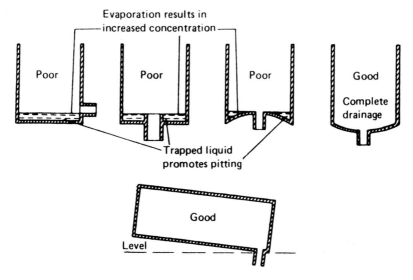

Figure 2-7 Poor and good designs for vessel drainage

tion to accumulate on walls, producing fouling and potential corrosion sites.
20. With side inlets and outlets on vessels, protrusions should be avoided that can cause additional turbulence.
21. Tanks designed to hold hygroscopic corrodants should be well sealed to prevent their breathing damp air.
22. Proper vessel designs should avoid discharges from high-positioned coolers directed down a pipe. This situation reduces pressure in coolers via siphon action.
23. Partially filled reactors and storage vessels containing vapors of corrosive constituents should be vented or provided with either vacuum removal or with a condenser return to the system.

GLOSSARY OF CORROSION TERMS

The following terms are widely used in the corrosion engineer's vocabulary. This subsection is included for the newcomer and defines terms used throughout the remainder of the text.

Active—a free-corroding condition.
Aluminizing—a process for impregnating a metal's surface with aluminum to provide protection against oxidation and corrosion.
Anchor pattern (surface profile)—the shape and amplitude of the profile of blast-cleaned or grooved steel, which influences the bond between metallic or paint films and the substrate.
Anion—negatively charged ions that migrate toward the anode of a galvanic cell.
Anode—the electrode at which oxidation of the surface or some component of the solution is occurring.
Anode polarization—the difference between the potential of an anode passing current and the steady-state or equilibrium potential of the electrode with the same electrode reaction.
Anodic inhibitor—a chemical constituent that reduces the rate of anodic or oxidation reaction.
Anodic metallic coating—a special coating usually comprised, either entirely or in part, of an anodic metal, which is electrically positive to the substrate to which it is applied.
Anodic protection—a technique for reducing corrosion of a metal surface via passing sufficient anodic current to it to cause its electrode potential to enter into the passive state.
Anodizing—the formation of a hard, corrosion-resistant oxide film on metals via anodic oxidation of the metal in an electrolytic solution.
Base potential—the potential toward the negative end of a scale of electrode potentials.

Glossary of Corrosion Terms

Blast peening—a treatment for relieving tensile stress via inducing beneficial compressive stress in the surface by kinetic energy of rounded abrasive particles.

Breakaway corrosion—a sudden increase in corrosion rate, particularly under conditions of high-temperature dry oxidation.

Cathode—the electrode of an electrolytic cell where reduction takes place. During corrosion, this is the area at which metal ions do not enter the solution. During cathodic reactions, cations take up electrons and discharge them, hence, reducing oxygen. That is, there is a reduction from a higher to a lower state of valency.

Cathodic inhibitor—a chemical constituent that reduces the rate of cathodic reaction.

Cathodic protection—a means of reducing the corrosion rate of a metal surface by passing sufficient cathodic current to it to cause its dissolution rate to become very low.

Cation—positively charged ions that migrate to the cathode in a galvanic cell.

Caustic embrittlement—a form of stress-corrosion cracking that occurs in steel exposed to alkaline solutions.

Composite plate—an electro deposit that consists of two or more layers of metals deposited separately.

Corrosion fatigue limit—the maximum stress that a metal can endure without failure. This is determined in a stated number of stress applications under defined conditions of stressing and corrosion.

Corrosion potential—the potential of a corroding surface in an electrolyte relative to some reference electrode.

Corrosion rate—the rate at which corrosion occurs. It is usually reported in units of inches of penetration per year (ipy), mils of penetration per year (mpy), milligrams of weight loss per square decimeter per day (mdd); microns per year (μm/yr), or millimeters per year (mmpy). Note that 1 μm = 0.0395 mils.

Couple—an electrical contact made between two dissimilar metals.

Critical humidity—the relative humidity (RH) at and above which the atmospheric corrosion rate of a metal increases significantly.

Current density—the average current flowing in an electrolyte. Common units are amperes per square foot (A/ft^2), amperes per square decimeter (A/dm^2), amperes per square centimeter (A/cm^2), or milliamperes per square centimeter (mA/cm^2) of either cathode or anode surface.

Deactivation—in corrosion control refers to the removal of a constituent of a liquid that is active in promoting corrosion.

Deposit attack—to localized corrosion under, and resulting from, a deposit on a metal surface.

Dielectric strength—the magnitude of electrical nonconductance of a material.

Differential aeration—the stimulation of corrosion at a localized area by differences in oxygen concentration in the electrolytic solution that is in contact with the metal surface.

Diffusion coating—application of a metallic coating. The chemical composition of the metal is modified by diffusing the coating into the substrate at the metal's melting temperature.

Electrogalvanizing—the process of galvanizing by electroplating.

Electrolysis—a reaction in which chemical change results in an electrolyte being produced from the passage of electric current.

Electrolyte—chemical constituent, usually a liquid, containing ions that migrate in an electric field.

Electrolytic cleaning—method of degreasing/descaling metal surfaces via electrolysis. The metal is utilized as an electrode.

Electrophoretic plating—the production of a layer of deposit as a result of discharge of colloidal particles in solution onto an electrode.

Electroplating—the process of electrode position onto a metallic substrate of a thin adherent layer of a metal or alloy having desirable chemical, physical, and/or mechanical properties.

Exfoliation—also called *lamination*, refers to the falling away of metal in layers.

Filiform corrosion—type of corrosion that takes place under a film in the form of randomly distributed hairlines.

Flame plating—the deposition of a hard metal coating onto a substrate via application of molten metal at supersonic velocities.

Flash corrosion—light surface oxidation of cleaned metals that are exposed to the environment for short times.

Fouling—deposition of scale materials on metal surfaces.

Galvanizing—the method of coating iron or steel with zinc by immersion of the metal in a bath of molten zinc.

Green rot—a corrosion product particular to nickel alloys and greenish in color that normally results from carburization and oxidation of certain nickel alloys at temperatures around 1,000°C (1,832°F).

Hermetic seal—an impervious seal made by the fusion of metals of ceramics, which prevents the passage of gas or moisture. The seal can be achieved by brazing, soldering, welding, fusing glass or ceramics.

Ion erosion—the deterioration of materials caused by ion impact.

Iron rot—the deterioration of wood caused by contact with iron.

Laminar scale—rust formation in heavy layers.

Localized attack—corrosion in which one area of the metal surface is primarily anodic and another predominantly cathodic.

Metal cladding—the combination of two or more metal compounds bonded metallurgically face to face.

Metallic coatings—coatings that consist fully or partially of metal applied to metals or nonmetals for the purpose of protection or to improve certain properties.

Metallizing—also called *metal spraying;* refers to the application of a metal coating to a surface (either metallic or nonmetallic) by means of a spray of molten particles.

Mill scale—an oxide layer on metals produced by metal rolling, hot forming, welding, or heat treatment.
Noble—positive direction of the electrode potential.
Noble potential—a potential that is more cathodic (that is, positive) than the standard hydrogen potential.
Oxidation—the loss of electrons by a constituent in a chemical reaction.
Parting—the selective attack of one or more constituents of a solid solution alloy.
Passivation—a reduction of the anodic reaction rate of an electrode involved in an electrochemical reaction, such as corrosion.
Passivity—a condition of a metal or alloy in which the material is normally thermodynamically unstable in a given electrolytic solution but remains visibly unchanged for a prolonged period. The electrode potential of a passive metal is always appreciably more noble than its potential in the active state.
Peen plating—the deposition of the coating metal (in powder form) on the substrate via a tumbling action in the presence of peening shot.
Pickling—a form of chemical and electrolytic removal of mill scale and corrosion products from the surfaces of metals in an acidic solution. Electrolytic pickling may be anodic or cathodic, depending on the polarization of the metal in the solution.
Plasma plating—deposition on critical areas of metal coatings resistant to wear and abrasion; normally this is done by means of a high-velocity and high-temperature ionized inert gas jet.
Rash rusting—also called *peak spotting;* refers to a local corrosion due to inadequate coating of the peaks of a rough surface.
Reduction—the reverse of oxidation; a chemical change of state in which one constituent gains electrons.
Rust—a corrosion product consisting mainly of hydrated iron oxide; the term is used to describe the corrosion products of iron and ferrous ions.
Rust creep—also called *underfilm corrosion;* refers to corrosive action that results in damaged or uncoated areas and extends subsequently under the surrounding inert protective coating.
Scaling—the formation of thick corrosion products as layers on a metal surface; in piping systems it is usually the deposition of water-insoluble constituents on a metal surface.
Season cracking—stress-corrosion cracking of brass.
Sherardizing—the process of coating iron or steel with zinc by heating the product to be coated in zinc powder at a temperature below the melting point of zinc.
Stress accelerated corrosion—increased corrosion rate caused by applied stresses.
Surface preparation—the cleaning of a surface prior to treatment.
Surface treatment—any suitable means of cleaning and treating a surface that

produces a desired surface profile that has required coating characteristics.

Tuberculation—the formation of localized corrosion products scattered over the surface in the form of knoblike tiles.

Vacuum deposition—also *vapor deposition* or *gas plating;* the deposition of metal coatings by means of precipitation (sometimes in vacuum) of metal vapor onto a treated surface. The vapor may be produced by thermal decomposition, cathode sputtering, or evaporation of the molten metal in air or an inert gas.

Weld decay—localized corrosion of weld metal.

3

Material Properties—Metals

Proper material selection for chemical and process equipment is one of the first important problems encountered by the designer. Among the many parameters that must be considered are structural strength specifications, heat resistance, corrosion resistance, physical properties, fabrication characteristics, composition and structure of material, and cost.

The properties that materials must have for a particular application depend largely on the environment in which they are to be used. Material selection begins with the determination of equipment, operating conditions, temperature, pressure, and various components in the process.

No one material meets all requirements. For example, good heat conductivity is a desirable property for the fabrication of heat-exchanger surfaces, but not for insulation purposes. Obviously, both positive and negative properties can coexist in a single material. A corrosion-resistant material may be insufficient for heat resistance or mechanical strength. Strong materials may be too brittle (for example, ferro silicon). Also, materials that have good mechanical and chemical properties may be too expensive or cost prohibitive.

The initial cost of a material does not provide the entire economic picture. At first, strong materials that are expensive may be more favorable than less expensive ones. The cost of processing inexpensive materials is sometimes high, thus resulting in prohibitively high fabrication costs. For example, the cost of a ton of granite is a dozen times cheaper than that of nickel chromium/

molybdenum steel. However, granite absorption towers would be more expensive than steel towers of the same volume because of the high costs associated with processing granite. Furthermore, granite towers are much heavier than the steel ones; therefore, they require stronger, and thus more expensive, foundations.

Because any material may be characterized by some desirable and nondesirable properties with respect to a specific application, the selection of materials is reduced to a reasonable compromise. In so doing, one strives to select materials so that properties correspond to the basic demands determined by the function and operating conditions of the equipment, tolerating some of the undesirable properties.

The basic requirements for materials intended for fabricating chemical apparatuses is mostly corrosion resistance because this determines the durability of the equipment. Often, corrosion data are reported as a weight loss per unit of surface area per unit of time. It is easy to transfer from such data to the penetration rate using the following relation:

$$V = 8.76pG \text{ mm/yr}$$

where G = weight loss at uniform corrosion (kg/m² hr)

V = corrosion rate (mm/yr)

p = density of material (kg/dm³)

Materials must have high chemical resistance as well as durability. For example, if the equipment material dissolves in the product, the product quality may deteriorate, or materials may act as catalysts promoting side reactions and thus decreasing the yield of the primary product. Usually there are several materials suitable for use under required process conditions. In such cases, the material is selected by additional considerations. For example, if a vessel must be equipped with a sight glass, the material for fabricating this item must be transparent and safe. In this case, plexiglas may be used if the vessel operates at low temperatures. For higher temperatures, glass is used; however, glass is brittle and very sensitive to drastic temperature changes. Therefore, the accessories must be designed so that the glass could not be broken occasionally and the poisonous or aggressive fluids allowed to escape. In this application, double glasses or valves must be provided for an emergency to shut off the accessory from the working space of the vessel. Consequently, the poor construction property of glass may cause additional complications in the design. At very high temperatures sight glasses are made from mica. For high-pressure drops they are made from rock crystal (an excellent but very expensive material). An example is shown in Figure 3-1. Gauges used for measuring the liquid level in vessels may be of semitransparent and even nontransparent materials.

Let us now consider the basic materials used in the fabrication of chemical equipment from the point of view of a designer. The principal construction materials for welded, forged, and cast chemical vessels are: cast irons, gray cast iron, white cast iron, malleable cast irons, nodular cast iron, austenitic cast iron,

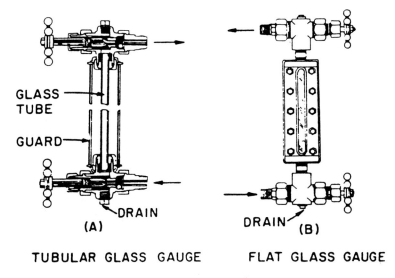

Figure 3-1 Typical glass sight gauges

high-silicon cast iron, low-carbon steels (mild steel), high-carbon steels, low-carbon/low-alloy steels, high-carbon/low-alloy steels, high-alloy steels (corrosion resistant, heat resistant, and high temperature), nickel, and nickel alloys. Descriptions of each of these follow.

PROPERTIES OF CAST IRONS

Three main factors that determine the properties of cast iron are:

- the chemical composition of the cast iron
- the rate of cooling of the casting in the mold
- the type of graphite formed

Most commercial cast irons contain 2.5 percent to 4 percent carbon, and it is the occurrence of some of this carbon as free graphite in the matrix that is the characteristic feature of thin material. About 0.8 percent to 0.9 percent carbon is in a bound form as cementite (iron carbide).

The cast irons usually have a ferrite-pearlite structure, which determines its mechanical properties. The ferrite content determines the cast iron's viscosity, while the pearlite content determines its rigidity and strength.

Because cast iron has a carbon content approximately equivalent to its eutectic composition, it can be cast at lower temperatures than steel and flows more readily than steel because of its much narrower temperature solidification range. The presence of the graphite flakes in cast iron decreases its shrinkage on

TABLE 3-1 TYPICAL MECHANICAL PROPERTIES OF VARIOUS TYPES OF CAST IRON

Material	Specification	Tensile strength (tonf/in.^{2})	Tensile strength (N/mm^2)	Elongation (%)
Gray Cast Iron	BS1452 Grade 10	10	155	—
	14	14	215	—
	26	26	400	—
Nodular Cast Iron	BS2789 SNG 24/17	24	370	17
	32/7	32	500	7
	47/2	47	730	2
Malleable Cast Iron				
Blackheart	BS310 B290/6		290	6
	B340/12		340	12
Whiteheart	BS309 W340/3		340	3
	W410/4		410	4
Pearlite	BS3333 P440/7		440	7
	P540/5		540	5
	P690/2		690	2

solidification much less than that of steel. These factors contribute to the fabrication of cast iron as sound castings in complex shapes and with accurate dimensions at low cost.

The physical properties of cast irons are characterized by the following data:

- density $p = 7.25$ kg/dm^3
- melting temperature $t_m = 1{,}250\text{--}1{,}280°$C
- heat capacity $C_p = 0.13$ kcal/kg°C
- heat conductivity $\lambda = 22\text{--}28$ kcal/m°C hr
- coefficient of linear expansion $\alpha = 11 \times 10^{-6}$

The cast irons do not possess ductility. They cannot be pressed or forged even while heated; however, their machining properties are considered good. Typical mechanical properties of various types of cast iron are given in Table 3-1.

Gray Cast Iron

Gray cast iron is the least expensive. It is the easiest to cast and machine. The tensile strength of gray cast iron ranges from 155 to 400 N/mm^2 (10 to 26 tonf/in.^{2}). The tensile modulus ranges from 70 to 140 kN/mm^2 and the hardness from 130 to 300 DPN.

In nearly all standards of gray cast iron, the grades are designated according to the tensile strength, not composition. In the British standard BS1452, for example, there are seven grades from 155 to 400 N/mm^2 (10 to 26 tonf/in.^{2}). This is the tensile strength measured on a test bar having a diameter of approximately

Properties of Cast Irons

TABLE 3-2 TYPICAL DATA SHOWING THE EFFECT OF STRENGTH ON GRAY IRON CASTINGS

Gray iron to BS1452	Tensile strength, N/mm², of casting with section thickness of:				
	10 mm	20 mm	75 mm	100 mm	150 mm
Grade 20	350	280	280	230	220
14	230	200	150	140	120
12	200	170	120	110	110
10	170	140	110	90	75

30 mm (1.2 in.). The actual strength of a casting will differ from that of the test bar according to the cross-sectional area. (See Table 3-2.) Castings are designed to be loaded in compression because the compressive strength of gray iron is about three times that of its tensile strength.

The recommended maximum design stress in tension is one quarter the ultimate tensile strength. For cast irons, this may be a value up to 185 N/mm² (12 tonf/in.²). The fatigue strength is one half the tensile strength. Notched specimens show the same value as unnotched specimens. For 200 N/mm² (14 tonf/in.²) grades and above, the fatigue strength of unnotched specimens is approximately one third the tensile strength. There is some notch sensitivity, although much less than is found in steel.

White Cast Iron

White cast iron is very hard (from 400 to 600 DPN) and brittle. All white cast irons are very difficult to machine and usually are finished by grinding. Table 3-3 gives properties of the four principal types of white cast irons.

Malleable Cast Irons

This type of cast iron is made by high-temperature heat treatment of white iron castings. The mechanical properties of malleable cast irons are given in Table

TABLE 3-3 PROPERTIES OF WHITE IRON

	Unalloyed white iron	Low-alloy white iron	Martensitic white iron (ni-hard)	High-carbon, high-chromium, white iron
Composition (%)				
Carbon	3.5	2.6	3.0	2.8
Silicon	0.5	1.0	0.5	0.8
Nickel			3.5	—
Chromium		1.0	2.0	27
Hardness (DPN)	600	400	600	500
Tensile strength, (N/mm²)	270	300	330	420

3-1; usually they are applied to the fabrication of conveyor chain links, pipe fittings, and gears.

Nodular Cast Iron

Nodular cast iron (also referred to as ductile cast iron) is manufactured by inoculating the molten metal with magnesium or cesium. It is characterized by a homogeneous structure, higher than usual abrasion resistance and strength for dynamic loads, and by easy machining. A wide variety of grades is available, with typical tensile strengths ranging from 380 to 700 N/mm^2 (25 to 40 tonf/in.^{2}), elongations from 17 to 2 percent, and hardness from 150 to 300 DPN (see Table 3-1). The tensile modulus is approximately 170 kN/mm^2. The design stress is half the 0.1 percent proof stress, and the fatigue design stress is one third the fatigue limit. The nodular cast iron is used for many applications such as valves in pipelines for petroleum products, underground pipelines, and so on.

Austenitic Cast Iron

Austenitic cast irons (either flake graphite irons or nodular graphite irons) are produced by mixing in nickel from 13 percent to 30 percent, chromium from 1 percent to 5 percent and copper from 0.5 percent to 7.5 percent (to lower nickel-containing grades to augment the corrosion resistance at lower cost). The main advantages of austenitic cast irons are corrosion and heat resistance. For corrosion resistance, the flake and nodular are similar, but the mechanical properties of nodular cast irons are superior.

APPLICATION REQUIREMENTS OF CAST IRONS

Abrasion Resistance

The white cast irons and their low alloys have good abrasion resistance properties. White cast irons are used for grinding balls, segments for mill liners, and slurry pumps. In the ceramic industry they are used for muller tyres and augers, in the pulp and paper industry for attrition mill plates and chip feeders, and in the paint industry for balls for grinding pigments.

Corrosion Resistance

The corrosion resistance of unalloyed and low-alloy flake, nodular, malleable, and white cast iron is comparable to mild- and low-alloy steel. However, these cast irons have a major advantage over steel; namely, the ability for greater cross section or wall thickness than steel. Consequently, they have a longer life, although they corrode at the same rates. For example, the corrosion rate for boiling concentrated sulfuric acid in large cast-iron pots is very high, but a reasonable life is obtained by making the bottom of the pots 3 inches thick. Secondly, although the matrix of a cast-iron pipe may rust, for example, the graphite network prevents disintegration of the pipe and permits its duty for a longer time than a steel pipe. Austenitic cast irons are in widespread use in many industries (food, pharmaceutical, petroleum, chemical, petrochemical, pulp and paper, and others) in mildly corrosive and erosive situations where the life of unalloyed or low-alloy cast iron or steel is short, but the high cost of stainless steel and nonferrous alloys cannot be justified. Other austenitic cast-iron applications can be found in food and dairy production, where the metallic contamination of the product must be eliminated.

Temperature Resistance

The persistent increase in volume of cast-iron items in high-temperature situations becomes the limiting factor in the use of unalloyed cast irons, especially in flake graphite castings. The addition in a casting of about 1 percent of chromium can control the growth in the temperature range from 400 to 600°C. Above 600°C, scaling due to surface oxidation becomes an undesirable phenomenon in the use of unalloyed cast irons. For achieving dimensional stability and long life, silal cast irons (containing 5.5 percent to 7 percent silicon) may be used for temperatures up to 800°C, in cases where it is not subjected to thermal shock. In thermal cycling and thermal-shock situations for temperatures up to 950°C, the 30 percent nickel austenitic cast irons are preferred; and above this temperature, where there is no thermal shock, the 28 percent chromium cast iron is recommended.

Gray cast irons do not have the abrupt ductile to brittle fraction transition down to $-40°C$ as takes place in steels. Special austenitic nodular cast iron similar to the AUS 203 grade, but with a higher manganese content of about 4

TABLE 3-4 RODS AND ELECTRODES FOR FUSION-WELDING CAST IRON

Filler rod or electrode	Welding method	Suitability	Comments
Cast Iron (e.g., BS1453)	Oxyacetylene	Gray irons but not malleable	Matching color Not machinable
Mild Steel	Metal arc	Cavity filling of thin castings	Not machinable
Nickel Iron (55/45)	Metal arc	For high-strength gray and nodular irons	Reasonable machinability
Nickel Copper (70/30)	Metal arc	For all gray irons	Easy to machine
Brass (60 Cu/40 Zn)	Oxyacetylene	Gray and malleable irons	Often called "bronze" welding
Tin Bronze (7 Sn/93 Cu)	Oxyacetylene	Gray and malleable irons	

percent, has been obtained for cryogenic purposes for temperatures down to −253°C.

Welding Cast Iron

Welding is sometimes used to repair broken and defective castings. This process is more difficult than welding steel because the high-carbon content in cast iron may lead to brittle structures on cooling, thus causing cracking. However, special techniques have been developed for fusion and nonfusion welding. Some of these are outlined in Table 3-4. The engineer should consider carefully whether the properties of varieties of cast irons will suit the demands before specifying more expensive materials.

PROPERTIES OF STEELS

The second group of structural materials in the iron-base category is steels. They have obtained an exclusive importance because of their strength, viscosity, and their ability to withstand dynamic loads. Also, they are beneficial for producing castings, forgings, stamping, rolling, welding, machining, and heat-treatment works. Steels change their properties over a wide range depending on their composition, heat treatment, and machining. Most steels have a carbon content of 0.1 percent to 1 percent, but in structural steels this does not exceed 0.7 percent. With higher carbon contents, steel increases in strength but decreases in plasticity and weldability. In the carbon steels designed for welding, the carbon content must not exceed 0.3 percent; in the alloy steels it must not exceed 0.2 percent. When the carbon content in the steels exceeds the preceding value, they are susceptible to air hardening. Hence, high stresses may be created and hardening fractures in welding zones may be formed. The steels with low-carbon content (below 0.2 percent) are well stamped and stretched, well cemented and nitrated, but badly machined. The physical properties of low-carbon, low-alloy steels are characterized in the following data:

- density $\rho = 7.85$ kg/dm^3
- heat capacity $C_p =$ dill kcal/m°C
- melting temperature $t_m = 1{,}400$–$1{,}500$°C
- thermal conductivity $\lambda = 40$–50 kcal/m°C hr

Low-Carbon Steels (Mild Steel)

Mild steel (<0.25 percent carbon) is the most commonly used, readily welded construction material, and has the following typical mechanical properties (Grade 43A in BS4360; weldable structural steel):

- tensile strength, 430 N/mm²
- yield strength, 230 N/m²
- elongation, 20 percent
- tensile modulus, 210 kN/mm²
- hardness, 130 DPN

No one steel exceeds the tensile modulus of mild steel. Therefore, in applications in which rigidity is a limiting factor for design (for example, for storage tanks and distillation columns), high-strength steels have no advantage over mild steel. Stress concentrations in mild-steel structures are relieved by plastic flow and are not as critical in other less ductile steels.

Low-carbon plate and sheet are made in three qualities: fully killed with silicon and aluminum, semikilled (or balanced), and rimmed steel. Fully killed steels are used for pressure vessels. Most general-purpose structural mild steels are semikilled steels. Rimming steels have minimum amounts of deoxidation and are used mainly as thin sheet for consumer applications.

The strength of mild steel can be improved by adding small amounts (not exceeding 0.1 percent) of niobium, which permits the manufacture of semikilled steels with yield points up to 280 N/mm². By increasing the manganese content to about 1.5 percent, the yield point can be increased up to 400 N/mm². This provides better retention of strength at elevated temperatures and better toughness at low temperatures.

Corrosion Resistance

Equipment from mild steel usually is suitable for handling organic solvents, with the exception of those that are chlorinated, cold alkaline solutions (even when concentrated), sulfuric acid at concentrations greater than 88 percent, and nitric acid at concentrations greater than 65 percent at ambient temperatures.

Mild steels are rapidly corroded by mineral acids even when they are very dilute (pH less than 5). However, it is often more economical to use mild steel and include a considerable corrosion allowance on the thickness of the apparatus. Mild steel is not acceptable in situations in which metallic contamination of the product is not permissible.

Heat Resistance

The maximum temperature at which mild steel can be used is 550°C. Above this temperature, the formation of iron oxides and rapid scaling makes the use of mild steels uneconomical. For equipment subjected to high loadings at elevated temperatures, it is not economical to use carbon steel in cases above 450°C because of its poor creep strength. (Creep strength is time dependent, with strain occurring under stress.)

Low Temperatures

At temperatures below 10°C the mild steels may lose ductility, causing failure by brittle fracture at points of stress concentrations (especially at welds). The temperatures at which the transition occurs from ductile to brittle fraction depends not only on the steel composition, but also on thickness.

Stress relieving at 600–700°C for steels decreases operation at temperatures some 20°C lower. Unfortunately, suitable furnaces generally are not available, and local stress relieving of welds, and so on is often not successful because further stresses develop on cooling.

High-Carbon Steels

High-carbon steels containing more than 0.3 percent are difficult to weld, and nearly all production of this steel is as bar and forgings for such items as shafts, bolts, and so on. These items can be fabricated without welding. These steels are heat treated by quenching and tempering to obtain optimum properties up to 1,000 N/mm^2 tensile strength.

Low-Carbon, Low-Alloy Steels

Low-carbon, low-alloy steels are in widespread use for fabrication-welded and forged-pressure vessels. The carbon content of these steels is usually below 0.2 percent, and the alloying elements that do not exceed 12 percent are nickel, chromium, molybdenum, vanadium, boron, and copper. The principal application of these steels are given in Table 3-5.

Properties

The maximum permissible loading of low-alloy steels according to the ASME code for pressure vessels is based on proof stress (or yield point), which is applicably superior to those of carbon steels. The cost of a pressure vessel in alloy steel may be more expensive than in carbon steel. However, consideration

TABLE 3-5 APPLICATIONS OF LOW-CARBON, LOW-ALLOY STEELS

0.5 Mo
1.25 CrMo
2.25 CrMo
6 to 12 CrMoVW
	High creep strength for:
	  1. pressure vessels such as boilers operating at elevated temperatures; and
	  2. oil refinery vessels such as crackers and reformers with high hydrogen pressures.

5 to 9% Cr	For oil refinery applications involving high-sulfur process streams, e.g., pipe stills.
CuCr (Corten)	Rust-resisting steels for structural applications.
2 to 9% Ni	For cryogenic applications.

should be given to other cost savings resulting from thinner-walled vessels, which provide fabrication savings on weldings, stress relieving, transportation, erection, and foundation.

The corrosion resistance of low-alloy steels is not significantly better than that of mild steel for aqueous solutions of acids, salts, and so on. The addition of 0.5 percent copper forms a rust-colored film preventing further steel deterioration; small amounts of chromium (1 percent) and nickel (0.5 percent) increase the rust resistance of copper steels still further. Low-alloy steels have good resistance to corrosion by crude oils containing sulfur.

In operations involving hydrogen at partial pressures greater than 35 kgf/cm^2 and temperatures greater than 250°C, carbon steels are decarborized and fissured internally by hydrogen. Small additions of molybdenum prevent hydrogen attack at temperatures up to 350°C and pressure up to 56 kgf/cm^2. For higher temperatures and pressures chromium/molybdenum steels (2.25 Cr, 0.5 Mo) are used.

Oxidation Resistance and Creep Strength

Chromium is the most effective alloying element for promoting resistance to oxidation. In atmospheres contaminated with sulfur, lower maximum temperatures are necessary. In fractionation columns for petroleum products, where the oxygen content is restricted, higher temperatures can be used without excessive waste of the metal. The creep strength of steels is a factor limiting the maximum temperatures for such high-pressure equipment as shells and stirrers of high-temperature reactors.

Low-Temperature Ductility

Nickel is the alloying element used for improving low-temperature ductility. The addition of 1.5 percent nickel to 0.25 percent Cr/0.25 percent Mo steels provides satisfactory application for moderately low temperatures down to about −50°C. Heat treatment by quenching and tempering improves the low-temperature ductility of steels such as 0.5 Cr, 0.5 percent Mo, 1 percent Ni Type V. For lower-temperature application (below −196°C), up to 9 percent nickel is used as the sole alloying element.

High-Carbon, Low-Alloy Steels

High-carbon (about 0.4 percent) low-alloy steels that are not weldable usually are produced as bars and forging for such items as shafting, high-temperature bolts and gears, and ball-bearing components. These steels can be less drastically quenched and tempered to obtain tensile strengths of at least 1,500 N/mm^2, thus minimizing the danger of cracking.

HIGH-ALLOY STEELS

Stainless and heat-resisting steels containing at least 18 percent by weight chromium and 8 percent nickel are in widespread use in industry. The structure of these steels is changed from magnetic body-centered cubic or ferritic crystal structure to a nonmagnetic, face-centered cubic or austenitic crystal structure.

Chromium Steels (400 Series), Low-Carbon Ferritic (Type 405)

The main use of this type steel (12 percent to 13 percent chromium) is for situations in which the process material may not be corrosive to mild steel, yet contamination due to rusting is not tolerable and temperatures or conditions are unsuitable for aluminum. However, prolonged use of these steels in the temperature range of 450–550°C causes low-temperature embrittlement of most ferritic steels with more than 12 percent chromium.

Medium Carbon Martensitic (Types 403, 410, 414, 416, 420, 431, 440)

These steels (13 percent to 17 percent chromium) resist oxidation scaling up to 825°C but are difficult to weld and, thus, are used mainly for items that do not involve welded joints. They are thermally hardened and useful for items that require cutting edges and abrasion resistance in mildly corrosive situations. However, they should not be tempered in the temperature range of 450–650°C. This reduces the hardness and wear resistance and also lowers the corrosion resistance because of the depletion of chromium in solution through the formation of chromium carbides.

Medium Carbon Ferritic (Types 430 and 446)

The 17 percent ferritic steels are easier to fabricate than the martensitic grades. They are used extensively in equipment for nitric acid production. The oxygen- and sulfur-resistant 30 percent chromium steel can be used at temperatures up to 1,150°C but only for lightly loaded and well-supported furnace items because of its poor creep and brittlement properties when equipment is down to ambient temperatures.

Chromium/Nickel Austenitic Steels (300 Series)

The excellent corrosion resistance over a wide range of operating conditions and readily available methods of fabrication by welding and other means of shaping metals make these steels the most extensively used throughout the chemical and allied industries.

The formation of a layer of metal oxide on the surface of this steel provides better corrosion resistance in oxidizing environments than under reducing conditions. Common steels 304, 304L, 347, 316, and 316L are used for equipment exposed to aqueous solutions of acids and other low-temperature corrosive conditions. For high-temperature regimes involving oxidation, carborization, and so on, the 309 and 310 compositions may be recommended because of their higher chromium content and, thus, better resistance to oxidation.

Type 304, 19/10 (chromium nickel), provides a stable austenitic structure under all conditions of fabrication. Carbon (0.08 percent max.) is sufficient to have reasonable corrosion resistance without subsequent corrosion resistance for welded joints. Type 304 is used for food, dairy and brewery equipment, and for chemical plants of moderate corrosive duties.

Type 304L is used for applications involving the welding of plates thicker than about 6.5 mm.

Type 321 is an 18/10 steel that is stabilized with titanium to prevent weld decay or intergranular corrosion. It has similar corrosion resistance to types 304 and 304L but a slightly higher strength than 304L; also, it is more advantageous for use at elevated temperatures than 304L.

Type 340 is an 18/11 steel that is stabilized with niobium for welding. In nitric acid it is better than type 321; otherwise, it has similar corrosion resistance. Type 316 has a composition of 17/12/2.5 chromium/nickel/molybdenum. The addition of molybdenum greatly improves the resistance to reducing conditions such as dilute sulfuric acid solutions and solutions containing halides (such as brine and sea water).

Type 316L is the low-carbon (0.03 percent max.) version of type 316 that should be used where the heat input during fabrication exceeds the incubation period of the 316 (0.08 percent carbon) grade. For example, it is used for welding plates thicker than 1 cm.

Type 309 is a 23/14 steel with greater oxidation resistance than 18/10 steels because of its higher chromium content.

Type 315 has a composition that provides a similar oxidation resistance to type 309 but has less liability to embrittlement due to sigma formation if used for long periods in the range of 425–815°C. (Sigma phase is the hard and brittle intermetallic compound FeCr formed in chromium-rich alloys when used for long periods in the temperature range of 650–850°C.)

Alloy 20 has a composition of 20 percent chromium, 25 percent nickel, 4 percent molybdenum, and 2 percent copper. This steel is superior to type 316 for severely reducing solutions such as hot, dilute sulfuric acid.

Precipitation Hardening Stainless Steel

Properties are higher than those of austenitic steels and they retain a general level of corrosion resistance considerably better than that of chromium martensitic steels. They can be supplied as forgings, castings, plate, bar, and sheets and can be readily welded and formed before hardening. A typical application is for gears in pumps used for metering chemicals where their hardness prevents wear and galling in contact with type 316 bodies.

Chromium/Nickel/Ferrite/Austenite Steels

These steels can be welded successfully because they are not predisposed to excess grain growth at elevated temperatures. However, the general level of their corrosion resistance is usually inferior to that of austenitic steels, although they have good resistance to stress-corrosion cracking. For example, using austenitic steels in hot, slightly acid solutions containing chlorides causes rapid cracking in a few weeks, whereas the ferrite/austenite steels may last many years.

Maraging Steels

For corrosion resistance, these steels (18 percent nickel, 9 percent cobalt, 3 percent molybdenum, 0.2 percent titanium, and 0.02 percent carbon) are similar to the 13 percent chromium steels and, therefore, are suitable for mildly corrosive situations. Because of their very high strength after heat treatment (yield strength—1,390 N/mm^2, elongation—15 percent, impact strength) maraging steels find some use in very high-pressure equipment.

APPLICATIONS OF HIGH-ALLOY STEELS

With austenitic stainless steels a high-carbon content may cause the formation of chromium carbides at grain boundaries, consequently producing intergranular corrosion. This is most likely to occur during welding (called *weld decay*). This phenomenon may be avoided by using either a low-carbon steel, grade L (that is, less than 0.03 percent carbon), or a steel containing titanium or niobium, such as types 321 and 347.

Intergranular corrosion depends on the length of time the steel is exposed to the sensitizing temperature (500–750°C), even if made from low-carbon or titanium- or niobium-stabilized steel.

Equipment fabricated from such a steel may undergo corrosion by condensation of even mild corrosives unless it is possible to keep it above the dew point or to neutralize acidic condensates. This kind of corrosion can be prevented by a preliminary heat treatment at temperatures of 815–915°C. The niobium-stabilized steels respond best to this treatment.

Stress-corrosion cracking, usually occurring at temperatures above 80°C, takes place in equipment made from austenitic stainless steel but does not affect ferritic steels in this way.

Stress cracking most often occurs in solutions of chlorides. Concentrations of a few parts per million can cause severe cracking, even in a medium that would not be considered corrosive, for example, in water main lines. Stress-corrosion cracking can be caused by some thermal-insulating materials, but can be prevented by cladding the insulation with aluminum. This eliminates rain from washing chlorides into contact with the steel.

Residual stresses occur from welding and other fabrication techniques even at very low stress values. Unfortunately, stress relief of equipment is not usually a reliable or practical solution. Careful design of equipment can eliminate crevices or splash zones in which chlorides can concentrate. The use of high-nickel stainless steel alloy 825 (40 percent nickel, 21 percent chromium, 3 percent molybdenum, and 2 percent copper) or the ferritic/austenitic steels would solve this problem.

Oxidation Resistance

The ferritic chromium steels (chromium is the principal alloying element) are the most economical for very lightly loaded high-temperature situations. However, they are inadequate when creep must be accounted for. Austenitic steels are often recommended for such conditions. The 17 percent chromium alloys (type 430) resist scaling up to 800°C; and 25 percent alloy (type 446) up to 1,100°C.

Mechanical Properties at Elevated Temperatures

The austenitic steels containing nickel are used for load-bearing applications, pressure vessels operating above 550°C, as well as for light-load cyclic operation because they have a more adherent scale than chromium steels and generally do not become brittle under high-temperature service. The 18/10 alloys are suitable for use up to 800°C in air; the 25/10 type 310 alloys are suitable for use up to 1,100°C. When using type 316 alloy at high temperatures, care should be taken that the atmosphere is not stagnant, as catastrophic oxidation of molybdenum may occur.

For high-pressure, high-temperature situations where steels are required with certified creep-strength properties, the AISI austenitic steels are given the suffix H (for example, 347H, 316H, and so on).

Below creep-range temperatures, economies can be made by using nitrogen-containing (for example, BS1501 Part 6, Grades 304S65 and 316S66) or worm-worked grades, as these have higher proof strength than ordinary grades.

Mechanical Properties at Low Temperatures

The austenitic steels can be used at very low temperatures (low-alloy ferritic steels containing 9 percent nickel down to −196°C) without the risk of brittle fracture.

CORROSION-RESISTANT NICKEL AND NICKEL ALLOYS

Nickel alloys have two main properties: good resistance to corrosion and high-temperature strength. There are alloys for medium- and low-temperature applications and for high-temperature conditions in which creep resistance is of main importance.

The standard quality of commercially pure nickel (nickel 99 percent minimum, carbon 0.15 percent maximum; nickel 200/201) can be readily welded and fabricated in all wrought forms and as castings. However, it is restricted to special applications for which nickel alloys are not adequate (for example, for equipment used in the production of caustic soda where it is not subject to stress-corrosion cracking in hot caustic soda solutions).

Unalloyed nickel is used where it is necessary to eliminate iron and copper contamination (nickel 200 up to 300°C and nickel 201 above 300°C).

Nickel/Copper (Alloy 400)

Alloy 400 has good mechanical properties and is easy to fabricate in all wrought forms and castings. K-500 is a modified version of this alloy and can be thermally treated and is suitable for items requiring strength, as well as corrosion resistance. Alloy 400 has immunity to stress-corrosion cracking and pitting in chlorides and caustic alkali solutions. Alloy 400 is also adequate for equipment processing of dry halogen gases and chlorinated hydrocarbons and can be used in reducing environments.

Nickel/Molybdenum

This alloy has a nominal composition of 65 percent nickel, 28 percent molybdenum, and 6 percent iron. It is generally used in reducing conditions. It is intended to work in very severely corrosive situations after postweld heat treatment to prevent intergranular corrosion. These alloys have outstanding resistance to all concentrations of hydrochloric acid up to boiling-point temperatures and in boiling sulfuric acid solutions up to 60 percent concentration.

Nickel/Molybdenum/Chromium

The composition of this alloy (54 percent nickel, 15 percent molybdenum, 15 percent chromium, 5 percent tungsten, and 5 percent iron) is less susceptible to intergranular corrosion at welds. The presence of chromium in this alloy gives it better resistance to oxidizing conditions than the nickel/molybdenum alloy, particularly for durability in wet chlorine and concentrated hypochlorite solutions, and has many applications in chlorination processes. In cases in which hydrochloric and sulfuric acid solutions contain oxidizing agents such as ferric and cupric ions, it is better to use the nickel/molybdenum/chromium alloy than the nickel/molybdenum alloy.

Nickel/Chromium/Molybdenum/Iron

Because the composition of this alloy (47 percent nickel, 22 percent chromium, 7 percent molybdenum, and 17 percent iron) has a higher iron content, it cannot withstand such aggressive corrosion conditions as nickel/molybdenum and nickel/molybdenum/chromium alloys. It is, however, less expensive. The nickel makes these alloys immune to stress-corrosion cracking and also superior to stainless steels with respect to pitting in chloride solutions. Because of these properties, their greater cost over stainless steel is justified.

Nickel/Chromium/Molybdenum/Copper

These alloys (50/60 percent nickel, 20/30 percent chromium, 5/8 percent molybdenum, and 5/7 percent copper) have very good resistance to hot sulfuric acid solutions and similar environments. They are only available as castings but can be hardened by heat treatment. The castings are suitable for parts requiring cutting edges and good wear resistance under corrosion conditions, but should not be used in contact with halogens, halogen acids, and halogen salt solutions.

Nickel/Silicon

Nickel/silicon alloy (10 percent silicon, 3 percent copper, and 87 percent nickel) is fabricated only as castings and is rather brittle, although it is superior to the iron/silicon alloy with respect to strength and resistance to thermal and mechanical shock. It is comparable to the iron/silicon alloy in corrosion resistance to boiling sulfuric acid solutions at concentrations above 60 percent. Therefore, it is chosen for this and other arduous duties where its resistance to thermal shock justifies its much higher price compared with iron/silicon alloys.

HEAT-RESISTANT NICKEL ALLOYS

Nickel/Chromium

The high-chromium casting alloys (50 percent nickel, 50 percent chromium and 40 percent nickel, 60 percent chromium) are designated for use at temperatures up to 900°C in furnaces and boilers fired by fuels containing vanadium, sulfur, and sodium compounds (for example, residual petroleum products). Alloys with lower chromium contents cannot be used with residual fuel oils at temperatures above 650°C because the nickel reacts with the vanadium, sulfur, and sodium impurities to form compounds that are molten above 650°C.

Nickel/Chromium/Iron

Alloy 800 (32 percent nickel, 20 percent chromium, and 46 percent iron) is used for furnace equipment such as muffles, trays, and radiant tubes and in oil and petrochemical plants as furnace coils for the reforming and pyrolysis of hydrocarbons. Higher-strength versions of alloy 800 were developed to meet this situation (802 has a higher carbon content; alloy 807 has a higher hot strength by adding cobalt and tungsten). For 807, the stress to produce rupture in 100,000 hr at 900°C is 13.8 N/mm² compared with 8.5 N/mm² for alloy 800.

COPPER AND COPPER ALLOYS

The outstanding properties of copper-base materials are high electrical and thermal conductivity, good durability in mildly corrosive chemical environments, and excellent ductility for forming complex shapes. As a relatively weak material, copper is often alloyed with zinc (brasses), tin (bronzes), aluminum, and nickel to improve its mechanical properties and corrosion resistance. The classification system used in the United States for copper and copper alloys is given in Table 3-6.

Different grades of copper are described in Table 3-7. The specific gravity of the soft pure metal is 8.94. Additional properties are:

- heat capacity C_p = 0.093 kcal/kg°C
- melting temperature t_m = 1,083°C
- thermal conductivity λ = 334 kcal/m°C hr
- linear expansion coefficient α = 1.65 × 10⁻⁵
- Young's modulus E = 1,080,000 kg/cm²
- molding temperature 1,150°C

The approximate tensile strength is 14 ton/in.² at ordinary temperatures, and its strength decreases with increasing temperature. Copper retains high impact strength and increases its tensile strength under low temperatures, including cryogenic applications.

Along with high mechanical properties, copper improves its conductivity in the range of lower temperatures (at −160°C ∼400, −190°C ∼450, and −252°C ∼1,600 kcal/m°C hr). It softens in the temperature range of 200–220°C. The casting properties of copper are rather fair, but copper can be readily stretched, flattened, rolled, welded, and brazed. For chemical plant work, welded or brazed joints have become almost universal.

TABLE 3-6 CLASSIFICATION USED FOR COPPER ALLOYS IN THE UNITED STATES

Series	Constituents
100	Not less than 99.4% copper
200	50–99% copper plus zinc and minor elements
300	Zinc and lead alloys
400	Zinc and tin alloys
500	Tin and phosphorus or phosphorus and zinc alloys
600	Aluminum, aluminum and zinc, or zinc and manganese alloys
700	Nickel, nickel and zinc, or zinc and lead

TABLE 3-7 VARIOUS GRADES OF COPPER

Alloy	Alloying elements (% by wt)	Thermal conductivity (Btu/ft²·hr °F/ft)	Ultimate tensile strength (ton/in.²)	
			(annealed)	(hard worked)
Tough-Pitch High-Conductivity (B.S. 1036)	0.04 oxygen	229	14	20
Tough-Pitch (B.S. 1038, 9)	0.10 oxygen	228	14	20
Tough-Pitch Arsenical (B.S. 1173)	0.10 oxygen 0.3–0.5 As	102	14	21–23
Deoxidized (B.S. 1172)	0.02–0.08 P	126–175	14	20
Deoxidized Arsenical (B.S. 1171)	0.02–0.08 P 0.3–0.5 As	102	14	21–23
Oxygen-Free High-Conductivity	Trace	229	14	20
Silver-bearing	0.05 Ag 0.04 oxygen	228	14	22
Chromium Copper	0.5 Cr	100	15	29
Tellurium Copper	0.75 Te	208	15	
Beryllium Copper	About 2% Be with other elements (Co, Ni, Cr)	50	33	75

Copper does not form protective oxide films. Therefore, its corrosion resistance is poor against most acids and salts. Many gases such as haloids, sulfurous anhydride, sulfur vapors, hydrogen sulfide, carbon dioxide, and ammonium destroy copper. However, copper is highly corrosion resistant to alkali solutions.

Brasses

These are alloys containing more than 50 percent of copper used to overcome the softness, low tensile strength, and high casting temperature of the pure metal. The compositions and properties of commonly used brasses are presented in Table 3-8.

These annealed brasses are used for fabrication of pressure vessels. They are characterized by the following physical properties:

- density = 8.5 kg/dm^3
- melting temperature t_m = 940°C
- heat capacity C_p = 0.092 kcal/kg°C
- heat conductivity λ = 90–100 kcal/m°C
- temperature elongation α = 2 × 10^{-5}

The strength and ductility of brasses are well maintained over a range of 300°C––180°C, and castings are easy to make as well as to machine. Brass behaves similarly to copper in chemical plant environments, with somewhat greater rates of attack.

Tin Bronzes

This is the name given to copper-tin alloys containing additional alloying elements (Table 3-9). Small amounts of phosphorus are added to deoxidize the metal and in residual amounts to harden the finished alloy. Mixtures treated in this way are referred to as phosphor-bronzes. These have the best corrosion resistance of the alloys listed in Table 3-9 and are used in applications involving contact with dilute acid solutions where bronzes containing zinc (as an alternative to phosphorus, that is, the gunmetals) would not be sufficiently durable. The phosphor-bronzes have a low coefficient of friction and good resistance to wear. They are most often used for gears and bearings. Lead-bearing alloys corrode more rapidly than those containing only tin and copper; however, apart from this, all bronze alloys can be used with confidence wherever copper can resist corrosion.

Aluminum and Manganese Bronzes

The aluminum-bearing (5 percent to 12 percent of aluminum) alloys retain high strength, good corrosion resistance, and good oxidation resistance at temperatures up to 400°C. The aluminum manganese bronzes are noted for high

TABLE 3-8 PROPERTIES OF COMMON BRASSES

Material	Nominal composition, %					Mechanical properties		
	Cu	Zn	Sn	Al	Others	Hardness (DPN)	0.1% Proof stress (N/mm^2)	UTS (N/mm^2)
70/30 Brass	70	30			Arsenic 0.02/0.06	78	80	325
Aluminum Brass	76	22		2	Arsenic 0.02/0.06	75	90	355
60/40 Brass	60	40				90	115	360
Naval Brass	62	37	1			95	135	390
High-Tensile Brass (manganese bronze)	59	38			Iron 1 Manganese 0.5 Aluminum 0.2	125	145	470

TABLE 3-9 PROPERTIES OF TIN BRONZES AND GUNMETALS

Materials	Nominal composition, %				Mechanical properties		
	Cu	Sn	Zn	Pb	Others	Hardness (DPN)	0.1% Proof stress (N/mm^2)	UTS (N/mm^2)
5% Phosphor Bronze, Wrought	95	5			Phosphorous	75	125	340
Admiralty Gunmetal, Castings	88	10	2			85	125	300
Leaded Gunmetal 85/5/5/5, Castings	85	5	5	5		65	95	220
Nickel Gunmetal	86	7	2		Nickel 5	80	155	300
Phosphor-Bronze, General-Purpose Castings	90	9.5			Phosphorus 0.1/0.4	85	125	270
Bronze for Bearings	89	10			Phosphorus 0.5			

strength and good corrosion resistance at temperatures on the order of 400°C. These bronzes are available only as castings. They have good machining qualities combined with easy welding. With regard to corrosion resistance, they appear to behave at least as well as the true bronzes.

Silicon Bronzes

Containing up to 3 percent silicon, silicon bronzes are characterized by high mechanical and antifriction properties. They are made in all wrought forms, such as plates, sheets, and castings. The silicon bronzes are well molded, cold- and hot-pressure shaped (rollings, forging, stamping, and so on) and welded. These alloys have corrosion resistance similar to that of copper, with mechanical properties equivalent to mild steel. Because silicon bronzes do not generate sparks under shocks, they can be used in the fabrication of explosion-proof equipment. Compared to tin bronzes, the tinless bronzes have a higher shrinkage (1.7 percent to 2.5 percent against 1.3 percent to 1.5 percent of tin bronzes) and less fluid flow, which is an important consideration in designing.

Cupro-nickels

The cupro-nickel alloys (5 percent to 30 percent of nickel) are perhaps the best of all for strength and resistance to corrosion.

Corrosion Resistance

Copper-base alloys perform best under reducing conditions and in the absence of aeration. Copper and its alloys are resistant to dilute solutions of several mineral acids such as sulfuric and hydrochloric, and to a wide range of organic acids such as acetic and formic. Aluminum bronze is suitable in slightly oxidizing situations. Copper-base alloys are resistant to most alkaline solutions but never should be exposed to strong oxidizing acids such as nitric and chromic, as well as aqueous ammonia. Copper-base alloys are also resistant to most neutral salts, except to those forming soluble complexes.

MECHANICAL PROPERTIES OF LEAD AND LEAD ALLOYS

Lead is the softest and most easily worked metal used in plant construction. The main difficulty in design is that the metal has a very low creep stress, even at ordinary temperatures, with or without work-hardening effects. In the form used for chemical plants, the purity of the metal is almost complete; small amounts of alloying additions in lead are intended to improve its mechanical properties without any significant decrease in corrosion resistance.

There are three standard leads available in the United States for process plant construction. These are described in Table 3-10.

TABLE 3-10 STANDARD U.S. LEADS

Analysis (%)	Chemical lead	Acid lead	Copper lead
Silver: max.	0.020	0.002	0.020
min.	0.002		
Copper: max.	0.080	0.080	0.080
min.	0.040	0.040	0.040
Arsenic, antimony and tin, max.	0.002	0.002	0.015
Zinc, max.	0.001	0.001	0.002
Bismith, max.	0.005	0.025	0.100
Lead, min.	99.900	99.900	99.850

Lead has the following physical properties:

- density $\rho = 11.35$ kg/dm^3
- melting point $t_m = 327°C$
- heat capacity $C_p = 0.031$ kcal/kg°C
- thermal conductivity $\lambda = 30$ kcal/m°C hr
- temperature elongation $= 3.9 \times 10^{-7}$

The mechanical properties of lead are given in Tables 3-11 and 3-12. Lead alloys have higher strength and lower melting points than pure lead and, therefore, have a lower service temperature (less than 100°C).

Dispersion-strengthened lead (DSL), obtained by a uniform dispersion of lead oxide through the lead particle matrix, has the traditional corrosion resistance of lead but much greater stiffness. DSL is fabricated as pipe and other extruded items but has a limited application for process plant construction because the welding technique does not provide adequate strength in joints.

The recommended maximum design stresses for a life of five to ten years based on long-time creep tests are given in Table 3-13.

Another important factor in the selection of a lead alloy is fatigue strength, which may arise from high-frequency vibration from pumps and stirrers or from differential expansion from heat and cooling cycles. The marked increase of fatigue strength obtained by alloying with copper, silver, and tellurium can be seen from Table 3-14.

TABLE 3-11 MECHANICAL PROPERTIES OF SHEET LEAD

Ultimate Tensile Strength (kg/cm^2)	130–180
Elongation (%)	40–50
Necking (%)	100
Brinell Hardness (HB)	4–4.6

TABLE 3-12 MECHANICAL PROPERTIES OF ANNEALED LEAD VS TEMPERATURE

	Temperature (°C)				
	20	80	150	200	265
Ultimate Tensile Strength (kg/cm^2)	135	80	50	40	20
Elongation (%)	31	24	23	20	18
Necking (%)	100	100	100	100	100

Corrosion Resistance

The corrosion resistance of lead is due to the formation of a thin surface film of an insoluble lead salt that protects the metal from sulfuric acid and related compounds of any strength at ordinary temperatures. Even when the temperature increases to nearly 100°C, the rates of corrosion are still low. However, strong, hot sulfuric acid attacks lead rapidly, especially if the acid is flowing.

Nitric acid in any concentration attacks lead steadily, but mixtures of nitric and sulfuric—nitration acids—are not as active and can be handled in lead.

Phosphoric acid made by the *wet process,* in which phosphate rock is treated with sulfuric acid, is highly inert toward lead in any concentration for temperatures up to 150°C. However, in the *dry process,* where hydrogen phosphate (H_3PO_4) is made directly from phosphorus or phosphorus pentoxide (P_2O_5), a chemical reaction with lead occurs.

Lead chloride is freely soluble in hot aqueous solutions, but lead fluoride is almost insoluble in dilute HF solutions. When the HF concentration reaches about 40 percent, steel is preferred.

Organic chlorinations are handled in lead where the presence of iron might produce catalyst substitution in an undesirable position. Hence, lead is the material most frequently specified for chlorinators.

Chromic acid and its salts normally are prepared in lead. Lead is especially suitable for organic oxidations because its inertness avoids any interference from reactions.

TABLE 3-13 MAXIMUM STRESSES IN PIPE WALL OF LEAD ALLOYS

Temperature (°C)	Maximum stress, S, N/mm^2, in pipe wall			
	99.99% Lead	Copper, tellurium, and silver leads	8% Antimonial lead	DSL
20	2.21	2.42	3.50	10.34
60	1.24	1.38	1.24	10.34
100	0.66	0.86	0.76	9.62
150		0.52		3.50

TABLE 3-14 FATIGUE-STRENGTH DATA OF LEAD ALLOYS

Lead		Endurance limit, $+N/mm^2$, for 20×10^6 cycles	
		20°C	80°C
99.99		3.17	2.10
99.99%	+0.06 copper	4.06	3.00
99.99%	+0.005% silver +0.005% copper	4.17	3.05
99.99%	+0.06% copper +0.04 tellurium	7.70	5.10
DSL		13.80	12.50

Neutral or weak acid-salt solutions usually can be handled in lead plants, with the exception of those few heavy metals that may form lead alloys by substitution. The alums and sulfates generally have little action.

ALUMINUM AND ALUMINUM ALLOYS

The main criteria in the selection of aluminum and its alloys for chemical plants are corrosion resistance, ease of fabrication, and price. High-quality aluminum grades are used for chemical and process plant applications.

Physical properties of aluminum are characterized by the following:

- density $\rho = 2.7 \text{ k/dm}^3$
- melting point $t_m = 657°C$
- heat capacity $C_p = 0.218 \text{ kcal/kg°C}$
- thermal conductivity $\lambda = 188 \text{ kcal/m°C hr}$
- thermal elongation coefficient $\alpha = 2.4 \times 10^{-5}$

The positive properties of aluminum are its high heat conductivity (4.5 times higher than that of steel), low specific gravity, high ductility providing good rolling, and cold and hot stamping. The negative properties are its poor

TABLE 3-15 MECHANICAL PROPERTIES OF ALUMINUM

	Mild, annealed aluminum	Hardened aluminum
Ultimate Tensile Strength (kg/cm^2)	700–1,000	1,500–2,000
Yield Strength (kg/cm^2)	300–400	1,400–1,800
Elongation (%)	30–40	4–8
Necking (%)	70–90	50–60
Brinell Hardness (HB)	15–25	40–55

TABLE 3-16 MECHANICAL PROPERTIES OF ALUMINUM ANNEALED AT 370°C

	Temperature, °C						
	20	75	135	310	400	510	600
Ultimate Tensile Strength (kg/cm^2)	1,160	1,000	765	260	125	55	35
Elongation (%)	19	24	32	39	42	45	48
Necking (%)	79	83	88	97	99	99	100

castability, poor cutting, and low strength. The most important specifications of aluminum as a structural material are given in Tables 3-15 to 3-17.

The low strength of aluminum can be considerably improved by alloying with magnesium, silicon, manganese, copper, and others. However, the alloys have substantially the same modulus of elasticity (70 kN/mm^2).

Aluminum alloys can be divided into three classes: aluminum of various degrees of purity, wrought alloys, and casting alloys.

Aluminum of commercial 99 percent minimum aluminum purity is used widely for chemical and process plant applications where its resistance to corrosion and high-thermal conductivity are desirable characteristics. The strength of aluminum can be increased by cold working. The addition of alloying elements to the commercially pure metal results in an increase in its strength and usually has some favorable effects on other characteristics.

Nonheat-treatable magnesium and manganese alloys represent the best compromise between corrosion resistance and strength. It is the most useful class of alloys for chemical and process plant construction.

Heat-Treatable Alloys

These are produced by adding small amounts of copper, magnesium, and/or silicon, which can increase their strength much more by heat treatment than by cold working. Differences in composition require different temperatures for the

TABLE 3-17 ALLOWABLE TENSILE AND COMPRESSION STRESSES FOR MILD ALUMINUM (ANNEALED) VERSUS METAL OPERATING TEMPERATURE

Aluminum temperature (°C)	Allowable tensile stress (kg/cm^2)	Allowable bending stress (kg/cm^2)
30	150	250
31–60	140	225
61–80	130	200
81–100	120	175
101–120	105	150
121–140	90	125
141–160	75	100
161–180	60	75
181–200	45	50

high-temperature solution treatment, as well as variations both in time and temperature of aging. Heat-treatable plate and sheet alloys are not widely used for process plant construction because heat treatment has to be applied after welding to restore the mechanical properties.

Casting Alloys

These are used as corrosion-resistant materials. Some can be strengthened by heat treatment. The alloys containing substantial amounts of silicon have the best foundry characteristics and a high resistance to corrosion but are not readily machined.

Temperature effects

Tensile strength diminishes rapidly with increasing temperatures above 200°C. The high-magnesium alloys should not be used above 65°C because higher temperatures make them susceptible to stress-corrosion cracking.

Aluminum and its alloys are excellent for low temperatures as well as for cryogenic applications because their tensile strength and ductility are increased at low temperatures.

They are widely employed in the manufacture, storage, and distribution of liquified gases, particularly on sea and road tankers. The most popular alloy for cryogenic applications is 4.5 percent magnesium alloy. Table 3-18 gives the boiling points of the most common cryogenic liquids and the minimum temperatures at which various materials can be used.

Corrosion Resistance

Clean metallic aluminum is extremely reactive. Even exposure to air at ordinary temperatures is sufficient to promote immediate oxidation. This reactivity is self-inhibiting, however, which determines the general corrosion behavior of

TABLE 3-18 ALUMINUM ALLOYS RECOMMENDED FOR CRYOGENIC APPLICATIONS

Cryogenic liquid	Boiling point (°C)	Material	Minimum temperature of use (°C)
Propane	−42	Carbon steels	−50
Carbon dioxide	−78	2.25% nickel steel	−65
Acetylene	−84	3.5% nickel	−100
Ethylene	−104	Aluminum/magnesium alloys	−270
Methane	−161	Austenitic stainless steel	−270
Oxygen	−182	Nickel alloys	−270
Argon	−186	Copper alloys	−270
Nitrogen	−196		
Hydrogen	−253		

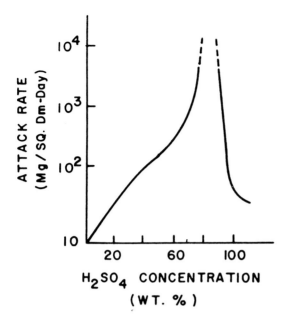

Figure 3-2 Effect of sulfuric acid on aluminum

aluminum and its alloys due to the formation of a thin, inert, adherent oxide film. In view of the great importance of the surface film, it can be thickened by anodizing in a bath of 15 percent sulfuric acid (H_2SO_4) solution or by cladding with a thin layer of an aluminum alloy containing 1 percent zinc.

Aluminum is suitable for contact with cold dilute sulfuric acid solutions, especially if an anodizing treatment has been applied. As the H_2SO_4 concentration increases, rates of attack also increase. This is illustrated in Figure 3-2. Aluminum may be used for handling oleum; however, the passivity of the metal with sulfuric acid is soon destroyed as the temperature rises, and the boiling solutions of an acid of any concentration attack rapidly.

Aluminum and stainless steel are used almost interchangeably for any strength of nitric acid. Figure 3-3 compares the rate of attack of cold nitric acid on stainless steel and aluminum. Figure 3-3 shows that higher rates of aluminum corrosion occur up to about 80 percent nitric acid (HNO_3), but aluminum is still to be preferred over stainless steel for any concentration above 80 percent.

Aluminum does not have the mechanical reliability of stainless steel, especially at higher temperatures. However, it is not nearly as susceptible to weld decay or localized attack around the welds.

Organic Acids

These are the second great field of application for aluminum alloys, with the exception of aluminum magnesium alloys. One restriction always applies in their use—the rapid increase in corrosion rate when perfectly anhydrous acids

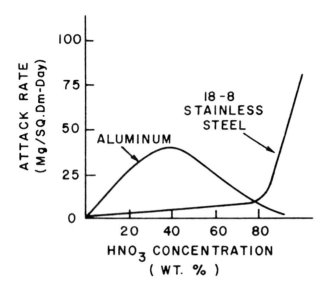

Figure 3-3 Effect of nitric acid on stainless steel and aluminum

are being handled at high temperatures. The resistance to most acid-reacting organic compounds increases with the acid concentration.

It is possible to carry out such oxidation processes as the conversion of acetaldehyde to acetic acid, or methyl alcohol to formaldehyde in aluminum plants, thus avoiding boiling anhydrous acids. The metal is especially valuable for handling delicate chemicals, which must not acquire metallic taste or color. For these reasons, aluminum has found extensive use in the food, dairy, brewing, and fishing industries.

Neutral salts and aqueous solutions of various acids generally follow the acid action. Aluminum has no apparent action or microbiological processes (that is, the production of antibiotics by deep-vessel fermentation). Fermentation tanks, as well as various absorbing and extracting units, can be made from aluminum.

Since aluminum is not attacked by hydrogen sulfide (H_2S) solutions, it is used widely as a material in refineries for the handling of hydrocarbons made from "sour" crudes. In the strongly oxidizing conditions of manufacturing hydrogen peroxide, aluminum is one of the few materials that does not undergo decomposition.

Steam-heated aluminum castings are used for the melt spinning of nylon and polyester fibers and have been used for the storage of raw materials during manufacturing, as well as for the storage of acetic acid in cellulose acetate plants.

MISCELLANEOUS PRECIOUS METALS

Titanium, tantalum, and zirconium are used for construction in process plants. The principal physical and mechanical properties of these three metals are given in Table 3-19.

TABLE 3-19 PROPERTIES OF TITANIUM, TANTALUM, AND ZIRCONIUM

	Density (g/cm^2)	Melting point (°C)	Coefficient of expansion × 10^{-6} (°C)	Thermal conductivity (W/m °C)	Yield strength (N/mm^2)	Tensile modulus (N/mm^2)	Hardness (DPN)
Titanium	4.5	1,668	9.0	15	345	103,000	150
Tantalum	16.6	2,996	6.5	55	240	185,000	170
Zirconium	6.5	1,852	7.2	17	290	80,000	180

Titanium

Titanium is a white metal and, when cold, is brittle and may be powdered. At a red heat, it may be forged and drawn. The tensile strength of titanium is almost the same as that of steel, while its specific gravity (4.5) is almost two times less than that of steel. Hence, its specific strength (tensile strength/specific gravity) is 1,000, which is considerably higher than that of 18/8 steel, which has a value of 700. Titanium is now available as plate, sheet, and tube, and its use in chemical plant construction is considered common.

The mechanical properties of titanium are greatly affected by small amounts of oxygen and nitrogen. The properties of the commercially pure grade metal and its alloys are given in Table 3-20.

The alloys with aluminum, vanadium, and tin have considerably greater strength but lower corrosion resistance. These alloys are used as rotating components in centrifuges, where the strength-to-weight ratio is important. About

TABLE 3-20 MECHANICAL PROPERTIES OF TITANIUM AND ALLOYS (ASTM B265/337/338)

Grade	Alloying element (% max.)	UTS (N/mm^2 min.)	Elongation (%)	Hardness (DPN)
1	Oxygen 0.18	240	24	150
2	0.25	345	20	180
4	0.40	550	15	260
5	Aluminum 6 Vanadium 4	900	10	
6	Aluminum 5 Tin 2.5	820	10	
7	Palladium 0.15/.25 Oxygen 0.25	345	20	180
8	Palladium 0.15/0.25 Oxygen 0.35	450	18	210

TABLE 3-21 EFFECT OF ELEVATED TEMPERATURES ON STRENGTH OF TITANIUM AND ALLOYS

Material ASTM grade	Tensile strength, N/mm^2, at test temperature of					
	Room temperature	100°C	200°C	300°C	400°C	500°C
1	310	295	220	170	130	
2	480	395	295	205	185	
3	545	460	325	250	200	
4	760	585	425	310	275	
5	1,030	890	830	760	690	655
6	960	820	690	620	535	460
7	480	395	295	205	185	
8	545	460	325	250	200	

0.2 percent palladium alloy gives better corrosion resistance than the first four grades in Table 3-20.

According to the ASME code (Section VIII, Div. 1), titanium may be used up to 300°C. The fatigue strength of the metal is about half the tensile strength. Typical values for the tensile strength of titanium and its alloys at temperatures up to 500°C are given in Table 3-21. The corrosion resistance of all grades of commercial pure titanium is similar. This protection relies on a surface film of metallic oxide. Therefore, titanium is most useful in oxidizing environments.

Titanium is generally suitable for use in boiling nitric acid, aqua regia, nitrites, nitrates, chlorides, sulfides, phosphoric acid, chromic acid, and organic acids. The main advantage of titanium over stainless steel is that it is not affected by pitting or stress-corrosion cracking in solutions containing chloride ions and has better resistance to erosion. It can be more easily protected anodically (less than 50 W for a surface equal to 100 m^2).

The corrosion resistance of unalloyed titanium in hydrochloric or sulfuric acids can be increased significantly by anodic protection, which maintains the oxide film so that the corrosion will be negligible even in severely reducing conditions.

If the metal is exposed to highly oxidizing conditions in the complete absence of water, a violent reaction may occur (for example, in completely dry chlorine). In this case, 0.015 percent water is added as the minimum for passivation of titanium.

Tantalum

Tantalum is a light bluish metal which is ductile, malleable, and resembles platinum when polished. The metal is characterized by high strength and infusibility. Its melting point is 3,000°C. The metal has high ductility, good forging,

Miscellaneous Precious Metals

flattening, and stamping. Excellent welds can be made by the TIG process; however, as tantalum reacts with oxygen and nitrogen at temperatures above 300°, careful shielding with argon of all areas likely to exceed this temperature is vital for success.

Tantalum has a degree of corrosion resistance similar to that of glass; therefore, it can be used in environments for which glass is required but without the risk of fracture and for purposes of heat transfer. The thermal conductivity of the metal is similar to that of nickel and nickel alloys.

In tantalum equipment very high flow rates can be experienced before erosion and cavitation occur, and a much higher thermal flux can be achieved. Therefore, the higher cost of tantalum can be justified for special uses when warranted.

The same volume of metal tantalum is 30 times more expensive than titanium, but it has the range of corrosion resistance more comparable with the precious, rather than the base, metals. It is only a fraction of the cost of platinum and gold.

In many applications tantalum can be substituted for platinum and gold, and there are some environments in which tantalum is more corrosion resistant than platinum. Table 3-22 lists the main chemicals for which tantalum is not a suitable substitute for platinum and, conversely, those for which tantalum is better than platinum. Tantalum is rapidly embrittled by nascent hydrogen even at room temperature. Therefore, it is very important to avoid the formation of galvanic couples between tantalum and other metals.

TABLE 3-22 COMPARATIVE CORROSION RESISTANCE OF TANTALUM AND PLATINUM

Chemical	Tantalum	Platinum
Acetylene	G^a	NR
Alkalis	NR^b	G
Bromine (wet or dry)	G	NR
Bromic Acid	G	NR
Cyanides	G	NR
Fluorine	NR	G
Compounds Containing Fluorine	NR	G
Ethylene	G	NR
Lead Salts	G	NR
Lead Oxide	NR	G
Metals (molten)	G	NR
Mercury	G	NR
Mercury Compounds	G	NR
Oleum	NR	G
Phosphoric Acid	NR	G
Sulfur Trioxide	NR	G

[a] G = good.
[b] NR = not recommended.

Zirconium

Of high purity, zirconium is a white, soft, ductile, and malleable metal. At 99 percent purity, when obtained at high temperatures it is hard and brittle. The rapid development of production techniques of zirconium has resulted because of its suitability for nuclear engineering equipment.

Zirconium has outstanding resistance to hydrochloric acid and is a cheaper alternative to titanium for this duty. It is superior to titanium in resistance to sulfuric acid. Zirconium has excellent resistance to caustic alkalies in all concentrations and is superior to both titanium and tantalum in this respect.

Precious Metals

The precious metals are many times the cost of the base metals and, therefore, are limited to specialized applications or to those in which process conditions are highly demanding (for example, where conditions are too corrosive for base metals and temperatures are too high for plastics; where base metal contamination must be avoided, as in the food and pharmaceutical industries; or where plastics cannot be used because of heat-transfer requirements; and for special applications such as bursting discs in pressure vessels). The physical and mechanical properties of precious metals and their alloys used in process plants are given in Table 3-23.

Silver

Silver is a white metal; it is softer than copper and harder than gold. One use of the pure metal (about 99.99 percent) is as a liner bonded to stronger or cheaper metals. The metallic bond is usually of high-thermal conductivity.

Both steel and copper vessels may be lined with thin silver sheets in the same way as for homogeneous lead lining. As silver is extremely resistant to

TABLE 3-23 PROPERTIES OF PRECIOUS METALS

	Platinum	Gold	Silver	10% Rh/Pt	20% Rh/Pt	10% Ir/Pt	20% Ir/Pt	70% Au 30% Pt 1% Rh
Density (g/cm^3)	21.45	19.3	10.5	20.0	18.8	21.6	21.7	20
Melting point (°C)	1,769	1,063	961	1,850	1,900	1,800	1,815	1,250
Thermal Conductivity (W/m °C)	70	290	418					
Young's Modulus (kN/mm^2)	170	70	70	195	215			
Tensile Strength Annealed (N/mm^2)	140	110	140	325	415	370	695	925
Hardness (DPN) Annealed	40	20	26	75	90	120	200	250

most organic acids at all concentrations and temperatures, it is used widely for handling foodstuffs and pharmaceutical products where nontoxicity and discoloration are essential. Silver is inert to hot alkaline solutions and very resistant to fused alkalies in the absence of oxidizing agents and to all neutral salt solutions.

Gold

Gold can be used only in very small portions or very thin coatings because of its cost. Most of the applications for which it was used in the past have now been accomplished with tantalum at a much lower cost. A gold/platinum/rhodium alloy is used in the manufacture of rayon-spinning jets in the production of rayon fibers. This alloy presents the combination of strength, corrosion resistance, and abrasion resistance necessary to prevent changes in hole dimensions.

Platinum

Platinum, plus other platinum group metals (Pt, Pd, Ir, Os, Rh, Ru) within the range of 99.8 percent to 99.99 percent platinum (Pt) content, are almost completely inert to chemical reagents under oxidizing conditions over a wide range of temperatures. At high temperatures under reducing conditions, however, it is attacked by all base metals, by molten silver and gold, and elemental silicon, boron, arsenic, phosphorus, bismuth, and sulfur. One particular case is the handling of molten glass in the manufacture of glass wool or glass threads for weaving. Again, platinum electrodes are used for electrolytic production of highly oxidizing materials, such as ammonium persulfate and chlorine from brine solutions.

METALLIC COATINGS

Metallic coating involves the deposition of metals and alloys onto other metals ranging in thickness from a few microns to several millimeters. This method allows for the possibility to obtain the properties of the coating at low cost, compared with making the items entirely from the coating composition. Coating permits the use of metals or alloys that are too brittle or too weak in the solid form (for example, chromium and zinc). The main reasons for applying a coating are prevention of corrosion, oxidation, and abrasion. Coatings are produced by four main methods: electrodeposition, spraying, dipping, and diffusion.

Electrodeposition

Nickel, chromium, and zinc are commonly used as electrodeposits. Chromium, the hardest of these coatings, is applied for abrasion resistance and low coefficient of friction. Nickel and zinc electrodeposits are used for resistance to corrosion, the latter for mildly corrosive conditions.

Dip Coating

Dip coating involves the immersion of steel or copper in a bath of molten coating metal (zinc, tin, and/or aluminum). Hot dip-galvanized (zinc-coated) steel should not be used in circuits containing copper equipment. This can result in galvanic corrosion at the copper/galvanized junctions, as well as cause overall galvanic corrosion of the zinc by copper redepositing from the water or process stream. Galvanized equipment is not recommended for use with liquors above 60°C. Above this temperature there is a reversal of the polarity of the zinc/steel couple, and the coating ceases to be protective where flaws appear in the coating. Impervious coatings of tin for mild corrosive conditions can be formed on steel and copper by dipping in a molten bath of tin.

Aluminum is the highest melting point metal (660°C) applied by hot dipping. Aluminized steel can be used at temperatures up to 550°C without appreciable oxidation. This steel has very good resistance to gases and vapors containing small quantities of sulfur dioxide and hydrogen sulfide.

Sprayed Coatings

Zinc, aluminum, nickel alloys, cobalt alloys, and tungsten carbide are applied for sprayed coatings, which are slightly porous. Flame-sprayed zinc coatings are used for corrosion protection of steel and provide similar properties for galvanized coatings.

Sprayed aluminum coatings used on steel for protection against atmospheric corrosion are preferred over zinc for use in areas with considerable contamination of the atmosphere by sulfur oxides. Sprayed aluminum also is used for the protection of steel at elevated temperatures up to 550°C. For temperatures of 550–900°C, aluminum is converted to a high melting point aluminum/iron compound by heating the coated equipment to 800/900°C and maintaining it at that temperature for 15 minutes. For protection up to 1,000°C, a sprayed coating of nickel chromium and nickel and cobalt alloys is applied. Nickel or cobalt alloys containing small amounts of boron or silicon can be deposited with very simple equipment, requiring very little heating of the base metal.

Diffusion Coatings

The purpose of diffusion coatings is not to produce a coating of another metal on the substrate, but to change the composition of the surface layers of the substrate by alloying with the diffusing metal chosen (zinc, aluminum, chromium, and silicon). The surface properties after such treatment depend not only on the metal diffused, but also on the composition of the substrate. The diffusion coating causes very little change in the dimensions of the piece being treated, which is important for items machined to fine limits, such as nuts and bolts.

Zinc diffusion is used for protection against atmospheric corrosion. Aluminum diffusion is used to improve the oxidation resistance of low-carbon steels.

Chromium diffusion applied to a low-carbon steel produces a surface that has the characteristics of ferritic stainless steel, such as AISI446 to a depth about 0.1 mm. When diffusion is applied to a high-carbon steel, a surface rich in chromium carbides is formed. This has a hardness greater than 1,000 VHN, which provides good resistance to abrasion. Nickel alloys and stainless steels such as AISI310 (25Cr/20Ni) diffusion treated with chromium enhance resistance to sulfur gases at high temperatures. The chromium-rich surface prevents the formation of nickel sulfide.

The use of equipment close to the temperature at which the material was diffusion treated will result in continuing diffusion of chromium, aluminum, and so on into the substrate, thus depleting chromium with a consequent loss in oxidation and corrosion resistance. For aluminum, this effect is noticeable above 700°C in steels, and above 900°C in nickel alloys. For chromium, the effect is pronounced above 850°C for steels and above 950°C for nickel alloys.

Silicon used for diffusion treatment of carbon steels enhances corrosion resistance to sulfuric acid. Such a treatment has the surface durability of iron/silicon alloys without their marked brittleness.

4

Carbon, Glass, and Plastics

CARBON AND GRAPHITE

Structural carbon shapes fabricated by heating coke with a mixture of tar and pitches are porous and are made impermeable by impregnation with a resin (usually a phenolic resin). Cashew nut shell liquid resin is used when resistance to alkalis and acids is required.

Graphite is used widely in process plants for its high-thermal conductivity (about six times that of stainless steel). Typical properties of impregnated carbon and graphite are given in Table 4-1.

Impregnated carbon and graphite can be used up to 180°C, and porous graphite can be used up to 400°C in oxidizing environments and 3,000°C in a reducing atmosphere. Carbon and graphite bricks and tiles are used for lining process vessels and are particularly suitable for applications involving severe thermal shock.

Tube and shell heat exchangers, small distillation columns, reactors, valves, pumps, and other items are available in impregnated graphite. Graphite can be joined only by cementing, which embrittles on aging. It is prone to mechanical damage, particularly when subjected to tensile stresses.

TABLE 4-1 PROPERTIES OF CARBON AND GRAPHITE

	Carbon	Graphite
Density (g/cm^3)	1.8	1.8
Tensile Strength (N/mm)	28	10
Compressive Strength (N/mm^2)	135	70
Tensile Modulus (kN/mm^2)	10	3
Thermal Conductivity (W/m °C)	4	70
Linear Coefficient of Expansion (°C^{-1})	3×10^6	4×10^6

Glass

By virtue of its chemical and thermal resistances, borosilicate glass has superior resistance to thermal stresses and shocks, and is used in the manufacture of a variety of items for process plants. Examples are pipe up to 60 cm in diameter and 300 cm long with wall thicknesses of 2–10 mm, pipe fittings, valves, distillation column sections, spherical and cylindrical vessels up to 400-liter capacity, centrifugal pumps with capacities up to 20,000 liters/hr, tubular heat exchangers with heat-transfer areas up to 8 m^2, maximum working pressure up to 275 kN/m^2, and heat-transfer coefficients of 270 kcal/hz/m^2°C.

Borosilicate glass has the following properties:

- density ρ = 2.7–31 kg/dm^3
- heat capacity C_p = 0.1–0.3 kcal/kg°C
- melting point t_m = 1,000–1,200°C
- thermal conductivity λ = 0.4–1.0 kcl/m°C hr
- linear coefficient of expansion $\alpha = 5 \times 10^{-6}$
- tensile strength 500–900 kg/cm^2
- compression strength 6,000–13,000 kg/cm^2
- modulus of elongation E = 620,000 kg/cm^2
- Poisson's ratio μ = 0.27–0.29

Because borosilicate is a brittle material, its design stress is restricted to less than 7 N/mm^2. Borosilicate glass is attacked by hydrofluoric acid even when a solution contains only a few parts per million of fluoride ions, and at elevated temperatures by strong solutions of phosphoric acid (85 percent acid at 100°C is the approximate limit). It is attacked also by strong bases such as sodium hydroxide and potassium hydroxide solutions, in which the effect is linear with time. As for all other materials, borosilicate represents almost the final resort in corrosion resistance. It has a very high degree of passivity to sensitive chemicals.

CEMENTS, BRICKS, AND TILES

Cements

Cements are used mainly for jointing brickworks, drainage gullies, pipes, and storage tanks. Portland cement has very poor resistance to acids but good resistance to alkaline liquors.

High-alumina cement is very quick setting, but its acid resistance is only slightly better than Portland cement, and it is rapidly attacked by alkalis. Supersulfated cement is used for liquors high in sulfates. It is resistant to acidic conditions down to a pH of 3.5 and has alkali resistance similar to Portland cement.

Bricks and Tiles

Lining vessels and equipment using this type of construction consists of a membrane, acid-resistant bricks or tiles, and chemical-resistant mortar for joining the bricks (tiles) together.

Membranes are applied directly to the metal or concrete surface to protect from corrosion by any liquor that penetrates the brick lining through pores and cracks. Membranes consist of sheet material bonded to the metal or concrete, for example, flexible PVC sheet, or it may be formed in situ (for instance, polyester resin reinforced with glass fiber, or synthetic rubber sheet, lead, polyisobutylene, polyethylene, and asphalt).

Brick lining (40–65 mm thick) is used for reducing the temperature at the membrane. This protects the membrane from deteriorating with free access of the process liquors and prevents its erosion and other mechanical damage.

Red and blue acid-resistant bricks are resistant to all inorganic and organic chemicals, except for hydrofluoric acid and hot concentrated caustic alkalis. Acid-resistant fireclay bricks are used for conditions involving elevating temperatures and corrosive condensates. Highly vitrified materials such as chemical stoneware, porcelain, and basalts are used for extremely severe duties or where contamination of the process liquors is undesirable.

Chemical-resistant cements are used for all acids up to temperatures of 170°C, but they are attacked rapidly by even dilute alkaline solutions.

Sulfur mortars (mixture of sulfur and inert fillers with small amounts of organic plasticizers) are used for shrinkage mitigation and for eliminating thermal shocks for temperatures up to 80°C. These mortars have poor resistance to alkalis and nonpolar organic solvents.

Phenolic mortars have excellent resistance to acids, particularly for dilute nitric acid (up to 50 percent) and sulfuric acid (60 percent to 90 percent) but can only tolerate very dilute alkaline solutions at low temperatures.

Furnace mortars are used over a very wide range of conditions. They are resistant to nonoxidizing acids, alkalis, and solvents up to 190°C. Carbon fillers should be used for conditions involving strong alkalis and compounds containing fluorine.

Plastic and Thermoplastic Materials

TABLE 4-2 CHEMICAL RESISTANCE OF BEDDING AND JOINTING CEMENTS

	Dilute acids	Concentrated acids	Caustic alkalis	Mineral oils	Animal and vegetable oils	Oxidizing acids
Rubber-Latex	G[a]	P[b]	G	F[c]	P	P
Silicate	VG[d]	VG	P	G	G	VG
Sulfur	G	G	P	G	P	G
Phenolic	VG	VG	P	G	G	F
Furane	VG	VG	P	G	G	P
Polyester	VG	G	VG	VG	VG	VG
Epoxy	VG	G	VG	VG	VG	F

[a] G = good.
[b] P = poor.
[c] F = fair.
[d] VG = very good.

Maximum chemical resistance of polyester mortars is obtained from cements made from isophthalic or biphenol resins. The biphenolic resins are preferred for alkaline and hypochlorite solutions.

Epoxy resin cements are specifically intended for resistance to caustic alkalis and organic solvents, but they also have fair acid resistance. They have excellent bond strength to other materials including ceramic and concrete. The corrosion resistances of the cements described previously are given in Table 4-2.

Bricks or tiles that line steel vessels have the tendency to crack when the vessel is heated because of the differences in the coefficients of thermal expansion between ceramic and steel.

PLASTIC AND THERMOPLASTIC MATERIALS

Plastics are highly resistant to a variety of chemicals. They have a high strength per unit weight of material; therefore, they are of prime importance to the designer of chemical process equipment. Their versatility in properties has provided new and innovative designs of equipment. They are excellent substitutes for expensive nonferrous metals.

Plastics are high molecular weight organic compounds of natural or mostly artificial origin. In fabrication, plastics are added with fillers, plasticizers, dyestuffs, and other additives, which are necessary to lower the price of the material, and give it the desired properties of strength, elasticity, color, point of softening, thermal conductivity, and so on.

Plastics are subdivided into two types: thermoplastic and thermosetting. The thermoplastics can be softened by heat and hardened again by subsequent cooling. This process is reversible and can be repeated many times. By contrast, the thermosetting resins are first softened and melted and, at subsequent heat-

TABLE 4-3 GENERAL PROPERTIES AND USES OF THERMOPLASTIC MATERIALS

Plastic material	First introduced	Strength	Electrical properties	Acids	Bases	Oxidizing agents	Common solvents	Product manufacturing methods	Common applications
Acetal Resins	1960	H^a		P^b	P	P	G^c	Injection blow, or extrusion molded	Plumbing, appliance, automotive industries
Acrylic Plastics	1931		G	P	P	F-P	F-P	Injection, compression, extrusion, or blow molded	Lenses, aircraft, and building glazing, lighting fixtures, coatings, textile fibers
Arc-Extinguishing Plastics	1964		E^d					Injection or compression molded and extruded	Fuse tubing, lightning arrestors, circuit breakers, panel boards
Cellulose Plastics Cellulose Acetate	1912	M^e					P	All conventional processes	Excellent vacuum-forming material for blister packages, and so on
Cellulose Acetate Butyrate		H				F	F	Molded with plasticizers	Excellent moisture-resistance metallized sheets and film, automobile industry
Cellulose Nitrate	1889	M						Cannot be molded	Little use today because of fire hazard
Cellulose Propionate		H						All conventional processes	Toys, pens, automotive parts, radio cases, toothbrushes, handles
Ethyl Cellulose		H+						All conventional processes	Military applications, refrigerator components, tool handles
Chlorinated Polyether	1959	M+	$A+^f$	VG	VG	VG	VG	Injection, compression, transfer, or extrusion molding	Bearing retainers, tanks, tank linings, pipe, valves, process equipment
Fluorocarbon (TFE)	1930	M	A	VG	VG	VG	VG	Molded by a sintering process following preforming	High-temperature wire and cable insulation, motor-lead insulation; chemical process equipment

Material	Year	Col1	Col2	Col3	Col4	Col5	Col6	Processing	Applications
Fluorinated Ethylene Propylene (FEP)		M+	A	G	G	G	G	Injection, blow molding, and extrusion and other conventional methods	Autoclavable laboratory ware and bottles
Glass-Bonded Mica	1919	M	G	VG	G	G	G	Moldable with inserts like the organic plastics	Arc chutes, radiation-generation equipment, vacuum tube components, thermocouples
Hydrogen Resins	1960	M	A	A	A	A	A	Molding with transfer and compression process, costing	Used as lamination resins for various industrial laminates
Methylpentene Polymers (TPX)	1965	M+	E	F	F	F	F	Most conventional processes	Used for electrical and mechanical applications
Parylene (poly-paraxylene)	1960							A monomer of the organic compound is vaporized and condensed on a surface to polymerize	Coating material for sensing probes
Phenoxy Plastics	1962	H-M	F	F	F	F	F	Injection, blow, and extrusion molding, coatings and adhesives	Adhesives for pipe-bonding compounds, bottles
Polyamide Plastics Nylon	1938	H-M	A	P	P	P	VG	Injection, blow, and extrusion molding	Mechanical components (gears, cams, bearings), wire insulation pipe fittings
Polycarbonate Plastics	1959	H-M	VG	G-F	G-F	F	F	All molding methods, thermoforming, fluidized-bed coating	Street light globes, centrifuge bottles, high-temperature lenses, hot dish handles

(*continued*)

TABLE 4-3 Continued

Plastic material	First introduced	Strength	Electrical properties	Acids	Bases	Oxidizing agents	Common solvents	Product manufacturing methods	Common applications
Polychlorotrifluoroethylene (CTFE)	1938	H	E	VG	VG	VG	VG	Molded by all conventional techniques	Wire insulation, chemical ware, pipe lining, pipe, process equipment lining
Polyester-Reinforced Urethane	1937	H		G	G	G	G	Compression molded over a wide temperature range	For heavy-duty leather applications—industrial applications
Polyimides	1964	H-M	E	G-VG	P	P	G-VG	Molded in a nitrogen atmosphere	Bearings, compressors, valves, piston rings
Polyolefin Plastics Ethlene Vinyl Acetate (EVA)	1940	H		G	G	G-F	G-F	Most conventional processes	Molded appliance and automotive parts, garden hose, vending machine tubing
Polyallomers	1962	H	G-VG	F	F	F	F	Molding processes, all thermoplastic processes	Chemical apparatus, typewriter cases, bags, luggage shells, auto trim
Polyethylene	1939	H	VG	G-VG	G-VG	P	P	Injection, blow, extrusion, and rotational	Pipe, pipe fittings, surgical implants, coatings, wire and cable insulation
Polypropylene Plastics	1954	H-M	VG	VG	VG	F	F	Same as PVC	Housewares, appliance parts, auto ducts and trim, pipe, rope, nets
Polyphenylene Oxide	1964	M	F-G	E	E	VG	VG	Extruded, injection molded, thermoformed and machined	Autoclavable surgical tools, coil forms, pump housings, valves, pipe
Polysulfone	1965	M	VG	VG	VG	F	P-F	Extrusion and injection molded	Hot-water pipes, lenses, iron handles, switches, circuit breakers

Material	Year	P1	P2	P3	P4	P5	P6	Molding	Uses
Polyvinylidene Fluoride (VF$_2$)	1961	H	VG+	G-VG		G-VG	G-VG	Molded by all processes, fluidized-bed coatings	High-temperature valve seats, chemical-resistant pipe, coated vessels, insulation
Styrene Plastics ABS Plastics	1933	M-H	VG+	G-VG	G-VG	F-G	F	Thermoforming, injection, blow, rotational, and extrusion molds	Business machine and camera housings, blowers, bearings, gears, pump impellers
Polystyrene	1933	M-H	VG+	G	G	F	P-F	Most molding processes	Jewelry, light fixtures, toys, radio cabinets, housewares, lenses, insulators
Styrene Acrylonitrile (SAN)		H	VG	VG	G	G	G	Most molding processes	Lenses, dishes, food packages, some chemical apparatus, batteries, film
Urethane	1955	M-H+		G-VG	G	G	F-G	Extruded and molded	Foams for cushions, toys, gears, bushings, pulleys, shock mounts
Vinyl Plastics, Copolymers of Vinyl Acetate and Vinyl Chloride, Polyvinyl Acetate	1835 1912	M+		G	G-F	G	G	All molding processes	Floor products, noise insulators
Polyvinyl Acetate	1928	M		P-G	G	G	P	Coatings and adhesives	Adhesives, insulators, paints, sealer for cinder blocks
Polyvinyl Aldehyde	1940	H		VG	VG	VG	VG	Most molding processes	Used for coatings and magnet wire insulation, interlayer of safety glasses
Polyvinyl Chloride (PVC)	1940	M-H		VG	G	G	G	Extrusion, injection, rotational, slush, transfer, compression, blow mold	Pipe conduit and fittings, cable insulation, downspouts, bottles, film

(continued)

TABLE 4-3 *Continued*

Plastic material	First introduced	Strength	Electrical properties	Acids	Bases	Oxidizing agents	Common solvents	Product manufacturing methods	Common applications
PVC Plastisols	1940	M-H		VG	G	G	G	Slush and rotationally molded, foamed, extruded	Used in coating machines to cover paper, cloth and metal
Polyvinylidene Chloride	1940	H		VG	VG	VG	VG	Same as PVC	Auto seatcovers, film, bristles, pipe and pipe linings, paperboard coatings

[a] H = high.
[b] P = poor.
[c] G = good.
[d] E = excellent.
[e] M = moderate.
[f] A = average.

ing to a definite temperature, they are irreversibly hardened, becoming insoluble.

Plastics are particularly resistant to inorganic chemicals but are often inferior to metals in resistance to organic chemicals. Table 4-3 gives general resistance properties and typical uses of thermoplastics. The strength characteristics obtained from short-time tests are not suitable for design purposes because all plastics gradually elongate or creep when subjected to sustained loads for long periods, even at ambient temperatures.

The physical and mechanical properties of the principal thermoplastics of interest for process plant applications are listed in Table 4-4. Table 4-5 gives typical hydrostatic design stresses for different types of thermoplastic pipe. Brief descriptions of plastics widely employed in piping systems follow.

Polyolefins

Polyethylene and polypropylene are semitransparent plastics made by polymerization. They are produced from ethylene and propylene in a variety of grades. Their mechanical properties are determined mainly by density (degree of crystallinity) and molecular weight, characterized by the Melt Index (MI).

The effect of these two parameters on mechanical and physical properties of polyethylene and polypropylene are shown in Tables 4-6 and 4-7. The copolymer grade is usually propylene with a little ethylene (5 percent), which considerably improves the impact strength while causing only a slight loss in stiffness.

Low-density polyethylene, a tough, relatively flexible material, is used at temperatures up to 50°C and does not become brittle until the temperature falls to $-40°C$.

This material is resistant to most hot, highly oxidizing inorganic acids, alkalis, and aqueous solutions of inorganic salts. Even though polyethylene is not actually dissolved by any organic substances at temperatures below 50°C, it is not a practical material for use in contact with many solvents (among them are chlorinated solvents and aromatics) because of its marked swelling. Although polar liquids such as alcohols, esters, amines, and phenols are not solvents suitable for use with polyethylene and do not cause swelling, they may cause environmental stress cracking. This occurs when polyethylene is stressed by either external loads or internal molding strains in contact with these solvents.

High-density polyethylene ($0.94–0.96$ g/cm^3) has up to five times the stiffness of low-density polyethylene at ambient temperatures and can be used at much higher temperatures. Its chemical resistance is similar to that of the low-density grades, but the resistance to swelling by solvents is higher.

High-density polyethylene is very susceptible to environmental stress-corrosion cracking, especially if used at the high end of the temperature range.

The chemical resistance of polypropylene is similar to that of polyethylene. Its standard grade is suitable for use at temperatures up to 90°C. All commercial grades of polypropylene contain an antioxidant, whose concentration in the polymer determines the maximum working temperature of equipment made

TABLE 4-4 MECHANICAL PROPERTIES OF THERMOPLASTICS

	Polyethylene	Polypropylene	PVC[a]	ABS[a]	PTFE[a]	Acrylics	Nylon 66	Acetal
Specific Gravity	0.93	0.90	1.4	1.05	2.20	1.2	1.15	1.4
Tensile Strength (N/mm^2)	10	34	55	34	21	70	83	70
Elongation at Break (%)	500	300	15	50	300	5	100	70
Tensile Modulus (N/mm^2)	170	1,360	3,100	2,100	415	3,450	2,760	2,900
Thermal Expansion(10^{-3}/°C)	15	11	7	10	10	7	10	10

[a] PVC = polyvinylchloride; ABS = acrylonitrile butadiene styrene; PTFE = polytetrafluorenethylene.

TABLE 4-5 HYDROSTATIC DESIGN PRESSURES FOR THERMOPLASTIC PIPE FOR TEMPERATURES UP TO 130°C

	Maximum hydrostatic design pressures, N/mm^2, for life of 10 years						Minimum temperature of usage[a]
	20°C	40°C	60°C	80°C	100°C	120°C	°C
Low-Density Polyethylene	2.7	1.7					−40
High-Density Polyethylene	4.8	3.9					−20
Polypropylene							
Homopolymer	6.9	4.1	2.2	1.2	1.0		−1
Copolymer	5.9	3.2	1.7	0.7			−5
CPVC[b]	13.8		6.9	3.5	1.1		−1
High-Impact PVC[b]	6.2	3.1					−20
ABS[b]	11.0		5.5	1.6			−30
PVF[b]	9.0	7.7	5.9	4.8	3.4	2.4	

[a] Temperature below which the pipe is very brittle.
[b] PVC = polyvinylchloride; ABS = acrylonitrile butadiene styrene; CPVC = chlorinated PVC; PVF = polyvinylfluoride.

from its high grades. The stiffness of polypropylene can be improved by incorporation of asbestos and glass fibers. Compared with polyethylenes, polypropylene has the great advantage of not being subject to environmental stress cracking.

The copolymer is safer to use in most process plant equipment where creep strength is low. However, for long life under pressure, particularly at the higher end of the temperature range, the homopolymer is preferred.

TABLE 4-6 EFFECT OF DENSITY OF POLYETHYLENE POLYMERS

	Polyethylene density, g/cm^3					Polypropylene (0.90)	
	0.92	0.93	0.94	0.95	0.96	Copolymer	Homopolymer
Crystallinity (%)	55	62	70	77	85		
Tensile Modulus, (N/mm^2)	170	275	450	625	850	1300	1500
ASTM Type[a]	I		II		III		
Melting Point (°C)	110				135	165	175
Heat Distortion Temperature (°C)	40	50	60	70	85	100	105

[a] D1248: Polyethylene plastics-molding and extrusion materials.

TABLE 4-7 EFFECTS OF DEGREE OF CRYSTALLINITY AND MOLECULAR WEIGHT

Increase in density (crystallinity)	Decrease in melt index (MI)	Basic properties unaffected by density and MI

Increases
  Elastic modulus
  Stress at yield
  Surface hardness
  Softening and melting temperature
  Impermeability to gases
Decreases
  Toughness
  Liquid absorption
	Increases
	  Resistance to environmental stress cracking
	  Tensile strength
	  Elongation at break
	
	Decreases
	  Temperature at which brittleness occurs
		Corrosion resistance, thermal properties, electrical properties, refractive index, decomposition temperature

Polyvinyl Chloride (PVC)

PVC, the polymerization product of chlorine-substituted ethylene derivatives, is probably the most widely used plastic for process plant construction. It is available in four different types: rigid, high impact, high temperature, and plasticized.

Rigid PVC (UPVC)

This is used most often in process plants. It is a tough, low-cost material with probably the widest range of chemical resistance of any of the low-cost plastics. On a volume basis, PVC is more favorable than polypropylene because the modulus of PVC is considerably higher than that of polypropylene, so it will form more rigid structures when used at the same thickness. On a weight basis it is not as favorable as PVC because it has a specific gravity of 1.4 compared with 0.92 for polypropylene.

The maximum temperature at which PVC can be used unsupported is 60°C. However, it can be used up to 100°C by external reinforcement with glass-reinforced polyester or epoxy resin. At low temperatures, rigid PVC becomes brittle and, in this case, a high-impact PVC should be used. Rigid PVC not suffering from environmental stress cracking is very resistant to most nonhighly oxidizing inorganic acids and alkalis. It is attacked by halogen gases, particularly when they are damp. PVC is much better than polyolefins for resistance to mineral oils, petrol and paraffin, but absorbs and is swollen by aromatic hydrocarbons, chlorinated solvents, esters, and ketones. This plastic is used widely in extruded forms such as sheet, pipe, and rod, and as injection moldings such as pipe fittings.

High-Impact PVC

These compounds are particularly advantageous for equipment exposed to low temperatures. This plastic containing chlorinated ethylene does not differ significantly in chemical resistance from rigid PVC, but does have a slightly lower creep strength.

Chlorinated PVC (CPVC)

This is used at temperatures up to 100°C and has similar chemical resistance impact. Presumably it will supersede normal PVC because of its superior tolerance to higher temperatures.

Plastic PVC

Produced from PVC compounds containing plasticizers such as dioctyl phthalate, this is a flexible sheeting suitable for lining tanks made from steel and concrete. The maximum temperature at which this lining can be used is 60°C.

Some solvents, including aromatic and chlorinated hydrocarbons, ketones, and ethers, will soften the sheet by acting as additional plasticizers.

Acrylonitrile-Butadiene-Styrene (ABS)

ABS, a tough rigid plastic, is widely used as sheet, pipe, and pipe fittings. ABS's chemical resistance is slightly inferior to rigid PVC. It has good impact and low-temperature properties similar to those of high-impact PVC, but has better resistance than rigid PVC to high temperatures (up to 80°C).

Fluorinated Plastics

Fluorinated polymers stand out sharply against other construction materials for their excellent corrosion resistance and high-temperature stability. In this respect they are not only superior to other plastics but also to platinum, gold, glass, enamel, and special alloys. The fluorinated plastics used in process plants are polytetrafluorethylene (PTFE), fluorinated ethylene/propylene (FEP), polytrifluoromonochlorethylene (PTFCE), and polyvinyl fluoride (PVF). They are much more expensive than other polymers and so are only economical in special situations.

In addition to excellent corrosion resistance at temperatures up to 250°C, PTFE has a very low coefficient of friction. Therefore, it is widely used for metal lining, self-lubricating bearings, seals, and so on. PTFE has a very low mechanical strength and so should be loaded only lightly. Its creep strength can be increased by using fillers (glass and carbon fibers). PTFE can be used in contact with all chemicals except molten alkali metals, fluorine, and chlorine trifluoride. Unlike PTFE, fluorinated ethylene propylene (FEP) can be extruded and injection molded. Therefore, FEP is available in a greater variety of forms than PTFE and can be used up to 200°C. The PTFE and FEP coatings do not protect metal from corrosion as they are porous. PTFCE (or CTFC) can be applied as a porous-free coating where the highest corrosion resistance is required.

Polyvinyl Fluoride (PVF)

PVF is a rigid plastic with corrosion resistance over a much wider range of conditions (it can be used with most chlorinated solvents up to 100°C) and temperatures ($-60°$ to 150°C).

Acrylics

Acrylics are used in applications in which transparency is necessary and breakage is likely to occur. It is tougher than glass and can be used in the temperature range from $-40°C$ to $+85°C$.

Acrylics are chemically resistant at room temperature to dilute acids, except hydrofluoric and hydrocyanic, all alkalis, and mineral oils. They are attacked by chlorinated solvents, aromatic hydrocarbons, ketones, alcohols, ethers, and esters.

Chlorinated Polyether

Chlorinated polyether used at temperatures up to 120°C in forms of coatings, extruded pipe, and sheet linings is an intermediate between the low-cost commodity polymers such as PVC and polyethylene, and fluorinated polymers.

Nylon (Polyamide)

Nylons, long-chain synthetic polymeric arnides, are fabricated in many different types. The most commonly used forms are described in Table 4-8.

High stiffness, fair resistance to corrosion, and good frictional properties sometimes make nylons more attractive than metals. The stiffness of the polymer may be increased by glass filling. Nylons containing molybdenum disulfide or graphite can be used over the temperature range of $-50°C$ to $+75°C$. Nylons are resistant at room temperature to very dilute acids, alkalis up to 207°C concentration, esters, alkyl halides, mercaptans, glycols, aldehydes (except formaldehyde), mineral oils, and petroleum fuels. They are attacked by halogens and halogenated compounds, as well as phenols and chesols.

Plastic and Thermoplastic Materials

TABLE 4-8 PROPERTIES OF DIFFERENT NYLONS

	Type nylon				
	66	Glass filled 66	8	610	11
Density (g/cm^3)	1.14	1.4	1.13	1.09	1.04
Melting point (°C)	260		220	220	190
Moisture Absorption (%)					
Equilibrium at 100% RH	8	5.5	11	3.5	1.5
Elastic Modulus (N/mm^2)					
Dry	3,000	7,500	2,750	1,750	1,250
Wet	1,200	2,500	700	1,100	900
Tensile Strength (N/mm^2)	85	170	80	60	60
Elongation at Break (%)	60	3	100	100	100
Heat Distortion Temperature (°C) (204 psi ASTM D648)	75	245	70	55	50

Miscellaneous Engineering Plastics

To overcome the disadvantage of nylon as an engineering material—high water absorption and poor creep strength at elevated temperatures—many newer polymers were developed. Table 4-9 lists polymers that are among the most commercially important: acetal, polycarbonate, polyphenylene oxide, and polysulfone.

TABLE 4-9 PROPERTIES OF DIFFERENT ENGINEERING PLASTICS

	Acetal	Polycarbonate	Polyphenylene oxide	Polysulfone	Nylon 66
Tensile yield stress (N/mm^2)	70	60	60	70	70
Elongation at fracture (%)	60	80	25	75	60
Elastic Modulus (dry) (N/mm^2)	2,800	2,500	2,500	2,500	3,000
Heat Deflection Temperature (°C) at 1.8 N/mm^2 (264 psi)	110	135	130	175	75
Water Absorption, 24 hr at 100% RH	0.22	0.15	0.07	0.22	1.5
Equilibrium Immersion	0.8	0.35	0.15		8
UL Temperature Index (°C)	80	115	105	140	65

Acetal Resin

This has the same stiffness as dry nylon but absorbs less than one fifth the amount of water and retains its strength and shape in humid conditions at considerably higher temperatures. Acetal has poor resistance to aqueous solutions of strong acids and should not be used at a pH below 3. The acetal's copolymers are satisfactory in most alkaline conditions. All grades have very good resistance to mineral oils and petrols and to most organic solvents including chlorinated compounds.

Polycarbonate

This has a very high resistance to impact damage, even at subzero temperatures. It has good creep strength in dry conditions up to 115°C but degrades by continuous exposures to water hotter than 65°C. It is resistant to aqueous solutions of acids, aliphatic hydrocarbons, paraffins, alcohols (except methanol), animal and vegetable fats and oils, but is attacked by alkalis, ammonia, aromatic and chlorinated hydrocarbons.

Polyphenylene Oxide

A stable material in humid conditions at temperatures up to 105°C, polyphenylene oxide is resistant to most aqueous solutions of acids and alkalis but is attacked by many organic solvents, particularly by aromatics and chlorinated aliphatics.

Polysulfone

Used at temperatures up to 140°C, polysulfone has good resistance to aqueous solutions of acids and alkalis; it is satisfactory with aliphatic solvents but is swollen by aromatics and stress cracked by several organic solvents, including acetone, ethyl acetate, trichlorethylene, and carbon tetrachloride.

THERMOSETTING PLASTICS

The thermosetting plastics—phenolics, polyesters, and epoxies—are used at higher temperatures (about 150°C) and pressures than thermoplastics. They are finding ever-increasing applications for process plant equipment.

Thermosetting resins are available as powders and liquids. The powders can be molded in a similar manner to thermoplastics and are used in process plants where large numbers of small items are required (for example, packing rings).

Liquid resins are usually reinforced with fibers (glass, asbestos) because of their brittleness. They are almost always used for process plant construction.

As liquid resins they can be catalyzed to cure at room temperature and low pressures. Relatively cheap wooden molds are required to build quite large items such as tanks and ducting on a one-off basis. Descriptions of the principal materials in this group of plastics follow.

Phenolic Resins

Phenoplasts manufactured on the basis of thermoreactive phenolformaldehyde resin harden on heating up to 120–170°C and then become insoluble. They are usually reinforced with asbestos fiber.

The phenolic/asbestos laminates (used up to 200°C) have excellent resistance to most mineral and organic acids but are attacked by strong oxidizing agents such as nitric and concentrated sulfuric acids and strong alkalis such as sodium and potassium hydroxide. Tanks, scrubbers, columns, pumps, pipes, and so on are fabricated from phenolic/asbestos laminates.

Polyester Resins

Polyester resins, reinforced with glass fibers, are used widely in the construction of process equipment. Some physical and mechanical properties are presented in Table 4-10. Table 4-11 lists various materials used as filler and the properties they impart to different plastics.

There are four types of polyester resins available: orthophthaleic, isophthaleic, biphenol, and chlorinated. The general-purpose orthophthaleic resins do not have adequate resistance to corrosion for most process plant requirements. Isophthaleic resins can be used up to 60°C. They are corrosion resistant to most organic and mineral (up to 25 percent) acids, salts that are not strongly alkaline, aliphatic solvents, and petroleum products, but not to aromatics, strongly oxidizing acids, hot distilled water, and strong alkalis.

The biphenol polyesters, being considerably more expensive than isophthaleics, have better chemical resistance to strong alkaline solutions and oxidizing solutions. For about the same price as the biphenols, the chlorinated resins are nonflammable. The addition of fire-retardant compounds—chlorine or bromine—can reduce the flammability of other polyester resins.

TABLE 4-10 VARIOUS PROPERTIES OF FIBERGLASS RESINS

Property	Polyester/ glass mat	Polyester/ woven glass cloth	Epoxy/ woven glass cloth	Epoxy/filament wound glass rovings
Glass (%w/w)	25	60	60	70
Specific Gravity	1.5	1.7	1.7	1.8
Thermal Expansion, $\times 10^{-6}$ °C	25	12	12	11
Modulus of Elasticity (kN/mm^2)	7	15	18	20
Flexural Strength (N/mm^2)	125	345	415	70

TABLE 4-11 VARIOUS FILLER MATERIALS AND THEIR PROPERTY CONTRIBUTIONS TO PLASTICS

Filler material	Chemical resistance	Heat resistance	Electrical insulation	Impact strength	Tensile strength	Dimensional stability	Stiffness	Hardness	Electrical conductivity	Thermal conductivity	Moisture resistance	Handleability
Alumina Powder									X	X		
Alumina Trihydrate	X	X	X								X	X
Asbestos			X	X		X	X	X		X		
Bronze									X			X
Calcium Carbonate		X				X	X	X				
Calcium Silicate		X				X	X	X				
Carbon Black		X				X	X	X	X	X		X
Carbon Fiber									X	X		
Cellulose				X	X	X	X	X				
Alpha Cellulose	X		X		X	X						
Coal (powdered)											X	
Cotton (chopped fibers)			X	X	X	X	X	X				
Fibrous Glass	X	X	X	X	X	X	X	X			X	
Graphite	X				X	X	X	X	X	X		
Jute				X			X					
Kaolin	X	X				X	X	X			X	X
Mica	X	X	X			X	X	X			X	X
Molybdenum Disulfide							X	X			X	X
Nylon (chopped fibers)	X	X	X	X	X	X	X	X				
Orlon®	X	X	X	X	X	X	X	X		X	X	
Rayon			X	X	X		X	X				
Silica, Amorphous			X			X	X	X			X	X
TFE-Fluorocarbon						X						
Talc	X	X	X			X	X	X			X	X
Wood Flour			X		X	X						

Epoxy Resins

The fully glass-reinforced epoxy laminates have much higher strength and corrosion resistance to solvents and alkalis than do polyester laminates, but are more expensive. They are used in the process industries for manufacturing piping to be used for very high working pressures by filament winding techniques with glass rovings.

Furane Resins

Furane resins are superior to polyesters and epoxies for resistance to ketones, chlorinated solvents, and carbon disulfide. However, as they are very brittle resins, they are difficult to make into self-supporting laminates. Therefore, they are often used as an inner layer to glass-reinforced polyester laminates.

Rubber Linings

Rubber, natural and synthetic, has been used extensively for many years in chemical process plants. Rubber is a product obtained by thermal processing (vulcanization) of a mixture of raw natural synthetic caoutchouc with sulfur.

Soft rubber is obtained by adding 2 percent to 4 percent sulfur; by adding extra sulfur (25 percent to 40 percent), the rubber can be made into ebonite, which is a hard, brittle material, having a wider range of chemical resistance than soft rubber. Soft ordinary rubber is chemical and erosion resistant, but its thermal resistance is not high (about 80°C).

Neoprene, which is basically polychloroprene, has better resistance to heat than does natural rubber (up to 105°C), and has better resistance to mineral and vegetable oils and fatty acids. It is attacked by aromatic and halogenated hydrocarbons. Butyl rubber, basically polyisobutylene, is used as a heat-resistant, impermeable material. It is better than natural rubber when in contact with oxidizing acids, such as dilute solutions of nitric acid. It is attacked by free halogens, chlorinated and aromatic hydrocarbons, and petroleum oils.

Nitrile rubbers, copolymers of butadiene and acrylonitrile, are used for resistance to swelling by mineral oils and fuels enhanced by formulations with a high acrylonitrile/butadiene ratio. They have poor resilience and low-temperature properties. However, these rubbers should not be used with ketones, phenols, or aromatic hydrocarbons.

Hypalon, chlorosulfated polyethylene, is particularly noted for its resistance to strong oxidizing materials such as sodium hypochlorite, chromic and nitric acids. It has good resistance to mineral and vegetable oils but is not recommended for use with aromatic and chlorinated hydrocarbons.

Fluorinated rubbers, copolymers of hexafluoropropylene and vinylidene-fluorides, have excellent resistance to oils, fuels, and lubricants at temperatures up to 200°C. They have better resistance to aliphatic, aromatic, and chlorinated hydrocarbons and most mineral acids than other rubbers, but their high cost restricts their engineering applications. A glossary of terms concerned with fabrication and properties of plastics is given in this chapter.

ORGANIC COATINGS AND PAINTS

Organic coatings are applied mainly to mild-steel structures and equipment. They are also used on aluminum, zinc-sprayed, and galvanized steel, but to a lesser extent. The applications for organic coatings can be divided into three areas: corrosion by atmospheric pollution, protection from splash by process liquors, and linings for immersion in process liquors.

Application of protective paints consists of surface preparation of steel, priming coat, and finishing coats. Wherever possible, steel should be cleaned by blasting before painting. Primers thoroughly wet the metal to promote adhesion of finishing paints and carry inhibitive pigments. For example, red lead oxide will minimize the spread of rust on metal surfaces. The total thickness of finishing coats must be at least 0.125 mm for adequate protection and life. Four coats of paint usually are necessary to achieve this.

Paints based on phenolic resins are oil modified to permit drying at ambient temperatures. They are very suitable for most industrial atmospheres. Paints with a higher standard of chemical resistance are required for equipment that is splashed by corrosive process liquors.

Chlorinated rubber paints and vinyl paints have excellent resistance to high concentrations of acids and alkalis at temperatures up to 0°C. High-build chlorinated rubber paints, which give a thickness of 0.12 mm per coat, are commonly used for process plant equipment.

Epoxy resin paints, inferior to chlorinated rubber for resistance to strong acids, are excellent for dilute acids and strong alkalis. They produce a harder, more abrasion-resistant coating than does chlorinated rubber and are much better for resistance to fats, oils, and many organic solvents. Table 4-12 gives data on the chemical resistance of epoxy resin coatings to different materials.

Chemical-resistant finishing paints are frequently used under immersion conditions; however, they are not used in cases in which pinholes in the coating might lead to catastrophic corrosion of the underlying metal. Coatings for immersion require special equipment for application (ovens for high-temperature curing, tanks for dipping, or special guns for spraying), but they provide the solution of complete elimination of porosity.

Stoved phenolics have outstanding acid resistance (up to 200°C in dry conditions and up to 100°C in wet conditions), except to strong oxidizing acids. They are unsuitable for use with alkaline solutions above pH 10, wet chlorine or hypochlorite solutions. Phenolics/silicon formulations can be used for steam up to 180°C without a significant effect on heat-transfer rates.

Polyester/glass-flake linings can be applied on site because they cure at ambient temperature. Their corrosion resistance depends on the type of polyester resin used.

TABLE 4-12   CHEMICAL RESISTANCE OF EPOXY RESIN COATINGS

Material	Concentration (%)	Temperature (°C)							
		5	15	27	38	49	60	71	82
Acetic Acid	1–5	G[a]	G	F[b]	F	F	F	F	P[c]
	5–10	F	F	P	P	P	NR[d]	NR	NR
	10–50	NR	NR	NR	NR	NR	NR	NR	NR
Acetone	1–5	G	G	G	F	F	F	P	P
	10–20	F	F	F	P	P	NR	NR	NR
Alcohols (ethyl)	—[e]	X[f]	X	G	G	F	F	P	P
Alum Sulfate	—	X	X	X	X	X	X	X	X
Ammonium Chloride	—	X	X	X	X	X	X	X	X
Ammonium Fluoride	—	X	X	X	X	X	X	X	X
Aromatic Solvents	—	X	X	X	X	G	G	G	F
Beer	—	X	X	X	X	X	X	X	X
Black Liquor	—	X	X	X	X	X	X	X	X
Boric Acid	1–5	X	X	X	G	G	G	F	F
Calcium Chloride	1–50	X	X	X	X	X	X	G	G
Carbon Tetrachloride	—	X	X	G	G	G	G	F	F
Chromic Acid	1–5	F	F	NR	NR	NR	NR	NR	NR
Citric Acid	1–5	F	F	F	G	G	G	G	F
Cooking Oils	—	X	X	X	X	X	X	X	G
Copper Salts	—	X	X	X	X	X	X	X	X
Esters	—	X	X	X	X	X	X	X	X
Esters (ethyl ether)	—	X	X	X	X	X	G	G	G
Formaldehyde	1–35	X	X	X	X	X	X	X	X
Ferric Chloride	—	X	X	X	X	G	G	G	F
Ferrous Salts	—	X	X	X	X	X	X	X	X
Gasoline	—	X	X	X	X	X	X	X	G
Glycerin	—	X	X	X	X	X	X	X	X
Hydrochloric Acid	1–5	X	X	X	G	G	F	F	F
Hydrofluoric Acid	1–5	G	G	G	P	P	NR	NR	NR
Kerosene	—	X	X	X	X	X	X	X	G
Lactic Acid	1–10	X	X	X	G	G	G	F	F
Lead Acetate	—	X	X	X	X	X	X	X	G
Manganese Salt	—	X	X	X	X	X	X	X	G
Methyl Ethyl Ketone	1–5	G	G	G	F	F	NR	NR	NR
Mineral Spirits	—	X	X	X	X	G	G	G	F
Naptha	—	X	X	X	X	X	X	X	X
Nitric Acid	1–5	F	F	P	P	P	P	NR	NR
	10–20	P	P	P	NR	NR	NR	NR	NR
Oxalic Acid	Saturated	G	G	G	F	F	NR	NR	NR
Phosphoric Acid	20	F	F	P	P	NR	NR	NR	NR
Potassium Hydroxide	—	X	X	X	X	X	X	X	X
Salt Brine	—	X	X	X	X	X	X	X	G
Soaps	—	X	X	X	X	X	X	X	G
Detergents	—	X	X	X	X	X	X	X	G
Sodium Chromate	—	X	X	X	X	X	X	X	G
Sodium Dichromate	—	G	G	G	G	G	G	F	F
Sodium Fluoride	—	X	X	X	X	X	X	X	X

(continued)

TABLE 4-12 Continued

Material	Concentration (%)	Temperature (°C)							
		5	15	27	38	49	60	71	82
Sodium Hydroxide	1–10	X	X	X	X	X	X	X	G
	50	X	X	X	G	G	G	G	F
Sodium Hypochlorite	3	G	G	F	P	P	NR	NR	NR
Sodium Phosphate	—	X	X	X	X	X	X	X	X
Sodium Sulfate	—	X	X	X	X	X	X	X	X
Sodium Sulfite	—	X	X	X	X	X	X	X	X
Sodium Thiosulfate	—	X	X	X	X	X	X	X	X
Sulfite Liquor	—	X	X	X	X	X	X	X	X
Sulfuric Acid	1–5	X	X	X	G	G	F	F	P
	10–20	X	X	P	P	NR	NR	NR	NR
Vegetable Oils	—	X	X	X	X	X	X	X	X
Water (fresh)	—	X	X	X	X	X	X	X	X
Water (distilled)	—	X	X	X	X	X	G	G	G
White Liquor	—	X	X	X	X	X	X	X	G

[a] G = good.
[b] F = fair.
[c] P = poor.
[d] NR = not recommended.
[e] — = all conditions.
[f] X = excellent.

PVC plastisol coatings are tough, with an abrasion resistance similar to rubber. They are resistant to acids and alkalis but usually are not suitable for solvents because of extraction of the plasticizers in the coating.

Most plastics are now available as powders and can be applied as coatings by fluidized-bed or spraying techniques. Nylon 11 and polyethylene have proved most useful for chemical plant applications.

GLOSSARY OF FABRICATION AND PLASTIC TERMS

A-stage—Initial or early stage in the reaction of some thermosetting resins in which the material is still soluble in certain liquids and fusible; referred to as *resol*.

Acid-acceptor—Chemical that acts as a stabilizer by chemically combining with an acid that may be present initially in trace quantities in a plastic; also may be formed via decomposition of the resin.

Acrylic plastics—Group of plastics based on resins generated from the polymerization of acrylic monomers (for example, ethyl acrylate and methyl methacrylate).

Adherend—A component or body held to another body by an adhesive.

Adhesion—Condition in which two surfaces are bonded together by interfacial forces caused by valence forces or interlocking forces or both (see *mechanical adhesion* and *specific adhesion*).

Adhesion, mechanical—Bonding between two surfaces caused by the interlocking action of molecules.

Adhesion, specific—Adhesion between surfaces whereby valence forces predominate that are similar to those promoting cohesion.

Adhesive—Material that holds parts together by surface attachment. Examples include glue, mucilage, paste, and cement. Various forms of adhesives include liquid or tape adhesives (physical type) and silicate or resin adhesives (chemical type).

Adhesive, assembly—Adhesive for bonding materials together, for example, boat, airplane, furniture, and so on; term commonly used in wood chemistry to distinguish between joint glues and veneer glues. Term applied to adhesives employed in fabricating finished goods, differs from adhesives used in fabricating sheet materials such as laminates or plywood.

Aging—The effect of exposure of plastics to the environment for a length of time. The specific effect and degree depend on the moisture in, and temperature and composition of, the environment in addition to the length of exposure.

Alkyd plastics—Group of plastics composed of resins based on saturated polymeric esters whereby the recurring ester groups are an integral part of the primary polymer chain and the ester groups exist in cross-links that are present between chains.

Alkyl plastics—Group of plastics composed of resins formulated by additional polymerization of monomers containing alkyl groups (for example, dialkyl phthalate).

Amino plastics—Group of plastics generated by the condensation of amines (for example, urea and melamine with aldehydes).

Anneal—As applied to molded plastics, the process of heating material to a specified temperature and slowly cooling it to relieve stresses.

Assembly—The positioning or placing together in proper order layers of veneer or other materials, with adhesives, for purposes of pressing and bonding into a single sheet or unit.

Assembly time—Refers to the elapsed time after an adhesive is applied until pressure effects curing.

B-stage—Intermediate-stage reaction step for various thermosetting resins. During this stage the material swells when in contact with certain liquids and becomes soft when heat is applied. The material may not dissolve or fuse entirely. Resin in this stage is referred to as *resitol*.

Backpressure-relief port—Opening from an extrusion die used for excess material to overflow.

Binder—Part of adhesive composition responsible for adhesive forces.

Blanket—Veneers laid up on a flat table. Complete assembly is positioned in a mold at one time; used primarily on curved surfaces to be molded by the flexible bag process.

Blister—Elevation of the surface of a plastic caused by trapped air, moisture, solvent; can be caused by insufficient adhesive, inadequate curing time, excess temperature, or pressure.

Blocking—Adhesion between layers of plastic sheets in contact; condition arises during storage or use when components are under pressure.

Bloom—Visible exudation or efflorescence on the surface of a plastic; caused by plasticizer, lubricant, and so on.

Bolster—Spacer or filler material in a mold.

Bond—The attachment at the interface or exposed surfaces between an adhesive and an adherend; to attach materials together with adhesives.

Bulk density—Density of a molding material in loose form, such as granular, nodular, and so on, with units in g/cm^3 or lb/ft^3.

Bulk factor—Ratio of the volume of loose molding compound to the volume of the same amount in molded solid form; ratio of density of solid plastic component to apparent density of loose molding compound.

Case harden—Process of hardening the surface of a piece of steel to a relatively shallow depth.

Cast film—Film generated by depositing a layer of liquid plastic onto a surface and stabilizing by evaporating the solvent, by fusing after deposition or by cooling. Cast films are generated from solutions or dispersions.

Catalyst—Material used to activate resins to promote hardening. For polyesters, organic peroxides are used primarily. For epoxies, amines and anhydrides are used.

Cavity—Portion of a mold that forms the outer surface of the molded product.

Cell—Single cavity caused by gaseous displacement in a plastic.

Cellular plastic—A plastic that suffers a density decrease by the presence of numerous cells dispersed throughout the material.

Cellular striation—Layering of cells within a cellular plastic.

Cellulosic plastics—Group of plastics composed of cellulose compounds, for example, esters (such as cellulose acetate) and ethers (such as ethyl cellulose).

Centrifugal casting—Process in which tubular products are fabricated through the application of resin and glass-strand reinforcement to the inside of a mold that is rotated and heated. The process polymerizes the resin system.

Chalking—Dry, chalklike deposit on the surface of a plastic.

Chase—Main portion of the mold containing the molding cavity, mold pins, guide pins, and so on.

Chemically formed plastic—Cellular plastic whereby the material's structure is formed by gases; plastic generated from the chemical reaction between its constituents.

Clamping plate—Mold plate that matches the mold and is used to fasten the mold to the machine.

Closed-cell foam—Cellular plastic composed predominantly of noninterconnecting cells.

Cohesion—Forces binding or holding a single material together.
Cold flow—Creep: the dimensional change of a plastic under load with time followed by the instantaneous elastic or rapid deformation at room temperature; permanent deformation caused by prolonged application of stress below the elastic limit.
Cold molding—The fashioning of an unheated mixture in a mold under pressure. The article is then heated to effect curing.
Cold pressing—Bonding process whereby an assembly is subjected to pressure without applying heat.
Cold slug—Material to first enter an injection mold.
Cold slugwell—Section provides opposite sprue opening of the injection mold, used for trapping cold slug.
Condensation—Chemial reaction whereby two or more molecules combine and separate out water or other substance. When polymers are formed, it is referred to as *polycondensation*.
Consistency—Resistance of a material to flow or to undergo permanent deformation under applications of shearing stresses.
Contact molding—Process whereby layers of resin-impregnated fabrics are built up one layer at a time onto the mold surface, forming the product. Little or no pressure is required for laminate curing.
Copolymer—Formed from two or more monomers (see *polymer*).
Core—Portion of the mold that forms the inner surfaces of the molded product.
Core and separator—Center section of an extrusion die.
Core pin—Pin for molding a hole.
Core-pin plate—Plate that holds core pins.
Crazing—Tiny cracks that develop on a laminate's surface caused by mechanical or thermal stresses.
Creep—See *cold flow*.
Cross-linkage—Generation of chemical linkages between long-chain molecules; can be compared to two straight chains joined together by links. The rigidity of the material increases with the number of links. The function of a monomer is to provide these links.
C-stage—Final reaction stage of various thermosetting resins. In this stage material is insoluble and infusible. Resin in fully cured thermosetting molding is in this stage and is referred to as *resite*.
Cull—Remaining material in the transfer vessel after the mold has been filled.
Cure—Process in which the addition of heat, catalyst, or both, with or without pressure, causes the physical properties of the plastic to change through a chemical reaction. Reaction may be condensation, polymerization, or addition reactions.
C-veil—Thin, nonwoven fabric composed of randomly oriented and adhered glass fibers of a chemically resistant glass mixture.
Degradation—Deleterious change in a plastic's chemical structure.
Delamination—Separation of a laminate's layers.
Deterioration—Permanent adverse change in the physical properties of a plastic.

Diaphragm gate—Gate employed in molding tubular or annular products.
Die adaptor—Piece of an extrusion die that serves to hold die block.
Die block—Part of an extrusion die that holds the core and forming bushing.
Die body—Part of an extrusion die used to separate and form material.
Dilatant—Property of a fluid whose apparent viscosity increases with shear rate.
Dished—Displays a symmetrical distortion of a flat or curved section; as viewed, it appears concave.
Dispersant—In an organosol, the liquid constituent that displays solvating or peptizing action on the resin; subsequent action aids in dispersing and suspending the resin.
Dispersion—Heterogeneous mixture in which finely divided material is distributed throughout the matrix of another material. Distribution of finely divided solids in a liquid or a solid (for example, pigments, fillers).
Doping—Coating a mandrel or mold with a material that prevents the finished product from sticking to it.
Dowel—Pin that maintains alignment between the various sections of a mold.
Draft—Angle of clearance between the molded article and mold, allowing removal from the mold.
Dry spot—Incompleted area on laminated plastics; the region in which the interlayer and glass are not bonded.
Durometer hardness—A material's hardness as measured by the Shore Durometer.
Ejector pin—Pin or dowel used to eject molded articles from a mold.
Ejector-pin retainer plate—Receptacle into which ejector pins are assembled.
Elasticity—Property of materials whereby they tend to retain or recover original shape and size after undergoing deformation.
Elastomer—A material under ambient conditions which can be stretched and, on release or with applied stress, returns with force to its approximate original size and shape.
Epoxy plastics—Group of plastics composed of resins produced by reactions of epoxides or oxiranes with compounds such as amines, phenols, alcohols, carboxylic acids, acid anhydrides, and unsaturated compounds.
Ethylene plastics—Group of plastics formed by polymerization of ethylene or by the copolymerization of ethylene with various unsaturated compounds.
Evenomation—Softening, discoloration, mottling, crazing, and so on. Process of deterioration of a plastic's surface.
Exotherm—Indicates that heat is given from a reaction between a catalyst and a resin.
Expandable plastics—Plastics that can be transformed to cellular structures by chemical, thermal, or mechanical means.
Extender—A material which, when added to an adhesive, reduces the amount of primary binder necessary.
Extraction—Transfer of materials from plastics to liquids with which they are in contact.

Extrusion—Process in which heated or unheated plastic compound is forced through an orifice, forming a continuous article.

Filament winding—Process in which continuous strands of roving tape are wound, at a specified pitch and tension, onto the outside surface of a mandrel. Roving is saturated with liquid resin or is preimpregnated with partially cured resin. Application of heat may be required to promote polymerization.

Filler—Inert material that is added to a plastic to modify the finished product's strength, permanence, and various other properties; an extender.

Fin—Portion of the flash that adheres to the molded article.

Finishing—Removal of any defects from the surfaces of plastic products.

Fisheye—A clump or globular mass that does not blend completely into the surrounding plastic.

Flash—Excess material that builds up around the edges of a plastic article; usually trimmed off.

Foamed plastic—Cellular structured plastic.

Force plate—A plate used for holding plugs in place in compression molding.

Furane plastics—Group of plastics composed of resins in which the furane ring is an integral portion of the polymer chain; made from polymerization or polycondensation of furfural, furfural alcohol, and other compounds containing furane rings; also formed by reaction of furane compounds with an equal weight or less of other compounds.

Fusion—As applied to vinyl dispersions, the heating of a dispersion, forming a homogeneous mixture.

Fusion temperature—Fluxing temperature; temperature at which fusion occurs in vinyl dispersions.

Gel—State at which resin exists before becoming a hard solid. Resin material has the consistency of a gelatin in this state; initial jelly-like solid phase that develops during the formation of a resin from a liquid.

Gel coat—Specially formulated polyester resin that is pigmented and contains fillers. Provides a smooth, pore-free surface for the plastic article.

Gel point—Stage at which liquid begins to show pseudoelastic properties.

Gelation—Formation of a gel.

Glass—Inorganic product of a fusion reaction. Material forms on cooling to a rigid state without undergoing crystallization. Glass is typically hard and brittle and will fracture conchoidally.

Glass transition—Transition region or state in which an amorphous polymer changed from (or to) a viscous or rubbery condition to (or from) a hard and relatively brittle one. Transition occurs over a narrow temperature region; similar to solidification of a glassy state. This transformation causes hardness, brittleness, thermal expansibility, specific heat, and other properties to change dramatically.

Gum—Class of colloidal substances prepared from plants. Composed of complex carbohydrates and organic acids that swell in water. Also, a number of natural resins are gums.

Halocarbon plastics—Group of plastics composed of resins generated from the polymerization of monomers consisting of a carbon and a halogen or halogens.

Hardener—Compound or mixture that, when added to an adhesive, promotes curing.

Heat treat—Refers to annealing, hardening, tempering of metals. Hot soils having a resistivity of less than 1,000 ohm-cm; generally very corrosive to base steel.

Hydrocarbon plastics—Plastics composed of resins consisting of carbon and hydrogen only.

Inhibitor—Material that retards chemical reaction or curing.

Isocyanate plastics—Group of plastics produced by the condensation of organic isocyanates with other plastics. Examples are the urethane plastics.

Isotactic—Type of polymeric molecular structure that contains sequences of regularly spaced asymmetrical atoms that are arranged in similar configuration in the primary polymer chain. Materials having isotactic molecules are generally in a highly crystalline form.

Isotropic—Refers to materials whose properties are the same in all directions. Examples are metals and glass mats.

Laminate—Article fabricated by bonding together several layers of material or materials.

Laminated, cross—Laminate in which some of the layers of materials are oriented at right angles to the remaining layers. Orientation may be based on grain or strength direction considerations.

Laminated, parallel—Laminate in which all layers of materials are oriented in parallel with respect to grain or strongest direction in tension.

Lignin plastics—Group of plastics composed of resins formulated from the treatment of lignin with heat or by reaction with chemicals.

Line pipe—Pipeline used for transportation of gas, oil, or water; utility distribution pipeline system ranging in sizes 1/8–42 in. o.d. inclusive. Fabricated to American Petroleum Institute (API) and American Water Works Association (AWWA) specifications.

Lyophilic—Referring to vinyl dispersions, having affinity for the dispersing medium.

Lyophobic—Referring to vinyl dispersions, no affinity or attraction for dispersing medium.

Mechanical tubing—Welded or seamless tubing manufactured in a large range of sizes of varied chemical compositions (sizes range 3/16–0 3/4 in. o.d. inclusive for carbon and alloy material); usually not fabricated to meet any specification other than application requirements; fabricated to meet exact outside diameter and decimal wall thickness.

Mechanically foamed—Cellular plastic whose structure is fabricated by plastic incorporated gases.

Melamine plastics—Group of plastics whose resins are formed by the condensation of melamine and aldehydes.

Metastable—Unstable state of plastic as evidenced by changes in physical properties not caused by the surroundings. Example is the temporary flexible condition some plastics display after molding.

Mold base—Assembly of all parts making up an injection mold, excluding cavity, cores, and pins.

Molding, bag—Process of molding or laminating in which fluid pressure is applied, usually by means of water, steam, air, or vacuum, to a flexible film or bag that transmits the pressure to the material being molded.

Molding, blow—Method of forming plastic articles by inflating masses of plastic material with compressed gas.

Molding, compression—Process of shaping plastic articles by placing material in a confining mold cavity and applying pressure and usually heat.

Molding, contact pressure—Method of molding or laminating whereby pressure used is slightly greater than is necessary to bind materials together during molding stage (pressures generally less than 10 psi).

Molding, high pressure—Molding or laminating with pressures in excess of 200 psi.

Molding, injection—Process of making plastic articles from powdered or granular plastics by fusing the material in a chamber under pressure with heat and forcing part of the mass into a cooler cavity where it solidifies; used primarily on thermoplastics.

Molding, low pressure—Molding or laminating at pressures below 200 psi.

Molding, transfer—Process of molding plastic articles from powdered, granular, or preformed plastics by fusing the material in a chamber with heat and forcing the mass into a hot chamber for solidification. Used primarily on thermosetting plastics.

Monomer—Reactive material that is compatible with the basic resin. Tends to lower the viscosity of the resin.

Nonrigid plastic—Plastic whose apparent modulus of elasticity is not greater than 10,000 psi at room temperature in accordance with the Standard Method of Test for Stiffness in Flexure of Plastics (ASTM Designation: D747).

Novolak—Phenolic-aldehyde resin that remains permanently thermoplastic unless methylene groups are added.

Nylon plastics—Group of plastics comprised of resins that are primarily long-chain synthetic polymeric amides. These have recurring amide groups as an integral part of the principal polymer chain.

Organosol—Suspension of finely divided resin in a volatile organic slurry.

Phenolic plastics—Group of plastics whose resins are derived from the condensation of phenols (for example, phenol and cresol, with aldehydes).

Piling pipe—Round-welded or seamless pipe for use as foundation piles where pipe cylinder acts as a permanent load-carrying member; usually filled with concrete. Used below the ground in foundation working the construction industry for piers, docks, highways, bridges, and all types of buildings. Fabricated to ASTM piling specifications (ASTMA-252).

Plastic—According to ASTM, a material containing an organic substance of large molecular weight that is sold in its finished state and, at some stage, is manufactured into finished goods and can be shaped to flow.

Plastic, semirigid—Plastic having apparent modulus of elasticity in the range of 10,000–100,000 psi at 23°C, as determined by the Standard Method of Test for Stiffness in Flexure Plastics (ASTM Designation: D747).

Plastic welding—Joining of finished plastic components by fusing materials either with or without the addition of plastic from another source.

Plasticate—Softening by heating or kneading.

Plasticity—Property of plastics that permits the material to undergo deformation permanently and continuously without rupture from a force that exceeds the yield value of the material.

Plasticize—Softening by adding a plasticizer.

Plasticizer—Material added to a plastic to increase its workability and flexibility. Plasticizers tend to lower the melt viscosity, the glass transition temperature, and/or the elastic modulus.

Plastisol—Suspension of finely divided resin in a plasticizer.

Polyamide plastics—See *nylon plastics*.

Polyester plastics—Group of plastics composed of resins derived principally from polymeric esters that have recurring polyester groups in the main polymer chain. These polyester groups are cross-linked by carbon/carbon bonds.

Polyethylene—Plastic or resin made by the polymerization of ethylene as the sole monomer.

Polymer—Material produced by the reaction of relatively simple molecules with functional groups that allow their combination to proceed to high molecular weights under suitable conditions; formed by polymerization or polycondensation.

Polymerization—Chemical reaction that takes place when a resin is activated.

Polypropylene—Plastic or resin derived from the polymerization of propylene as the principal monomer.

Polystyrene—Plastic derived from a resin produced by the polymerization of styrene.

Polyvinyl acetate—Resin derived from the polymerization of vinyl acetate.

Polyvinyl alcohol—Polymer derived from the hydrolysis of polyvinylesters.

Polyvinyl chloride—Resin derived from the polymerization of vinylchloride.

Polyvinyl chloride acetate—Copolymer of vinyl chloride and vinyl acetate.

Pot life—Time period beginning once the resin is catalyzed and terminating when material is no longer workable; working life.

Preform—Coherent block of granular plastic molding compound or of fibrous mixture with or without resin. Prepared by sufficiently compressing material, forming a block that can be handled readily.

Prepolymer—An intermediate chemical structure between that of a monomer and the final resin.

Pressure tubing—Tubing used to convey fluids at elevated temperatures and/or pressures. Suitable for head applications, it is fabricated to exact o.d. and decimal wall thickness in sizes ranging from 1/2–6 in. o.d. inclusive and to ASTM specifications.

Prime—Coating that is applied to a surface before application of an adhesive, enamel, and so on. The purpose is to improve bonding.

Promoted resin—Resin with an accelerator added but not catalyst.

Reinforced plastic—According to ASTM, those plastics having superior properties over those consisting of the base resin, due to the presence of high-strength fillers embedded in the composition. Reinforcing fillers are fibers, fabrics, or mats made of fibers.

Resin—Highly reactive material which, in its initial stages, has fluid-like flow properties. When activation is initiated, material transforms into a solid state.

Roller—A serrated piece of aluminum used to work a plastic laminate. Purpose of device is to compact a laminate and to break up large air pockets to permit release of entrapped air.

Roving—Bundle of continuous, untwisted glass fibers. Glass fibers are wound onto a roll called a *roving package*.

Saran plastics—Group of plastics whose resins are derived from the polymerization of vinylidene chloride or the copolymerization of vinylidene chloride and other unsaturated compounds.

Shelf life—Period of time over which a material will remain usable during storage under specified conditions such as temperature and humidity.

Silicone plastics—Group of plastics whose resins consist of a main polymer chain with alternating silicone and oxygen atoms and with carbon-containing side groups.

Softening range—Temperature range in which a plastic transforms from a rigid solid to a soft state.

Solvation—Process of swelling of a resin or plastic. Can be caused by interaction between a resin and a solvent or plasticizer.

Standard pipe—Pipe used for low-pressure applications such as transporting air, steam, gas, water, oil, and so on. Employed in machinery, buildings, sprinkler, irrigation systems, and water wells but not in utility distribution systems; can transport fluids at elevated temperatures and pressures not subjected to external heat applications. Fabricated in standard diameters and wall thicknesses to ASTM specifications, its diameters range from 1/8–42 in. o.d.

Stress crack—Internal or external defect in a plastic caused by tensile stresses below its short-time mechanical strength.

Structural pipe—Welded or seamless pipe used for structural or load-bearing applications in above-ground installations. Fabricated in nominal wall thicknesses and sizes to ASTM specifications in round, square, rectangular, and other cross-sectional shapes.

Structural shapes—Rolled flanged sections, sections welded from plates, and specialty sections with one or more dimensions of their cross section greater than 3 in. They include beams, channels and tees, if depth dimensions exceed 3 in.

Styrene plastics—Group of plastics whose resins are derived from the polymerization of styrene or the copolymerization of styrene with various unsaturated compounds.

Styrene-rubber plastics—Plastics that are composed of a minimum of 50 percent styrene plastic and the remainder rubber compounds.

Syneresis—Contraction of a gel, observed by the separation of a liquid from the gel.

Thermoelasticity—Rubber-like elasticity that a rigid plastic displays; caused by elevated temperatures.

Thermoforming—Forming or molding with heat.

Thermoplastic—Reverse of thermoset. Materials that can be reprocessed by applying heat.

Thermoset—Those plastics that harden on application of heat and cannot be reliquefied, resin state being infusible.

Thixotropy—Describes those fluids whose apparent viscosity decreases with time to an asymptotic value under conditions of constant shear rate. Thixotropic fluids undergo a decrease in apparent viscosity by applying a shearing force such as stirring. If shear is removed, the materials' apparent viscosity will increase back to or near its initial value at the onset of applying shear.

Tracer yarn—Strand of glass fiber colored differently from the remainder of the roving package. It allows a means of determining whether equipment used to chop and spray glass fibers is functioning properly and provides a check on quality and thickness control.

Urea plastics—Group of plastics whose resins are derived from the condensation of urea and aldehydes.

Urethane plastics—Group of plastics composed of resins derived from the condensation of organic isocyanates with compounds containing hydroxyl groups.

Vacuum forming—Fabrication process in which plastic sheets are transformed to desired shapes by inducing flow; accomplished by reducing the air pressure on one side of the sheet.

Vinyl acetate plastics—Group of plastics composed of resins derived from the polymerization of vinyl acetate with other saturated compounds.

Vinyl alcohol plastics—Group of plastics composed of resins derived from the hydrolysis of polyvinyl esters or copolymers of vinyl esters.

Vinyl chloride plastics—Group of plastics whose resins are derived from the polymerization of vinyl chloride and other unsaturated compounds.

Vinyl plastics—Group of plastics composed of resins derived from vinyl monomers, excluding those that are covered by other classifications (that is, acrylics and styrene plastics). Examples include PVC, polyvinyl acetate, polyvinyl butyral, and various copolymers of vinyl monomers with unsaturated compounds.

Vinylidene plastics—Group known as saran plastics.

Weathering—Exposure of a plastic to outdoor conditions.

Yield value—Also called *yield stress*; force necessary to initiate flow in a plastic.

Chemical Resistance of Thermoplastics

NOMENCLATURE

C_p heat capacity (kcal/kc-°C)
E modulus of elongation (kg/cm²)
G weight loss at uniform corrosion (kg/m²-hr)
t_m melting temperature (°C)
V corrosion rate (mm/yr)
λ thermal conductivity (kcal/m-°C-hr)
μ Poisson's ratio
ρ density (kg/dm³ or g/cm³)

CHEMICAL RESISTANCE OF THERMOPLASTICS

Chemical resistance of plastic piping and felting thermoplastic materials is given in Table 4-13.

The major advantages of thermoplastics in the process industries is economy of installation and maintenance and their outstanding chemical resistance and immunity to the corrosion that affects many metal piping systems.

Table 4-13 gives the chemical resistance of the four thermoplastic piping materials for a list of chemical and industrial fluids and gases.

TABLE 4-13 CHEMICAL RESISTANCE OF THERMOPLASTICS

CHEMICAL	PVC		CPVC			PP[3]				PVDF[4]		
	72°F	140°F	72°F	185°F	212°F	73°F	120°F	150°F	180°F‡	70°F	150°F	250°F
Acetaldehyde	NR	NR	NR	NR	NR	R	R			R	R	
Acetate Solvents, Crude	NR	NR	NR	NR	NR	NR	NR	NR	NR	R	F	NR
Acetate Solvents, Pure	NR	NR	NR	NR	NR	NR	NR	NR	NR	R	F	NR
Acetic Acid, 10%	R	R	R	F	NR	R	R	R	R	R	R	R
Acetic Acid, 20%	R	R	R	NR	NR	R	R	R	R	R	R	
Acetic Acid, 50%	F	NR	NR	NR	NR	R	R	R	R	R	R	R
Acetic Acid, 80%	NR	NR	NR	NR	NR	R	F	F		R	R	NR
Acetic Acid, Glacial	NR	NR	NR	NR	NR	R	R	F		R	R	NR
Acetic Anhydride	NR	NR	NR	NR	NR					F	NR	NR
Acetone	NR	NR	NR	NR	NR	R	R	F		NR	NR	NR
Acetophenone						F	F	F	NR	R	NR	NR
Acetyl Chloride	NR	NR	NR	NR	NR					R	R	
Acetylene	R	R	R	R		R				R		
Adipic Acid	R	R	R	R		R	R	R		R		
Air	R	R	R	R	R	R	R	R	R	R	R	R
Alcohol, Allyl	NR	NR	NR	NR	NR	R	R			R	R	
Alcohol, Amyl	NR	NR	R	NR	NR	R	F	F		R	R	R
Alcohol, Butyl	NR	NR	R	NR	NR	R	R	R	R	R	R	R
Alcohol, Ethyl	R	R	R	R	R	R	R	R	R	R	R	R
Alcohol, Methyl	R	R	R			R	R	R	R	R	R	R
Alcohol, Propyl	R	NR	R			R	R			R	NR	NR
Allyl Chloride	NR	NR	NR	NR	NR	R				R	R	
Alum	R	R	R	R	R	R	R	R	R	R	R	R
Alum, Ammonium	NR	NR	NR	NR	NR	R	R	R	R	R	R	R
Alum, Chrome	R	R	R	R	R	R	R	R	R	R		
Alum, Potassium	R	R	R	R	R	R	R	R	R	R	R	R
Aluminum Chloride	R	R	R	R	R	R	R	R	R	R	R	R
Aluminum Fluoride	R	R	R	R	R	R	R	R	R	R	R	R
Aluminum Hydroxide	R	R	R	R	R	R	R	R	R	R	R	R
Aluminum Nitrate	R	R	R	R	R	R	R	R	R	R	R	R

TABLE 4-13 Continued

CHEMICAL	PVC[1]		CPVC[2]			PP[3]				PVDF[4]		
	72°F	140°F	72°F	185°F	212°F	73°F	120°F	150°F	180°F‡	70°F	150°F	250°F
Aluminum Oxychloride	R	R	R	R		R	R	R	R			
Aluminum Sulfate	R	R	R	R	R	R	R	R	R	R	R	R
Ammonia, Gas	R	R	R	R	R	R	R	R		R	R	R
Ammonia, Aqua, 10%	R	R	R	R	R	R	R	R		R	R	R
Ammonia, (25% Aqueous Solution)	R	R	R	R		R	R	R		R	R	R
Ammonia (Concentrated)	NR	NR	NR	NR	NR	R	R	F		R	R	R
Ammonium Acetate	R	R	R	R	R	R	R	R		R	R	
Ammonium Bifluoride	R	R	R	R		R	R	R	R	R	R	R
Ammonium Carbonate	R	R	R	R		R	R	R	R	R	R	R
Ammonium Chloride	R	R	R	R		R	R	R	F	R	R	R
Ammonium Fluoride, 10%	R	R	R	R		R	R	R	R	R	R	R
Ammonium Fluoride, 25%	NR	NR	NR	NR	NR	R	R	R	R	R	R	R
Ammonium Hydroxide	R	R	R	R		R	R	R	R	R	R	R
Ammonium Metaphosphate	R	R	R	R		R	R	R	R			
Ammonium Nitrate	R	R	R	R		R	R	R	R	R	R	R
Ammonium Persulphate	R	R	R	R		NR	NR	NR	NR	R	R	R
Ammonium Phosphate	R	R	R	R		R	R	R	R	R	R	R
Ammonium Sulfate	R	R	R	R		R	R	R	R	R	R	R
Ammonium Sulfide	R	R	R	R		R	R	R	R	R	R	R
Ammonium Thiocyanate	R	R	R	R		R	R	R				
Amyl Acetate	NR	NR	NR	NR	NR	NR	NR	NR	NR	R	NR	NR
Amyl Chloride	NR	NR	NR	NR	NR	NR	NR	NR		R	R	R
Aniline	NR	NR	NR	NR	NR	R	R	R	F	R	NR	NR
Aniline Hydrochloride	NR	NR	NR	NR	NR	R				R		
Anthraquinone Sulfonic Acid	R	R	R	R		R				R		
Antimony Trichloride	R	NR	NR	NR	NR	R	R	NR	NR	R	NR	NR
Aqua Regia	NR	NR	NR	NR	NR	NR	NR	NR	NR	NR	NR	NR
Arsenic Acid	R	R	R	R		R				R	R	R
Aryl Sulfonic Acid	NR	NR	NR			NR				NR		
Barium Carbonate	R	R	R	R		R	R	R	R	R	R	R
Barium Chloride	R	R	R	R		R	R	R	R	R	R	R
Barium Hydroxide	R	R	R	R		R	R	R	R	R	R	R
Barium Sulfate	R	R	R	R	R	R	R	R	R	R	R	R
Barium Sulfide	R	F	R	R	R	R	R	R	R	R	R	R
Beer	R	R	R	R	R	R	R	R	R	R	R	R
Beet Sugar Liquors	R	R	R	R	R	R	R	R	R	R	R	R
Benzaldehyde, above 10%	NR	NR	NR	NR	NR	R	R	R		R	NR	NR
Benzene, Benzol	NR	NR	NR	NR	NR	NR	NR	NR	NR	R	NR	NR
Benzene Sulfonic Acid, 10%	R	R	R	R		R	R	R	R	R	NR	NR
Benzoic Acid	R	R	R	R		R	R	R		R	R	R
Benzyl Alcohol	NR	NR	NR	NR	NR	R	R	NR	NR	R	R	
Bismuth Carbonate	R	R	R	R	R	R	R	R	R	R	R	R
Black Liquor	R	R	R	R		R	R	R	NR	R	R	R
Bleach, 12.5% Active Cl,	R	R	R	R	R	R	NR	NR	NR	R	R	R
Borax	R	R	R	R		R	R	R	R	R	R	R
Boric Acid	R	R	R	R		R	R	R	R	R	R	R
Breeder Pellets, Deriv. Fish	R	R	R	R	R	R	R	R	R	R	R	R
Brine, Acid	R	R	R	R	R	R	R	R	R	R	R	R
Bromic Acid	R	R	R	R		NR				R	R	
Bromine, Liquid	NR	NR	NR	NR		NR	NR	NR	NR	R	R	
Bromine, Water	NR	NR	NR	NR	NR	NR	NR	NR	NR	R	R	
Bromine, Water, Saturated	NR	NR	NR	NR	NR	NR	NR	NR	NR	R	R	
Butadiene	NR	NR	R	R		R	R			R	R	R
Butane	R	NR	R	NR	NR	R	NR	NR	NR	R	NR	NR
Butanol, Primary	NR	NR	R	R	NR	R	R	R	R	R	R	R
Butanol, Secondary	NR	NR	R	R	NR	R	R	R	R	R	R	R
Butyl Acetate	NR	NR	NR	NR	NR	F	NR	NR	NR	R	NR	NR
Butyl Alcohol	NR	NR	R	NR	NR	R	R	R	R	R	R	R
Butyl Phthalate	NR	NR	NR	NR	NR	R	R	R	R	R	R	R
Butylene	R	NR	NR	NR	NR	NR	NR	NR	NR	R	R	R
Butyl Phenol	NR	NR	NR	NR	NR	R				R	R	

TABLE 4-13 Continued

CHEMICAL	PVC[1] 72°F	PVC[1] 140°F	CPVC[2] 72°F	CPVC[2] 185°F	CPVC[2] 212°F	PP[3] 73°F	PP[3] 120°F	PP[3] 150°F	PP[3] 180°F‡	PVDF[4] 70°F	PVDF[4] 150°F	PVDF[4] 250°F
Butyne Diol	NR	NR	R	NR	NR	R	R			R		
Butyric Acid	NR	NR	R	NR	NR	R	R	R	R	R	R	R
Calcium Bisulfide	R	R	R	R	R	R	R	R	R	R	R	R
Calcium Bisulfite	R	R	R	R	R	R	R	R	R	R	R	R
Calcium Carbonate	R	R	R	R	R	R	R	R	R	R	R	R
Calcium Chlorate	R	R	R	R	. .	R	R	R	R	R	R	R
Calcium Chloride	R	R	R	R	R	R	R	R	F	R	R	R
Calcium Hydroxide	R	R	R	R	R	R	R	R	R	R	R	R
Calcium Hypochlorite	R	F	R	R	. .	R	F	F	F	R	R	R
Calcium Nitrate	R	R	R	R	. .	R	R	R	R	R	R	R
Calcium Sulfate	R	R	R	R	R	R	R	R	R	R	R	R
Cane Sugar Liquors	R	R	R	R	R	R	R	R	R	R	R	R
Carbon Bisulfide	NR	NR	NR	NR	NR	NR	NR	NR	NR	R	NR	NR
Carbon Dioxide, wet	R	R	R	R	R	R	R	R		R	R	R
Carbon Dioxide, dry	R	R	R	R	R	R	R	R	. .	R	R	R
Carbon Disulfide	NR	NR	NR	NR	NR	NR	NR	NR	NR	R	NR	NR
Carbonic Acid	R	R	R	R	. .	R	R	R	R	. .	. .	. .
Carbon Monoxide	R	R	R	R	R	R	R	R	R	R	R	. .
Carbon Tetrachloride	NR	NR	NR	NR	NR	NR	NR	NR	NR	NR	NR	NR
Castor Oil	R	R	R	R	NR	R	R	NR	R	R	R	R
Caustic Potash	R	R	R	R	R	R	R	R	R	R	R	R
Caustic Soda	R	R	R	R	. .	R	R	R	R	R	R	. .
Cellosolve	R	F	R	F	. .	R	. .	. .	. .	R	R	R
Chloral Hydrate	R	R	R	R	. .	R	. .	. .	. .	R	. .	. .
Chloric Acid, 20%	R	R	R	R	. .	NR	. .	. .	. .	R	R	. .
Chlorine Gas, dry	NR	NR	NR	NR	NR	NR	NR	NR	NR	R	R	
Chlorine Gas, wet	NR	NR	NR	NR	NR	NR	NR	NR	NR	R	R	
Chlorine, Liquid	NR	NR	NR	NR	NR	NR	NR	NR	NR	R	R	
Chlorine Water, Saturated	R	R	R	R	R	R	R	R	R	R	R	
Chloroacetic Acid	NR	NR	R	F	NR	R	NR	NR	. .	R	R	R
Chlorobenzene	NR	NR	NR	NR	NR	NR	NR	NR	NR	R	R	NR
Chlorobenzyl Chloride	NR	NR	NR	NR	NR	NR	NR	NR	NR	R	. .	. .
Chloroform	NR	NR	NR	NR	NR	F	NR	NR	NR	R	R	
Chlorosulfonic Acid, 100%	F	NR	F	NR	NR	NR	NR	NR	NR	NR	NR	NR
Chlorox Bleach Solution, 5.5% Cl₂	R	R	R	R	R	F	NR	NR	NR	R	R	R
Chrome Alum	R	R	R	R	NR	R	R	NR	NR	R	NR	NR
Chromic Acid, 10%	NR	NR	NR	NR	NR	R	R	F	NR	R	R	
Chromic Acid, 30%	NR	NR	NR	NR	NR	R	R	F	NR	R	R	. .
Chromic Acid, 40%	NR	NR	NR	NR	NR	R	R	F	NR	R	R	. .
Chromic Acid, 50%	NR	NR	NR	NR	NR	R	R	F	NR	R	R	NR
Citric Acid Solution	R	R	R	R	R	R	R	R	. .	R	R	R
Citric Acid, 10%	R	R	R	R	R	R	R	R	NR	R	R	R
Coconut Oil	R	R	R	. .	. .	R	R	R	. .	R	R	. .
Coke Oven Gas	NR	NR	R	R	R	R	. .	. .	. .	R	. .	. .
Copper Carbonate	R	R	R	R	R	R	R	R	. .	R	R	R
Copper Chloride	R	R	R	R	R	R	R	R	. .	R	R	R
Copper Cranide	R	R	R	R	R	R	R	R	. .	R	R	R
Copper Fluoride	R	R	R	R	R	R	R	R	. .	R	R	R
Copper Nitrate	R	R	R	R	R	R	R	R	. .	R	R	R
Copper Salts	R	R	R	R	R	R	R	R	. .	R	R	R
Copper Sulfate	R	R	R	R	R	R	R	R	. .	R	R	R
Cottonseed Oil	R	R	R	R	R	R	R	R	. .	R	R	R
Cresol	NR	NR	F	NR	NR	R	NR	. .	. .	R	R	. .
Cresylic Acid, 50%	NR	NR	F	NR	NR	R	NR	. .	. .	R	R	. .
Croton Aldehyde	NR	NR	NR	NR	. .	R	. .	. .	. .	R	R	NR
Crude Oil	R	R	R	R	. .	R	. .	. .	. .	R	R	R
Cyclohexane	R	R	R	. .	. .	NR	NR	NR	NR	R	R	R
Cyclohexanol	NR	NR	NR	NR	. .	R	NR	NR	. .	R	R	NR
Cyclohexanone	NR	NR	NR	NR	. .	NR	NR	NR	NR	R	NR	NR
Decalin	NR	NR	NR	NR	NR	R	R	R	R	R	R	R

(continued)

TABLE 4-13 *Continued*

CHEMICAL	PVC[1]		CPVC[2]			PP[3]				PVDF[4]		
	72°F	140°F	72°F	185°F	212°F	73°F	120°F	150°F	180°F‡	70°F	150°F	250°F
Detergents	R	R	R	R		R	R	R	R	R		
Detergent Solution (Heavy Duty)	R	R	R	R		R	R	R	F	R	R	R
Dextrin	R	R	R	R		R	R	R		R	R	
Dextrose	R	R	R	R		R	R	R		R	R	R
Diacetone Alcohol	NR		NR			R	R			R	NR	
Diazo Salts	R	R	R	R		R	R			R		
Dibutyl Phthalate	NR	NR	NR	NR		R	F	NR		R		
Dichloroethylene	NR		NR			NR				R		
Diesel Fuels	R	F	R	F	NR	R	F			R	R	R
Diethyl Ether	NR	NR	NR	NR	NR	R	NR	NR	NR	R	NR	NR
Diglycolic Acid	R	R	R	R		R				R		
Dimethylamine, Aqueous	NR	NR	NR	NR		R	R			NR	NR	NR
Dimethyl Formamide	NR	NR	NR	NR		R		R	NR	R		
Dimethylamine	NR	NR	NR			R	R			NR	NR	NR
Dioctyl Phthalate	NR	NR	NR	NR	NR	NR	NR	NR	NR	R		
Dioxane	NR	NR	NR	NR	NR	R	NR	NR	NR	NR	NR	NR
Dioxane, 1,4	NR	NR	NR	NR	NR	R	NR	NR	NR	NR	NR	NR
Disodium Phosphate	R	R	R	R	R	R	R	R	R	R	R	R
Distilled Water	R	R	R	R	R	R	R	R	R	R	R	R
Divinylbenzene	NR	NR	NR	NR	NR	NR	NR	NR	NR	NR	NR	NR
Epichlorohydrin	NR	NR	NR	NR	NR	R	R			R		
Esters	NR	NR	NR	NR	NR	F	NR	NR	NR	R	NR	NR
Ethanol	R	R	R	R	R	R	R	R	R	R	R	R
Ethers	NR	NR	NR	NR		F	F	NR	NR	R		
Ethyl Acetate	NR	NR	NR	NR		F	F	F	NR	R	NR	NR
Ethyl Acetoacetate	NR	NR	NR	NR		NR				R	NR	NR
Ethyl Acrylate	NR	NR	NR	NR		NR				R	NR	NR
Ethyl Alcohol	R	R	R			R	R	R		R	R	R
Ethyl Chloride	NR	NR	NR	NR		NR	NR	NR	NR	R	R	R
Ethyl Ether	NR	NR	NR	NR	NR	R	NR	NR	NR	R	NR	NR
Ethylene Bromide	NR	NR	NR	NR		F				R	R	
Ethylene Chloride	NR	NR	NR	NR		F	NR	NR	NR	R	R	R
Ethylene Chlorohydrin	NR	NR	NR	NR		R				R	NR	NR
Ethylene Diamine	NR		NR			R				F	NR	NR
Ethylene Dichloride	NR	NR	NR	NR	NR	F	F	NR	NR	R	R	R
Ethylene Glycol	R	R	R	R		R	R	R	NR	R	R	R
Ethylene Oxide	NR	NR	NR	NR	NR	NR	NR	NR	NR	R	R	
Fatty Acids	R	R	R	R		R	R	R	R	R	R	R
Ferric Chloride	R	R	R	R		R	R	R	R	R	R	R
Ferric Nitrate	R	R	R	R		R	R	R	R	R	R	R
Ferric Sulfate	R	R	R	R		R	R	R	R	R	R	R
Ferrous Chloride	R	R	R	R		R	R	R	F	R	R	R
Ferrous Nitrate	R	R	R	R		R	R	R	R	R	R	R
Ferrous Sulfate	R	R	R	R		R	R	R	R	R	R	R
Fish Solubles	R	R	R	R		R	R	R	R	R	R	R
Fluorine Gas, wet	R	R	R			NR				R		
Fluoboric Acid	R	R	R	R		R				R		
Fluosilicic Acid	NR	NR	R	R		R	R			R	R	R
Formaldehyde, 35% Solution	R	R	R	NR		R	R		F	R	R	R
Formaldehyde, 37%	R	R	R	NR		R	R	R	F	R		
Formaldehyde, 50%	R	R	R	NR		R	R	R		R		
Formic Acid	R	NR	R	NR		R	R	F		R	R	R
Formic Acid (Anhydrous)	F	NR	R	NR	NR	R	R	R	NR	R	R	R
Freon F-11	R		R			R				R	R	
Freon F-12	NR		R			R				R	R	
Freon F-22	NR	NR	NR			R				R	R	
Fructose	R	R	R	R		R	R	R	R	R	R	R
Fruit Juices, Pulp	R	R	R	R		R	R	R	R	R	R	R
Furfural	NR	NR	NR	NR		F	F	F	NR	F	F	NR

Chemical Resistance of Thermoplastics

TABLE 4-13 *Continued*

CHEMICAL	PVC[1]		CPVC[2]			PP[3]				PVDF[4]		
	72°F	140°F	72°F	185°F	212°F	73°F	120°F	150°F	180°F‡	70°F	150°F	250°F
Gallic Acid	R	R	R	R		R				R	F	
Gas, Manufactured	R		R	R		R				R	R	R
Gas, Natural	R	R	R	R		R				R	R	R
Gasoline, Leaded	F	F	R	NR		NR	NR	NR	NR	R	R	R
Gasoline, Unleaded	F	F	R	NR		NR	NR	NR	NR	R	R	R
Gasoline, Sour	R	R	R	R		NR	NR	NR	NR	R	R	R
Gasoline, Refined	F	F	R	NR		NR	NR	NR	NR	R	R	R
Gelatin	R	R	R	R		R	R	R	R	R	R	R
Gin	R	R	R	R		R	R			R	R	R
Glucose	R	R	R	R		R	R	R	R	R	R	R
Glue	R	R	R	R		R	R			R	R	
Glycerine, Glycerol	R	R	R	R		R	R	R	R	R	R	R
Glycolic Acid	R	R	R	R		R	R	R	R	R	NR	NR
Glycols	R	R	R	R		R	R	NR	NR	R	R	R
Green Liquor	R	R	R	R		R				R	R	
Heptane	R	NR	R	R		R	NR	NR	NR	R	R	R
Hexane	NR	NR	R			R	NR	NR		R	R	R
Hexanol, Tertiary	R	R	R	R	NR	R				R	R	
Hydrobromic Acid, Dilute	R	R	R	R		R	R	R		R	R	R
Hydrobromic Acid, 20%	R	R	R	R		R	R	R	F	R	NR	NR
Hydrobromic Acid, 50%		NR				R	R	NR		R	R	R
Hydrochloric Acid, 35%	R	R	R	R	R	R	R	F		R	R	R
Hydrochloric Acid Conc., 38%	R	R	R			R	R	NR		R	R	R
Hydrochloric Acid, 50%	R	R	R	R								
Hydrochloric Acid, Dilute	R	R	R	R	R	R	R	R	R	R	R	R
Hydrocyanic Acid	R	R	R	R		R				R	R	R
Hydrocyanic Acid, 10%	R	R	R			R				R	R	R
Hydrofluoric Acid, Dilute	R	NR	NR		R	R	R	R	F	R	R	R
Hydrofluoric Acid, 30%	R	NR	NR			R	R	F	F	R	R	R
Hydrofluoric Acid, 40%	R	NR	NR			R	R	F	F	R	R	R
Hydrofluoric Acid, 50%	R	NR	NR			R				R	R	R
Hydrofluoric Acid, 100%												
Hydrofluosilicic Acid	NR	NR	NR	NR	NR	R				R	R	R
Hydrogen	R	R	R			R				R	R	R
Hydrogen Cyanide	R	R	R	R	R	R				R	R	
Hydrogen Fluoride	NR	NR	NR			R				R	R	
Hydrogen Peroxide Dilute	R	R	R			R	F	F		R		
Hydrogen Peroxide, 50%	R	R	R			R	NR			R		
Hydrogen Peroxide, 90%	NR	NR	NR			R	R	R		R		
Hydrogen Phosphide	NR	NR	R			R				R		
Hydrogen Sulfide, Dry	R	R	R	R		R	R	R		R	R	R
Hydrogen Sulfide, Aq. Sol.	R	R	R	R		R	R	R		R	R	R
Hydroquinone	R	R	R	R		R	R	R		R	R	
Hydroxylamine Sulfate	R	R	R	R		R	R			R	R	
Hypo. Sodium Thiosulfate	R	R	R	R	R	R	R	R	R	R	R	R
Hypochlorous Acid	R	R	R	R		R	R			R	R	R
Iodine	NR	NR	R	NR		R				R	R	
Iodine Solution	NR		R			R				R	R	
Iodine Solution, 10%	R		R			R				R	R	
Iodine in Alcohol	R		R			R				R	R	
Isopropyl Alcohol	R		R			R	R	R	R	R	R	
Isopropyl Ether	NR	NR	NR	NR	NR	F	NR	NR		R	R	
Isooctane	R	NR	R	NR	NR	R	F	NR	NR	R	R	R
Jet Fuel, JP-4	R	R	R			F	NR	NR		R	R	
Jet Fuel, JP-5	R	R	R			F	F	NR		R	R	
Kerosene	R	R	R	R		R	NR	NR	NR	R	R	R
Ketones	NR	NR	NR	NR	NR	R	F	F	NR	R	NR	NR
Kraft Liquor	R	R	R	R		R				R		

(continued)

TABLE 4-13 *Continued*

CHEMICAL	PVC[1]		CPVC[2]			PP[3]				PVDF[4]		
	72°F	140°F	72°F	185°F	212°F	73°F	120°F	150°F	180°F‡	70°F	150°F	250°F
Lactic Acid, 25%	R	R	R	R		R	R	R		R	NR	NR
Lactic Acid, 80%	R		R			R	R	R		R		
Lard Oil	R	R	R	R		R				R	R	R
Lauric Acid	R	R	R	R		R	R	R		R	R	
Lauryl Chloride	R	R	R			R				R	R	
Lead Acetate	R	R	R	R		R	R	R	R	R	R	R
Lemon Oil	R		R			F				R	R	R
Ligroine	NR		NR			R	R			R	NR	
Lime Sulfur	R	R	R	R		R	R			R	R	
Linoleic Acid	R	R	R	R		R	R	R	R	R	R	R
Linseed Oil	R	R	R			R	R	R		R	R	R
Linseed Oil, Blue	R		R			R	NR	NR	NR	R		
Liqueurs	R		R			R	R			R	R	R
Lubricating Oil	R	R	R	R		R	F	NR	NR	R	R	R
Lye Solutions	R	R	R	R		R	R			R	R	
Machine Oil	R	R	R	R		R	R	NR	NR	R	R	
Magnesium Carbonate	R	R	R	R		R	R	R	R	R	R	
Magnesium Chloride	R	R	R	R		R	R	R	R	R	R	R
Magnesium Hydroxide	R	R	R	R		R	R	R	R	R	R	R
Magnesium Nitrate	R	R	R	R		R	R	R	R	R	R	R
Magnesium Sulfate	R	R	R	R		R	R	R	R	R	R	R
Maleic Acid	R	R	R	R		R	R			R	R	
Malic Acid	R	R	R	R		R	F	F		R	R	R
Manganese Salts	R		R			R	R	R	R	R	R	
Mercuric Chloride	R	R	R	R		R	R	R	R	R	R	R
Mercuric Cyanide	R	R	R	R		R	R	R	R	R	R	R
Mercurous Nitrate	R	R	R	R		R	R			R	R	R
Mercury	R	R	R	R		R	R	R		R	R	R
Methane	R	R	R			R	R	F		R	R	R
Methyl Alcohol	R	R	R	R	R	R	R	R	R	R	R	R
Methyl Bromide	NR	NR	NR	NR		NR				R	R	R
Methyl Cellosolve	NR		NR			R				R	R	R
Methyl Chloride	NR	NR	NR	NR		NR				R	R	R
Methyl Chloroform	NR	NR	NR	NR	NR	F				R		
Methyl Ethyl Ketone	NR	NR	NR	NR		R	F	F		F	F	NR
Methyl Isobutyl Ketone	NR	NR	NR	NR		R	NR	NR	NR	R	F	NR
Methyl Salicylate	R		R			R				R		
Methyl Sulfate	R	NR	R	NR	NR	R				R		
Methyl Sulfuric Acid	R	R	R	R		R	R	F		R		
Methylene Chloride	NR	NR	NR	NR		NR	NR	NR	NR	R	NR	NR
Methylisobutyl Carbinol	R		R			R	R			R	R	
Milk	R	R	R	R		R	R	R	R	R	R	
Mineral Oil	R	R	R	R		R	F	NR		R	R	R
Molasses	R	R	R	R		R	R			R	R	
Monoethanolamine						R	R	R		F	NR	NR
Motor Oil	R	R	R	R		F	F	NR	NR	R		
Naphtha	R	R	R			R	R	F		R	R	R
Naphthalene	NR	NR	NR	NR		R	R	R	R	R	R	R
n-Heptane	R	NR	R	R		R	NR	NR	NR	R	R	R
Natural Gas	R		R			R				R	R	R
Nickel Chloride	R	R	R	R		R	R	R	R	R	R	R
Nickel Nitrate	R	R	R	R		R	R	R	R	R	R	R
Nickel Salt	R	R	R	R	R	R	R	R	R	R	R	R
Nickel Sulfate	R	R	R	R		R	R	R	R	R	R	R
Nicotine	R	R	R	R		NR				R	F	
Nicotinic Acid	R	R	R	R		R				R	R	R
Nitric Acid, 10%	R	NR	R			R	R	R		R		
Nitric Acid, 30%	R	NR	R			R	R			R		
Nitric Acid, 40%	R	NR	R			R	R			R		
Nitric Acid, 50%	R	NR	R			R	NR	NR	NR			

TABLE 4-13 Continued

CHEMICAL	PVC[1] 72°F	PVC[1] 140°F	CPVC[2] 72°F	CPVC[2] 185°F	CPVC[2] 212°F	PP[3] 73°F	PP[3] 120°F	PP[3] 150°F	PP[3] 180°F‡	PVDF[4] 70°F	PVDF[4] 150°F	PVDF[4] 250°F
Nitric Acid, 70%	NR	NR	NR	NR	...	NR	NR	NR	NR	NR		
Nitric Acid, 100%	NR	NR	NR	NR	...	NR	NR	NR	NR	NR		
Nitric Acid, Fuming	NR	NR	NR	NR	NR	NR	NR	NR	NR	NR		
Nitrobenzene	NR	NR	NR	NR	...	R	F	NR		R	NR	
Nitrous Acid, 10%	R	R	R	R	...	F				R	R	
Nitrous Oxide	R	R	R	...	...	R				R	R	
Ocenol	R	R	R	R	...	NR	...			R		
Oil and Fats	R	R	R	R	...	R	R			R	R	R
Oils, Vegetable	R	R	R	R	...	R	R			R	R	R
Oleic Acid	R	R	R	R	...	R	F	NR	NR	R	R	R
Oleum	NR	NR	NR	NR	...	NR	NR	NR	NR	NR	NR	NR
Oxalic Acid	R	R	R	R	...	R	F	F		R	F	
Oxalic Acid, 50%	R	R	R	R	...	R	R	R	...	R	F	
Oxygen Gas	R	R	R	R	...	NR	NR	NR	NR	R	R	R
Palmitic Acid, 10%	R	R	R	...	...	R	R	R	R	R	R	R
Palmitic Acid, 70%	NR	NR	R	...	...	R	R	R	R	R	R	R
Peracetic Acid, 40%	NR	NR	NR	NR	NR	NR	NR	NR	NR	R		
Perchloric Acid, 10%	R	F	R	R	...	R	R	R	R	R		
Perchloric Acid, 70%	NR	NR	NR	NR	...	R	R	R	R	R		
Petrolatum	R	R	R	R	R	R	R			R	R	R
Petroleum Oils, Sour	R	R	R	R	...	R	NR	NR	NR	R	R	R
Petroleum Oils, Refined	R	R	R	R	R	R	NR	NR	NR	R	R	R
Phenol	NR	NR	R		...	R	NR	NR	NR	R	R	F
Phenylhydrazine	NR	NR	NR	NR	...	NR				R	R	
Phenylhydrazine Hydrocholoride	R	NR	R	R	...	NR				R	R	
Phosgene, Liquid	NR	NR	NR	NR	NR	NR	NR	NR	NR	F		
Phosgene, Gas	NR	NR	NR	NR	NR	F	NR	NR	NR	R		
Phosphoric Acid, 10%	R	R	R	R	...	R	R	R	R	R	R	R
Phosphoric Acid, 25, 50%	R	R	R	R	...	R	R	R	R	R	R	R
Phosphoric Acid, 50, 85%	R	R	R	R	R	R	R	R	R	R	R	R
Phosphorus Trichloride	NR	NR	NR	NR	NR	R				R	R	
Photographic Solutions	R	R	R	R	...	R	R	R		R		
Picric Acid	NR	NR	NR	NR	...	R				R		
Plating Solutions, Brass	R	R	R	R	R	R	R			R		
Plating Solutions, Cadmium	R	R	R	R	R	R	R			R		
Plating Solutions, Chrome, 25%	R	R	R	R	F	R	R			R		
Plating Solutions, Chrome, 40%	R	R	R	R	F	R	R			R		
Plating Solutions, Copper	R	R	R	R	R	R	R	R	R	R	R	R
Plating Solutions, Gold	R	R	R	R	R	R	R			R		
Plating Solutions, Indium	R	R	R	R	R	R	R			R		
Plating Solutions, Lead	R	R	R	R	R	R	R			R		
Plating Solutions, Nickel	R	R	R	R	R	R	R	R	R	R	R	R
Plating Solutions, Rhodium	R	R	R	R	R	R	R			R		
Plating Solutions, Silver	R	R	R	R	F	R	R	R	R	R	R	R
Plating Solutions, Tin	R	R	R	R	F	R	R	R	R	R	R	R
Plating Solutions, Zinc	R	R	R	R	R	R	R	R	R	R	R	R
Potassium Aluminum Sulfate	R	R	R	R	...	R	R	R	R			
Potassium Bicarbonate	R	R	R	R	R	R	R	R	R	R	R	R
Potassium Bichromate	R	R	R	R	R	R	R	R		R	R	R
Potassium Borate	R	R	R	R	...	R	R					
Potassium Bromate	R	R	R	R	...	R	R					
Potassium Bromide	R	R	R	R	...	R	R	R	R	R	R	R
Potassium Carbonate	R	R	R	R	...	R	R	R	R	R	R	R
Potassium Chlorate, Aqueous	R	R	R	R	...	R	R	R	R	R	R	R
Potassium Chloride	R	R	R	R	...	R	R	R	R	R	R	R
Potassium Chromate	R	R	R	R	...	R	R	R	R	R	R	R
Potassium Cyanide	R	R	R	R	...	R	R	R	F	R	R	R
Potassium Dichromate	R	R	R	R	R	R	R	R	R	R	R	R
Potassium Ferricyanide	R	R	R	R	...	R	R	R	R	R	R	R
Potassium Ferrocyanide	R	R	R	R	...	R	R	R	R	R	R	R

(continued)

TABLE 4-13 Continued

CHEMICAL	PVC[1]		CPVC[2]			PP[3]				PVDF[4]		
	72°F	140°F	72°F	185°F	212°F	73°F	120°F	150°F	180°F‡	70°F	150°F	250°F
Potassium Fluoride	R	R	R	R		R	R	R	R	R	R	R
Potassium Hydroxide	R	R	R	R		R	R	R	R	R	R	
Potassium Hydroxide, 50%	R	R	R	R		R	R	R	R	R	R	
Potassium Hypochlorite	R	R	R	R	NR	R	F	NR	NR	R	R	
Potassium Iodide	R		R			R				R	R	R
Potassium Nitrate	R	R	R	R		R	R	R		R	R	R
Potassium Perborate	R	R	R	R		R	R	R		R	R	
Potassium Perchlorate	R	R	R	R		R	R	R		R	R	
Potassium Permanganate, 10%	R	R	R	R		R	R			R	R	R
Potassium Permanganate, 20%	NR	NR	R			R	F	F		R	R	R
Potassium Persulfate	R	R	R	R		R	R			R	R	
Potassium Sulfate	R	R	R	R		R	R	R	R	R	R	R
Propane	R		R			R				R	R	R
Propargyl Alcohol	R	NR	R	NR	NR	R	R			R	R	
Propyl Alcohol	R	NR	R	NR		R	R	R		R	R	
Propylene Dichloride	NR	NR	NR	NR	NR	F	NR			R	R	
Pyridine	NR	NR	NR	NR		R	R	R	R	F	NR	NR
Rayon Coagulating Bath	R	R	R	R		R				R		
Selenic Acid, Aqueous	R	R	R	R		R				R		
Sea Water	R	R	R	R		R	R	R	R	R	R	R
Silicone Oil	R	R	R	R		R	R	R		R		
Silver Cyanide	R	R	R	R		R	R	R	R	R	R	R
Silver Nitrate	R	R	R	R		R	R	R	R	R	R	R
Soaps	R	R	R	R		R	R	R	R	R	R	
Soap Solutions	R	R	R	R		R	R	R	R	R	R	
Soap Solution, 5%	R	R	R	R		R	R	R	R	R	R	
Sodium Acetate	R	R	R	R		R	R	R	R	R	R	R
Sodium Benzoate	R	R	R	R		R	R	R	R	R	R	R
Sodium Bicarbonate	R	R	R	R	R	R	R	R	R	R	R	R
Sodium Bisulfate	R	R	R	R	R	R	R	R	R	R	R	R
Sodium Bisulfite	R	R	R	R	R	R	R	R	R	R	R	R
Sodium Borate	R	R	R	R		R	R			R	R	R
Sodium Bromide	R	R	R	R		R	R	R	R	R	R	R
Sodium Carbonate	R	R	R	R		R	R	R	R	R	R	R
Sodium Chlorate	R	F	R	R		R	R	R	R	R	R	R
Sodium Chloride	R	R	R	R	R	R	R	R	R	R	R	R
Sodium Chlorite	NR		NR			R	R	R	F	R		
Sodium Cyanide	R	R	R	R		R	R	R		R	R	R
Sodium Dichromate	R	R	R	R		R	R	R		R	R	R
Sodium Ferricyanide	R	R	R	R		R	R	R		R	R	R
Sodium Ferrocyanide	R	R	R	R		R	R	R		R	R	R
Sodium Fluoride	R	R	R	R		R	R	R	R	R	R	R
Sodium Hydroxide, 15%	R	R	R	R		R	R	R	R	R	R	
Sodium Hydroxide, 30%	R	R	R	R		R	R	R	R	R	R	NR
Sodium Hydroxide, 50%	R	R	R	R	R	R	R	R	R	R	R	NR
Sodium Hydroxide, 70%	R	R	R	R		R	R	R	R	NR		
Sodium Hydroxide Solution	R	R	R	R		R	R	R	R	R		
Sodium Hypochlorite, Conc.	R	R	R	R		R	F	NR	NR	R	R	R
Sodium Hypochlorite	R	R	R	R		R	R	NR	NR	R	R	R
Sodium Metaphosphate	R	R	R	R		R	R	R		R	R	
Sodium Nitrate	R	R	R	R		R	R	R	R	R	R	R
Sodium Nitrite	R	R	R	R		R	R	R	R	R	R	R
Sodium Palmitrate Solution, 5%	R	R	R	R		R	R			R		
Sodium Perborate	R	R	R	R		R	R	R	R	R		
Sodium Phosphate, Alkaline	R	R	R	R		R	R	R	R	R	R	R
Sodium Phosphate, Acid	R	R	R	R		R	R	R	R	R	R	R
Sodium Phosphate, Neutral	R	R	R	R		R	R	R	R	R	R	R
Sodium Silicate	R	R	R	R		R	R	R	R	R	R	R
Sodium Sulfate	R	R	R	R		R	R	R	R	R	R	R

Chemical Resistance of Thermoplastics

TABLE 4-13 *Continued*

CHEMICAL	PVC¹		CPVC²			PP³				PVDF⁴		
	72°F	140°F	72°F	185°F	212°F	73°F	120°F	150°F	180°F‡	70°F	150°F	250°F
Sodium Sulfide	R	R	R	R		R	R	R	R	R	R	R
Sodium Sulfite	R	R	R	R		R	R	R	R	R	R	R
Sodium Thiosulfate	R	R	R			R	R	R	R	R	R	R
Sour Crude Oil	R	R	R	R		R	R			R	R	R
Stannic Chloride	R	R	R	R		R	R	R		R	R	R
Stannous Chloride	R	R	R	R		R	R	R		R	R	R
Stearic Acid	R	R	R	R		R	R			R	R	R
Stoddard's Solvent	NR	NR	NR	NR		R				R	R	R
Succinic Acid	R	R	R	R		R	R	R		R	R	
Sulfated Detergents	R	R	R	R		R	R			R	R	
Sulfate Liquors	R	R	R	R		R	R	R		R	R	
Sulfur Slurries	R	R	R	R		R	F	F	NR	R	R	R
Sulfur Chloride Solution	R	R	R	R		F	NR	NR	NR	R		
Sulfur Dioxide, Dry	R	R	R			R				R	R	
Sulfur Dioxide, Wet	NR	NR	R			R				R	R	
Sulfuric Acid, 10%	R	R	R	R	R	R	R	R	R	R	R	R
Sulfuric Acid, 30%	R	R	R	R		R	R	R	F	R	R	R
Sulfuric Acid, 50%	R	R	R	R	F	R	F	F	F	R	R	R
Sulfuric Acid, 60%	R	R	R	R	F	R				R	R	R
Sulfuric Acid, 70%	R	R	R	R	F	F				R	R	R
Sulfuric Acid, 80%	NR	NR	R	R	F	F				R	R	F
Sulfuric Acid, 90%	NR	NR	R	NR		NR				R	R	F
Sulfuric Acid, 93%	NR	NR	R	NR	NR	NR				R	R	F
Sulfuric Acid, 98%	NR	NR	NR	NR	NR	NR	NR	NR	NR	R	R	F
Sulfuric Acid, 100%	NR	NR	NR	NR	NR	NR	NR	NR	NR	R	R	F
Sulfuric Acid, 103%	NR	NR	NR	NR	NR	NR				R		
Sulfurous Acid	R	R	R	R		R	R	R		R	R	
Sulfur Trioxide	R	R	R	R		NR				NR	NR	
Tall Oil	R	R	R	R	R	R	R	R	R	R	R	R
Tannic Acid	R	R	R	R	R	R	R	R	R	R	R	R
Tanning Liquors	R	R	R	R		R				R		
Tartaric Acid	R	R	R	R	R	R	R	R		R	R	R
Tetraethyl Lead	R		R			R				R	R	R
Tetrahydrofuran	NR	NR	NR	NR		NR	NR	NR	NR	NR	NR	
Tetralin	NR	NR	NR	NR		NR	NR	NR	NR	R	R	
Thionyl Chloride	NR	NR	NR	NR		NR				NR		
Thread Cutting Oils	R	R	R	F		R	R			R	R	
Titanium Tetrachloride	NR	NR	NR	NR		NR				NR		
Toluene, Toluol	NR	NR	NR	NR		F	NR	NR	NR	R	R	
Toluene-Kerosene, 25-75%	NR	NR	NR	NR		F	NR	NR	NR	R	R	
Tomato Juice	R	R	R	R		NR	NR	NR	NR	R	R	
Toxaphene-Xylene, 90-10%	NR	NR	NR	NR		NR	NR	NR	NR	R	R	
Transformer Oil	R	R	R	R		R	F	NR	NR	R		
Transformer Oil DTE/30	R	R	R	R		R	R	R	R	R	R	
Tributyl Phosphate	NR	NR	NR	NR		F	F	NR		R	NR	NR
Trichloroacetic Acid	R		R			F				R	NR	NR
Trichloroacetic Acid 2N	R		R			R	R	R	NR	R		
Trichloroethylene	NR	NR	NR	NR		F	NR	NR	NR	R	R	R
Triethanolamine	R	NR	NR	F		R	R	R	R	R		
Triethylamine	R	R	R			NR				R	R	
Trimethylpropane	R		R			R				R	R	
Trisodium Phosphate	R	R	R	R		R	R	F		R	R	R
Turpentine	NR	NR	R			NR	NR	NR	NR	R	R	R
Urea	R	R	R	R		R	R	R	R	R	R	R
Urine	R	R	R	R		R	R	R	R	R	R	R
Vaseline	R	R	R	R		R	R	F	F	R	R	R
Vegetable Oil	R	R	R	R		R	R			R	R	R
Vinegar	R	F	R	R	R	R	R	R		R	R	R
Vinegar, White	R	F	R	R		R	R	R		R	R	R

(continued)

144 Carbon, Glass, and Plastics   Chap. 4

TABLE 4-13 Continued

CHEMICAL	PVC[1]		CPVC[2]			PP[3]				PVDF[4]		
	72°F	140°F	72°F	185°F	212°F	73°F	120°F	150°F	180°F‡	70°F	150°F	250°F
Vinyl Acetate	NR	NR	NR	NR						R	R	R
Water	R	R	R	R		R	R	R	R	R	R	R
Water, Acid Mine	R	R	R	R	NR	R	R	R		R	R	R
Water, Demineralized	R	R	R	R		R	R	R		R	R	R
Water, Distilled or Fresh	R	R	R	R		R	R	R	R	R	R	R
Water, Potable	R	R	R	R		R	R	R	R	R	R	R
Water, Salt	R	R	R	R	NR	R	R	R	R	R	R	R
Water, Sewage	R	R	R	R		R	R	R	R	R	R	R
Whiskey	R	R	R	R		R	R	R	R	R	R	R
White Liquor	R	R	R	R		R				R	R	R
Wines	R	R	R	R		R	R	R		R	R	R
Xylene or Xylol	NR	NR	NR	NR		NR	NR	NR	NR	R	R	
Zinc Chloride	R	R	R	R		R	R	R	R	R	R	R
Zinc Nitrate	R	R	R	R		R	R	R	R	R	R	R
Zinc Sulfate	R	R	R	R	R	R	R	R	R	R	R	R

[1] Polyvinyl Chloride Type 1 Grade 1   [2] Chlorinated Polyvinyl Chloride   [3] Polypropylene   [4] Polyvinylidene Fluoride   ‡ For drainage application only

This information may be considered as a basis for use but not as a guarantee, because of the widespread variation in actual service conditions, in many chemicals and industrial fluids with respect to the presence of contaminants, concentration, and so on. Pipe and fittings made of each of the materials listed should be tested under actual service conditions to determine suitability for a particular purpose.

In Table 4-13 the letter symbols signify:

R   The results of laboratory immersion tests of the plastic material in the chemical, its aqueous solution or slurry, indicate that no appreciable loss or gain in weight or serious loss of physical properties occurs. It may be expected that under normal circumstances pipe and fittings made of this plastic material will provide good long-term service in piping the chemical at the temperature indicated and appropriate recommended pressures.

F   Some grades of the chemical or industrial fluid in question have been shown in laboratory immersion tests to affect the plastic material appreciably. Considerable difference in actual chemistry analysis exists in many industrial fluids and different effects can be obtained with materials called by the same general name. Varying in degree from slight to severe, one or more of such effects on the plastic as swelling, embrittlement, gain or loss in weight, loss of tensile strength, softening, discoloration, and so on may have been noted. Nonetheless, plastic pipe and fittings have with some industrial chemicals in this category proven economically advantageous and serviceable in applications such as suction or intermittent drainage lines not involving continuous pressure. Field testing of the plastic piping under actual service conditions is recommended in these cases.

NR   These chemicals produce drastic deleterious effects on the plastics material in a relatively short period of time. Therefore, pipe and fittings made of the plastic material cannot be expected to convey these chemicals satisfactorily. However, in waste piping systems, these chemicals can be disposed of intermittently in reasonable quantities with continuous and abundant flushing with water.

Thermoplastics listed include:

- Polyvinyl chloride (PVC)
- Chlorinated polyvinyl chloride (CPVC)
- Polypropylene (PP)
- Polyvinylidene fluoride (PVDF)

5

Fabrication Methods

FABRICATION AND DESIGN PRINCIPLES

Properties of a structural material determine the method of fabrication of equipment and their components. Cast-iron apparatus is cast; steel components also may be cast but more often are fabricated from rolled steel; pipe or profile iron is made by cutting, bending, and welding half-finished articles; ceramic vessels are formed and cured. Pieces of equipment fabricated by different methods require different design criteria, even if they are to be used for the same application.

As an example, Figure 5-1 shows two alternative methods for shaping tray towers. One tower is made from sheet metal by welding. Its trays, furnished by a number of bubble caps, are fastened to the shell by welding, as are all the shell nozzles. The second tower is made from cast iron and consists of several flanged sections. The trays are cast separately and fastened between the flanged sections. Each tray has only one bubble cap. All nozzles are molded together with the tower.

Design variations between the two towers are established by the different methods of welding and casting. It is not possible to cast the tower in one piece; it must be made in several sections. A thicker shell than is needed for the application has to be fabricated because of the casting method. A decrease in the number of bubble caps was established by a need for simplifying the shape of a casting pattern, thus lowering its price. It is difficult both to cast a thin-walled

Fabrication and Design Principles

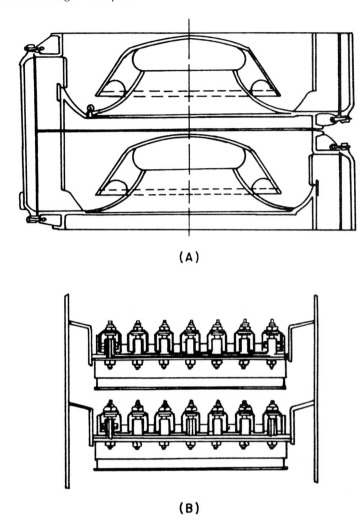

Figure 5-1 Schematic section through a bubble-cap tray tower: (A) cast-iron tower; (B) steel tower

nozzle with a thick-walled section of tower and to avoid its breaking during transportation.

As another example, assemblies for heating chambers of two vaporizers (from steel and cast iron) are shown in Figure 5-2A and B. The tubular plate of the steel vaporizer is welded to its shell and the tubes rolled in the plate. The design of a similar assembly of a cast-iron vaporizer is more complicated: The plate is fastened with bolts to the shell and the tubes are tightened with special profile nuts. Such complications are inevitable because cast iron cannot be welded; also the tubes cannot be rolled out because of the metal's brittleness.

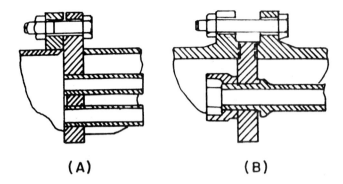

Figure 5-2 Tube joints: (A) steel heating chamber; (B) cast-iron heating chamber

Equipment fabricated from the same structural material but by different techniques also will differ from each other. Figure 5-3A shows the old design of a vessel head riveted to the shell of a receiver. The lower head was made concave to support the rivets during clinching (riveting). For the welded version, there is no need for a concave head; both heads are fastened with outside convexity. This design permits butt-welded joints that improve the final design (Figure 5-3B).

It follows from these examples that the design of apparatuses and their assemblies depend on the method of fabrication. Consequently, the designer must be well versed in the principal fabrication methods and consider all modern techniques and their limitations.

Assuming our reader is familiar with machine-building technology, only data and conclusions about this subject that are necessary for a design are presented next.

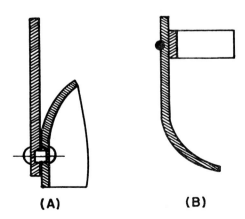

Figure 5-3 Riveted and welded head connections

DESIGN OF CAST EQUIPMENT

Casting processes permit a very wide latitude in chemical engineering because one can obtain castings of cast iron and steel that have very complicated shapes. The designer must be familiar with concepts and details before producing a successful design. There are many variables in castings: properties of the material, the simplicity of a pattern, ease of molding and subsequent machining of castings, cost considerations, and so on.

Castings must be designed so that the material of the article after cooling has the desired requirements in all its parts. Therefore, one of the most important criteria in casting designs is the ability to provide a uniform cooling rate. As a first approximation, it is possible to assume that the casting cooling rate depends on the ratio of the casting perimeter to the cross-section area. Care must be taken to ensure the same thickness of all outer walls of the casting to obtain equal cooling rates. The inside wall, wells, and ribs generally are harder to cool at the same rate as outside walls. To maintain constant rates, the thickness S_1 should be less than that of an outside wall, S':

$$S_1 = 0.7S' \text{ to } 0.8S'$$

Violating the "equal wall principle" can cause the formation of residual thermal stresses, cracks, pores, or even blow holes.

Thermal stresses originate in the following manner. In cooling a nonequal-walled piece, such as a tube (Figure 5-4), the hardening of its thin wall occurs before the thick wall. The thin-wall hardened metal will prevent the free shrinkage and size shortening of the thick wall. As a result, stresses will bend the article, as illustrated in Figure 5-4. If there is a great difference in wall thickness, the residual thermal stresses may exceed the material's ultimate strength and the casting during cooling.

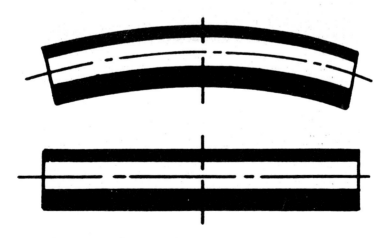

Figure 5-4 Deformation of nonequal-walled tube after cooling

TABLE 5-1 MINIMUM WALL THICKNESS IN SAND CASTINGS

Material	Minimum wall thickness (mm)		
	Small	Average	Large
Gray Cast Iron	3–5	8–10	12–15
Malleable Cast Iron	2.5–4	6–8	
Cast Steel	6	10–12	20
Alloys Nonferrous Metals	2–3	5–7	10–20

Local thickening and metal accumulation are extremely objectionable. They form pores and blow holes as a result of a discontinuous metal supply in the thickened part and a more rapid cooling of the thin wall.

It is desirable to specify a minimum wall thickness of the casting, which should be determined from strength considerations. Minimum wall thickness provides a good metal structure as well as economical efficiency of the casting. It is determined by metal flowability, its capability to fill up the mold, features of the molding, and the size of the casting. Recommended minimum wall thickness for sand castings are given in Table 5-1. Note, however, that sand casting does not permit as small a thickness as other casting processes (for example, die casting, casting in shell molds, chill casting, injection molding).

The wall thickness of a founding in sand castings may be determined by using the plot in Figure 5-5, where the wall thickness is given with respect to the casting overall size, N N (in meters) is calculated from the following formula:

$$N = \frac{2L + b + h}{3}$$

where L = casting length

b = casting width

h = casting height

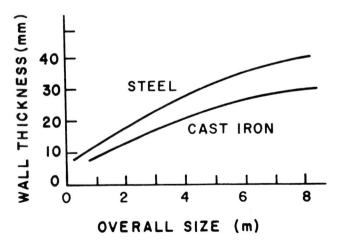

Figure 5-5 Wall thickness versus conditional overall size

Design of Cast Equipment

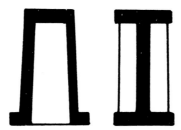

Figure 5-6 Ribbed and box sections

Increasing the wall thickness of castings more than 30–40 mm decreases their strength, porosity, and graphite grain growth. Fabricating vessels from cast iron with walls thicker than 50–60 mm is not recommended. However, they can be made from steel, with a considerably greater wall thickness, and their mechanical properties can be considerably improved by forging in simple shapes. Ribs and hollow or box sections are used to decrease the weight of cast apparatus and to make the construction either stronger or more rigid.

The hollow or box sections should not be combined with ribbed sections since this complicates the casting unnecessarily. In selection, it is necessary to account for ribbed constructions, which are better cast. Their cooling conditions are more favorable than box sections. However, box sections are more rigid and their shapes are streamlined, smooth, and provide a more agreeable shape (Figure 5-6).

In designing chemical equipment, it is impossible from a practical standpoint to make all parts of a casting the same thickness. Parts of equipment are loaded differently and because of full-strength considerations must be fabricated in different wall thicknesses. In this case, castings should be designed so that the casting massiveness and its wall thickness gradually increase in the direction of improving heat evacuation.

For obtaining a sound casting without residual stresses, it is necessary to maintain strictly a smooth change of sections (Figure 5-7) and to avoid wall connections without fillets (Figure 5-8A and B).

It is not desirable to design conjugating walls with a thickness ratio greater than 2:

$$\frac{S}{S_1} \leq 2$$

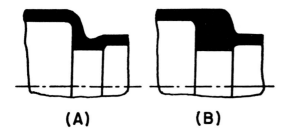

(A)   **(B)**

Figure 5-7 Smoothing the wall thickness for improving the casting: (A) good design; (B) poor design

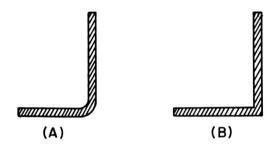

Figure 5-8 Conjugation of cast walls: (A) good design; (B) poor design

Junctions of sections of different thicknesses are achieved using a taper (Figure 5-9A) with the following ratio:

$$\frac{S_1 - S}{l} = \frac{1}{4} \text{ to } \frac{1}{5}$$

If $S/S_1 < 2$, it is possible to have a corner junction (Figure 5-9B) that can be combined with a paper junction (Figure 5-9C).

The selection of the curvature radius for junctions of walls at a right angle with different wall thicknesses as, for example, the junction of a pipe with a flange, is contradictory. On the one hand, increasing the radius decreases the stress concentration and reduces the probability of crack formation. A large fillet may produce a large accumulation of metal and a high probability of cavity formation.

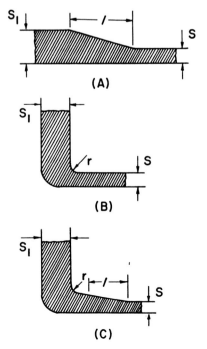

Figure 5-9 Design of junctions between walls of different thickness

Design of Cast Equipment

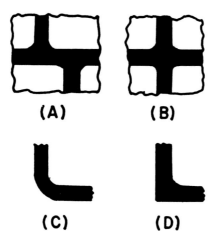

Figure 5-10 Treatment of intersections

If the junction radius ranges from one-sixth to one-third of the arithmetic mean of the wall thickness,

$$r = \frac{S_1 + S}{6} \text{ to } \frac{S_1 + S}{12}$$

then both problems are minimized. Greater-size radii for junctions at right angles and at sharp angles will cause unacceptable local wall thickening. Intersections should be designed so that the concentration of metal is at a minimum. Figures 5-10A and C illustrate good designs. Figures 5-10B and D illustrate poor design.

Particular attention must be given to junctures of sections of different thicknesses. An abrupt change from a thin section to a heavy one (Figure 5-11A) will produce a high-stress concentration at the juncture. This may result in shrinkage cracks that often occur during cooling or in cracking when a tensile load is applied. A large fillet (Figure 5-11B) is apt to produce large local accumulation of metal that will be still liquid when the outside fibers have solidified. This results in pore formation inside the casting, or even a blow hole.

A moderate fillet, as shown in Figure 5-11C (approximately $r = 0.5h'$), will eliminate excessive stress concentration and prevent the formation of a crack. A gradual taper, Figure 5-11D, will provide even better results.

When heavy sections, or bosses, are surrounded by light sections, it is important to secure proper feeding of the boss by placing ribs of sufficient cross

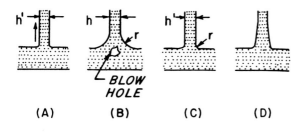

Figure 5-11 Junctures of different sections

(A) STRONG ; LOW RIGIDITY.

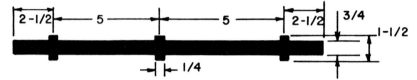
(B) HIGHER RIGIDITY; EXCESSIVE STRESS.

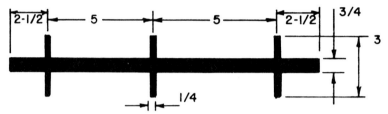
(C) VERY RIGID ; STRONG BUT HARD TO CAST.

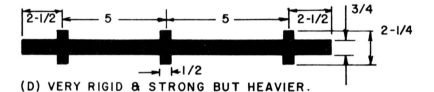
(D) VERY RIGID & STRONG BUT HEAVIER.

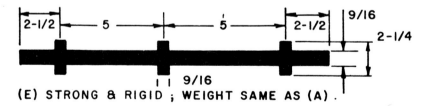
(E) STRONG & RIGID ; WEIGHT SAME AS (A).

Figure 5-12 Various ribs added to a plate

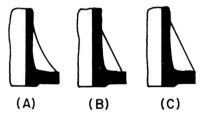

Figure 5-13 Cast ribs: (A) concave; (B) straight; (C) straight expanded to trim

Design of Cast Equipment

sections on the light parts. The main sections should be made sufficiently thick to permit the addition of a gate of a size that secures good feeding of all sections. The ribs improve evacuation of heat from a massive part of the casting and provide a more even load distribution over the entire volume of material. Ribs are added to make a construction either stronger or more rigid. While ribs always attain the latter objective, they may fail in regard to the former. The ribs strengthen the part if their addition lowers the maximum stress of it. If the maximum stress in the original design is lowered, but the stress in the added rib is equal to or higher than in the original design, the rib is either useless or weakens the part. A crack may develop in the high-stressed ribs that eventually will spread and cause the casting to fail.

Figure 5-12 shows various ribs added to a plate for stiffening. The shape of the rib has to be simple. A curved reinforced rib, Figure 5-13A, is better than a straight one, Figure 5-13B, as it is less rigid and causes less stress concentration. However, the maximum bending stress in rib A will be higher than in rib B. Care must be taken to put in sufficiently heavy ribs.

In the event of impact loading, the decrease of failure may be achieved by increasing the resilience of casting parts involved, which, in turn, requires greater deformations. Since ribs act in the opposite direction, it is obvious that they are to be avoided where impact loads are anticipated.

Fluctuating loads have a similar effect, although not as pronounced as shock loads; therefore, all conclusions about the latter relating to ribs apply to fluctuating loads as well. Ribs should be used chiefly for static loads. Wide and low ribs are safer than thin and high ones.

In casting design one must account for the fact that cast iron, and especially ferrosilicon and similar alloys, resist compression loads but act poorly where bending and tensile loads are applied. Gray iron resists four times greater compression loads than tensile ones. Therefore, the ribs have to be located at the side of compressed fibers, for example, in vessels under internal pressure (Figure 5-14A) and in vessels under external pressure (Figure 5-14B). Care should be taken to avoid loading the vessels with bending moments.

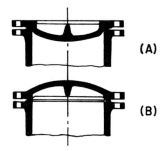

Figure 5-14 (A, B) Layout of ribs on the cast-iron head

Figure 5-15 Pilot wheel with conically situated spokes

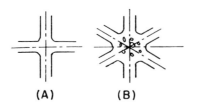

Figure 5-16 Junction of intersecting ribs

Figure 5-17 Junction of radial ribs by ring rib

Ribs that are allocated incorrectly may produce defective castings. For example, placing radial ribs on heads of large vessels will increase their rigidity and prevent free shrinkage during cooling. This may cause excessive stresses and shrinkage cracks. The radial allocation of wheel spokes, pulleys, and flywheels is apt to produce an unsound casting. To increase elasticity, it is desirable that all radial ribs, spokes, and so on be changed by tangent, curved, or conical shapes. Figure 5-15 illustrates the pilot wheel with conically situated spokes.

The intersection of ribs crosswise is less favorable than that of T- or V-shape. Intersecting ribs (Figure 5-16A) should be avoided as they are apt to produce unsound castings. Six ribs (Figure 5-16B) make a poor design, which can be improved materially by the addition of a ring rib (Figure 5-17), which permits the escape of metal accumulation.

The correctness and smoothness of junctures are checked by the inserted circle method, which is explained in Figure 5-18.

The smaller the difference in the circles, the better the casting. The diameter ratio in a good design must meet the following criterion:

$$\frac{D}{d} \leq 1.2$$

Figures 5-17 and 5-18 illustrate how minor changes in the casting shape can significantly decrease local accumulation of the metal.

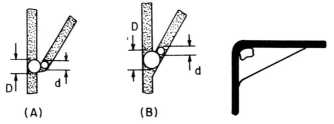

Figure 5-18 Checking the casting design by the inserted circle method: (A) good design; (B) poor design

Figure 5-19 Hole in the rib

Design of Cast Equipment

To decrease stress and metal accumulation, it is desirable not to bring the ribs inside of an apparatus up to its angles. That is, at the point of juncture of three walls, it should leave a hole not less than 30–50 mm in diameter (Figure 5-19).

Holes in medium and large castings are provided if their diameters are more than 50 mm. For the lesser holes, it is more economical to drill them in a finished casting. To lower the stress concentration around a hole, it must be reinforced by a one- or two-sided bead.

Rectangular or square openings, such as windows in compressor pedestals, must have well-rounded corners (Figure 5-20A) and should be reinforced with beads.

Too small a fillet will produce a stress concentration that will cause cracks at the corners under fluctuating loads (see Figure 5-20B). Long nozzles should not be cast together with thick-walled cylinders—they can be broken during cooling or transportation. Such nozzles must be provided with ribs or be changed by bosses. Their thickness should not exceed 50 percent to 60 percent of the wall thickness. Location of the parting plane must be selected in a manner such that cores are not required because they complicate and raise the price of the molding.

Height of the chamber H (Figure 5-21) formed as a result of metal displacement by the part of the pattern fixed on its base in the lower mold box may reach up to $H \leq D$ for hand molding and $H \approx D$ for machine molding. The height of a chamber formed by a part of the pattern that is suspended downward is $h \leq 0.3d$.

Standardization possibilities and machine molding of bosses in designing boss castings must be given careful consideration. Where possible, core marks must have a cross section equal to the core section. This relaxes requirements for exact and sure fixing of a core mark necessary for preventing displacement during casting and fabricating defective pieces as a result of casting unevenness. In all cases, one should avoid console cores fastening. The cores must be fixed on two sides, as shown in Figure 5-22A.

In casting design it is necessary to avoid foundry gaggers. They are one of the basic sources of an unsound casting because of pores and liquation. The

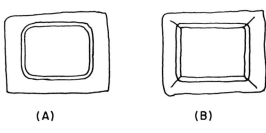

Figure 5-20 Shaping the rim of a rectangular opening

Figure 5-21 The sizes of the chamber in the mold without bosses

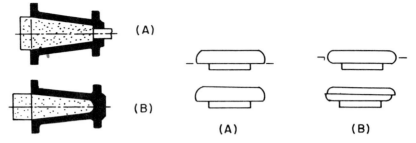

Figure 5-22 Core fixing: (A) on both sides; (B) console fixing

Figure 5-23 Selection of the location of a parting plane: (A) good design; (B) poor design. Lines show the parting plane

gaggers decrease the casting density and are not permissible in fabricating pressure vessels. The design of such vessels must provide the boss fixing in the mold without gaggers. They are not used with ferrosilicon and similar castings. Openings must be provided in the casting (no less than 50 mm in diameter) to withdraw bosses and frames.

Wherever possible, the parting plane should be flat and located so that the largest dimension of the part is in this plane. All mold cavities should be as shallow as possible. Also, the parting plane should be kept away from surfaces that are likely to be used for locating or clamping when the part is machined.

Location of the parting plane may considerably influence the cast finish. For example, at displacement of pattern parts during molding, thick seams may result that require trimming. It is possible to avoid seam formation by the location of other parting planes (Figure 5-23).

To lower the price of castings and decrease defective products, the designer should consider the fabrication features of patterns, core boxes and molding, and the casting process itself. Although casting provides fabrication parts of complicated shapes, it is always necessary to strive for simple configurations and shapes. The pattern fabrication (Figure 5-24A) with only two radii is easier than that with four radii of curvature (Figure 5-24B).

Patterns and bosses contoured by planes and surfaces of revolution are

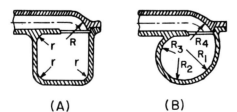

Figure 5-24 Two-pattern version of (A) simple form; (B) more complicated form

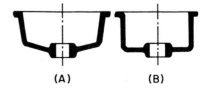

Figure 5-25 Bottoms: (A) casting with cone bottom; (B) casting with flat bottom

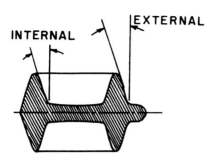

Figure 5-26 External and internal draft

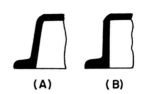

Figure 5-27 Castings with a draft: (A) good design; (B) poor design

more simple in fabrication and not expensive. Notice that large, flat surfaces are extremely unfavorable for castings. It is better to manufacture them with short ribs or pits and round bottoms as cones with a large central angle (Figure 5-25A), thus avoiding flat castings (Figure 5-25B) because blow holes may be formed in horizontal planes.

To permit easy withdrawal of a pattern, all surfaces perpendicular to the parting plane—walls, ribs, and so on—require draft (Figures 5-26 and 5-27). External draft is used at surfaces that, on cooling, will shrink away from the mold. Internal draft is used on surfaces that, on cooling, tend to tighten the mold. The pattern must have a shape that enables easy withdrawal from the mold (Figure 5-27A). It is not desirable to provide for ribs, partitions, projections, and notches (Figure 5-28). These make the molding process very difficult.

If ribs, projections, and so on are necessary, they should be arranged so that they do not prevent the withdrawal of the pattern. Figures 5-29A and C illustrate good designs, and Figures 5-29B and D illustrate poor designs.

Checking the pattern shape may be achieved if we imagine a beam directed parallel to the pattern motion (Figures 5-30A and B). The absence of shadows

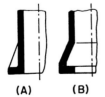

Figure 5-28 The edge of an apparatus of silicon cast iron: (A) easy withdrawal pattern; (B) difficult withdrawal pattern

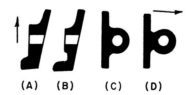

Figure 5-29 The notches do not permit easy withdrawal of the pattern: (A, C) good designs; (B, D) poor designs. The arrows show the direction of pattern withdrawal

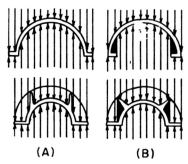
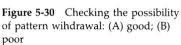
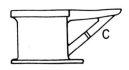

Figure 5-30 Checking the possibility of pattern wihdrawal: (A) good; (B) poor

Figure 5-31 Bed cast with extension

shows that the pattern shape favors the pattern withdrawal as illustrated in Figure 5-30A; and the shaded sections (Figure 5-30B) show that the pattern cannot be withdrawn if there is no detachable part.

To avoid scrapping of the heavy piece through breakage of the extension, the light and protruding extensions should not be cast in one piece with heavy sections or parts. Figure 5-31 shows an incorrect design; that is, brace c is out of proportion with the rest of the bed. Figure 5-32 shows the correct, two-piece design.

In most cases, straight lines look better than curved lines and are cheaper to obtain in pattern. However, the parabola is the proper shape for a heavy pedestal (Figure 5-33), which resists a bending moment, because its strength is more nearly uniform. As both the ultimate strength and yield point of cast iron in compression are several times greater than those in tension, cast-iron parts should be designed as much as possible to avoid their loading in tension. Sections subjected to tensile stress either should be reinforced or relieved of stress by appropriate tie rods (Figure 5-34), or clamp-shaped devices such as forged-steel bearing caps, b (Figure 5-35).

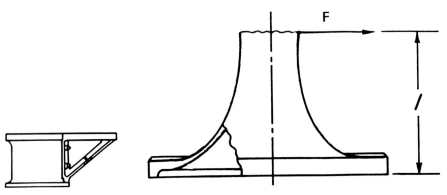

Figure 5-32 Bed with a separate extension

Figure 5-33 Pedestal loaded in bending

Design of Cast Equipment

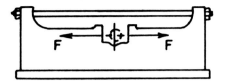

Figure 5-34 Bed plate with tie rod

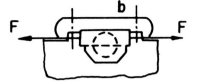

Figure 5-35 Steel bearing cap

For the same reason, heavy-loaded items, for example, a punch-box section frame, should be made with a thick wall (Figure 5-36). Stress concentration in section A may cause it to break, whereas section B, with a gradual change of wall thickness, gives a satisfactory frame.

Steel is not as fluid as cast iron, so complicated shapes, sharp corners, and thin sections cannot be obtained. The steel castings must be annealed to relieve internal stresses. Subsequent heat treatment can increase their strength.

Light alloys are more sensitive to stress concentrations produced by fillets and notches than is cast iron. Fillets should be made larger, notches should be avoided, and a uniform wall thickness provided. The strength of aluminum sections in tension can be increased by cast-in steel reinforcing rings or anchors.

In designing a casting part, care should be given to any machining that must be performed. Machining allowance (the amount of excess material) is required to permit a proper machining or finishing of the surfaces of the casting. In determining the machining allowance, there must be a tolerance of both the casting and the machining dimensions. Because different amounts of shrinkage are caused by variations in pouring temperature, incomplete closing of the molds and rapping the pattern to make it easier to remove it from the mold, castings require relatively large tolerances.

Machining allowance will vary with the size and shape of the casting. A cylinder 2 in. in diameter and 4 in. long, for instance, might be cast 2¼ in. in

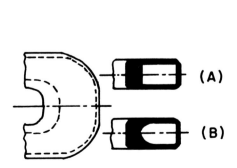

Figure 5-36 Punch-press frame

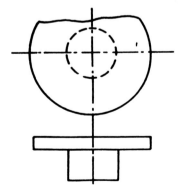

Figure 5-37 Finished or machined part for which a casting is to be made

diameter and 4¼ in. long to permit 1/8 in. to be removed in machining its surfaces [9]. Should the cylinder be 12 in. long, however, a cast diameter of 2 1/2 in. might be required because of the possibility of warpage of the casting. For the part illustrated in Figure 5-37, 1/4-in. allowance on all diameters and 1/8-in. allowance on lengths probably will suffice. Therefore, the diameters will become 10 1/4 in. and 4 1/4 in., and the lengths 1 1/8 in. and 3 in.

Most metals contract during cooling, so provision must be made for this change in dimension. The shrinkage allowance depends on the size of the casting and the material of which it is made, and is generally specified as a definite amount per foot of length or diameter. The usual allowance for cast iron is 1/8 in./ft; for steel, 1/4 in./ft; for aluminum, 3/16 in./ft. So that the casting may have the required dimensions, the 10 1/4-in. diameter must be increased to approximately 10 3/8 in.; the 4 1/4-in. diameter to 4 19/64 in.; the 3-in. length to 3 1/32 in.; and the 1 1/8-in. length to 1 9/64 in.

To remove the pattern without injury to the mold, it should be tapered in the direction in which it is drawn or removed from the mold. The necessary increase in size of the upper portion of the pattern (that is, the draft allowance) depends on the size of the casting and the manner in which the pattern is drawn from the mold. The pattern (Figure 5-38) will be drawn in a direction parallel to the axis of the component cylinders.

Therefore, it will be necessary to change these cylinders into a frusta of cones. Draft allowance is generally 1/4 in./ft of depth, so the upper diameters of the two component cones are 1/32 in. and 1/16 in. larger than the lower diameters. The final dimensions are given in Figure 5-38.

Somewhat controversial is the best casting to be used—either a heavy casting in one piece or its dismembering as several small ones. In most cases, it is more expeditious to make use of dismembering heavy and complicated castings in several small and simple ones and then to assemble them after casting and machining. As a result of such subdivision, the quality of castings increases and the amount of defective parts becomes lower. Each part can be cast from a more suitable material and made thin walled, which considerably decreases the casting weight and material consumption. The cost of machining and assembly increases in this case. The net cost of casting depends on material cost, plant

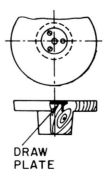

DRAW PLATE

Figure 5-38 Typical pattern

Design of Cast Equipment

equipment, casting technique, weight and complexity of casting, and number of parts. Assuming the cost of a ton of casting from gray iron as a unit, it is possible to obtain a relative net cost of castings from other metals.

Castings usually are divided into four groups: simple, mean complex, complex, and very complex. Most parts of chemical apparatus are of mean and complex groups. Table 5-2 presents the net cost relationship versus casting complexity.

The higher increase in net cost of steel castings with complexity compared to that of cast iron depends on poorer cast properties of steel—its lower fluidity and worse mold filling. On the other hand, the capability of bronze to fill extremely narrow channels and molds of complicated shapes makes the difference in net cost of items of various complexity insignificant.

Heavy castings are cheaper than light ones. Assuming the cost of one ton of material in a 3,000-kg casting as unity, we will obtain the following relationship for cost versus casting weight for mean and complex castings: 300 kg-1.00; 100-1,000 kg-1.05; 10-100 kg-1.20; 1-10 kg-1.40; 0.21-1 kg-1.80.

Correspondingly, for steel castings we have: all castings heavier than 3,000 kg-0.95; 1,000-3,000 kg-1.00; 500-1,000 kg-1.08; 200-500 kg-1.20; 50-200 kg-1.35; 10-50 kg-1.65; up to 10 kg-2.15.

The cost of patterns is usually 10 percent to 15 percent of the casting cost. For up to ten castings, the cost per casting is increased by 10 percent and by 5 percent for up to 25 pieces. For many parts the cost does not increase. The net cost of casting depends very much on the founding method:

- Sand mold casting-1
- Shell molding-2
- Casting by the lost-wax process-10

The last two methods are only economical in mass production and for precision casting, which eliminates aftertreatment.

Preceding data on net cost show that, wherever possible, the designer must use the cheapest material (for example, gray cast iron). Steel castings should be used only for cases in which it cannot be substituted by cast iron. Because of pattern rapping, misalignment of bosses, and various other reasons,

TABLE 5-2 RELATIONSHIP BETWEEN NET COST AND COMPLEXITY OF CASTINGS

Complexity	Castings of gray iron	Castings of steel	Castings of bronze
Simple	1.0	1.0	1.0
Mean Complex	1.15–1.20	1.30–1.40	1.0
Complex	1.40–1.60	1.70–1.80	1.05
Very Complex	1.70–2.00	2.20–2.30	1.10

TABLE 5-3 TYPICAL ALLOWANCES IN SIZES OF CASTINGS FROM STEEL AND CAST IRON WITH USE OF WOODEN PATTERNS

Maximum casting size (mm)	Nominal size, mm						
	Up to 50	50–120	120–260	260–500	500–800	800–1,250	1,250–2,000
Up to 500	1.0	1.5	2.0	2.5	—	—	—
500–1,250	1.2	1.8	2.2	3.0	4.0	5.0	—
1,250–3,150	1.5	2.0	2.5	3.5	5.0	6.0	7.0
3,150–6,300	1.8	2.2	3.0	4.0	5.5	6.5	8.0
6,300–10,000	2.0	2.5	3.5	4.5	6.0	7.5	9.0

the casting will never follow the drawing exactly. Hence, matching allowances must be made to produce a successful design. Reference surfaces should be noted in drawings for checking the castings and finished products. Typical allowances in sizes for steel and cast-iron castings used are given in Table 5-3.

DESIGN OF WELDED EQUIPMENT

Process equipment is fabricated by a number of well-established methods such as casting, welding, forging, machining, brazing and soldering, and sheet-metal forming. Each method has certain advantages for particular equipment. However, welding is the most important and is widely used for fabrication of steel apparatuses. This method of construction is virtually unlimited with regard to size and is applied extensively for fabrication and erection of large-sized process equipment in the field, which is often made by the method of subassembly [10–24].

Three factors infuence the economics of welding processes: (1) the size of the weldment required; (2) the number of weldments required; and (3) the relative complexity of the weldment. The size of welding has a marked effect on the process that is used.

The design may be broken down into a number of subassemblies of perhaps similar form if a large weldment is made from a thin section. An example of this kind of construction is an all-welded steel flat-bottomed cylindrical vessel. If the fabrication of the vessel is made in the field, then manual arc welding by highly skilled craftsmen is likely to be the major process. In these circumstances capital investment charges may be comparatively low, but overall production costs will be high because labor cost will be very high. If the vessel construction is streamlined by prefabrication of relatively small subassemblies in a comfortable workshop not subject to climatic conditions, and by employing manipulators and fixtures that enable less skilled operators to use rapid-production automatic and semiautomatic high-speed welding processes, then the skilled workers and slower manual processes need be used only for difficult parts of the final assembly. This means relatively large initial capital investment charges but comparatively low overall labor costs, so the total production cost tends to fall with increasing number of orders. If the component is large with relatively thick sections, then subassemblies are likely to be difficult to arrange. The problem is then one of adapting the process to the fabrication techniques most suited to the circumstances. If it is not possible to install some form of automatic arc-welding equipment or an electro-slag welding plant, then costly positioned manual arc welding may have to be used. The cost of the manual process is not limited to the deposition costs but may have to allow for more careful treatment and a greater risk of defective welding. Suitable large manipulating equipment is a great help for manual welding of larger components and can reduce labor costs greatly.

The greater the number of finished parts required, the more sophisticated the methods become and the lower the production cost per part. With the increasing number of parts, it is possible to devise fixtures to simplify and speed up the manual welding and then, possibly, to arrange an automatic process, employing an existing machine.

For complex shape components requiring expensive manufacturing techniques, it is worthwhile to use one of the more costly welding processes to aid in fabricating the shape in such a way that overall fabrication is simplified. In fabrication of process equipment from steel, the following welding methods are used:

- Automatic and semiautomatic hidden-arc welding.
- Manual electric-arc welding.
- Electro-slag welding.
- Automatic and semiautomatic carbon dioxide gas-shielded welding.
- Automatic, manual, and machine argon-arc welding.
- Acetylene welding.

Automatic and semiautomatic hidden-arc welding is now the basic method for the fabrication of process equipment from carbon, low-alloy, and high-alloy steels.

Compared to manual electric-arc welding, the automatic hidden-arc welding possesses a number of significant advantages; the major ones are:

- the higher productivity of welding—four to five times higher than manual electric-arc welding
- the higher equality of welded joints
- butt-welding without beveling, which considerably simplifies preliminary operations, saves weld metal, and decreases the cost of equipment

The hidden-arc welding process involves submerging the arc beneath a blanket of granulated mineral flux, where it generates heat for the electrode and deposits weld metal under a protective layer formed from a portion of melted flux. Several inches of weld metal can be deposited in one pass, which greatly decreases the welding time involved. As the arc is covered, there is no arc flash, and smaller amounts of smoke and obnoxious fumes are produced compared to other welding processes.

Hidden-arc welding is used primarily for the fabrication of straight and circumferential buttjoints of vessel shells from sheet steel of any thickness up to 3,200 mm in diameter, as well as corner welds of flanges and other connections for apparatuses of medium and large sizes, tee welds, and other standard part joints for chemical equipment.

The semiautomatic hidden-arc welding of steels is similar to automatic welding and differs by a manual motion of arc along the seam. Compared to automatic welding, semiautomatic welding has the following advantages:

- Due to the high passability and maneuverability of a semiautomatic device, it is possible to weld in locations that are inaccessible for automatic hidden-arc welding.
- One can butt-weld thin sheets (2–3 mm) and corner welds with a 3–4 mm fillet, which is inaccessible or difficult for automatic hidden-arc welding.
- The cost of welding is lower due to lower energy consumption (approximately 30 percent to 40 percent less), and there is less consumption (approximately 15 percent to 20 percent) of welding (filler) wire compared to automatic welding.

Semiautomatic hidden-arc welding is used for fabricating small-sized and thin-walled equipment and for welding to apparatus flanges, nozzles, tubes, and other parts where the use of automatic welding is either impossible or ineffective.

Manual electric-arc welding of steel is used widely in the fabrication of process equipment. This method is used preferentially for tube joints, accessories, internal appliances of apparatuses, supporting constructions, and different kinds of steel structures completing the process equipment.

Electro-slag welding is used widely in fabricating high-pressure, thick-walled apparatuses, thick and large flanges, and other items that have a large initial thickness. Compared to other welding methods, electro-slag welding has the following advantages:

- It consumes a small amount of energy, approximately 1.5–2.0 times less than for automatic hidden-arc welding and four times less than manual arc-welding.
- From 150–200 mm of weld metal can be deposited in one pass with the use of one electrode; when more electrodes are used, the welding of any metal thickness is possible.

- Welding of any metal thickness is possible without beveling; this makes preliminary operations easier and decreases the laboriousness and cost of apparatus fabrication.
- Better conditions for evacuation of gases and slags provides a high quality of welded joints.

The productivity and economy of electro-slag welding depends significantly on the thickness of welded parts. If their thickness is less than 50 mm, the laboriousness and cost of welded joints made by electro-slag welding is higher than with electric-arc welding; however, if the part thickness is more than 100 mm, electro-slag welding is considerably more economical.

Automatic and semiautomatic carbon dioxide gas-shielded welding is used widely for low-carbon, high-alloy, and low-alloy steels. This method provides a high quality of welded joints. Automatic, manual, and machined argon-arc welding is used mostly for welding apparatuses from high-alloy steels, as well as for welding tubes to the tube plate.

Acetylene welding of steels is used in fabricating process apparatuses in limited quantities because of its defects. The considerable heating of the main metal determined by a longer heat action results in higher deformation of parts to be welded and enables superheating and grain raise in the weld joint. Compared to other welding methods, acetylene welding is less productive and not economical. However, it may be applied for light gauges of metal (20 gauge or less) which are difficult to weld by the arc-welding process. In common with every other type of fabrication, selection of a particular welding method is a matter of balanced consideration of design requirements, metallurgy, and economics.

The influence of welding on the apparatus design depends primarily on the design of a welded joint. This is determined by the strength and tightness that can be obtained by the material in and around the joint, accessibility to the welding process, and accommodation to the side-welding effects.

Each of the preceding factors has to be considered in every application. Unfortunately, each will not have the same relative importance in every case. In one case, limited weld strength will determine location; in some cases, the process may be the determining influence; and in many cases incidental effects, such as residual stresses or deformations, could be the main influence.

The designer has a free hand in selecting joint locations if full parent metal strength is assured in a weld joint, provided that the selected position is accessible and side effects can be absorbed. In this case, the selection of joint positions is determined primarily by economics.

If full parent strength is not attainable in a weld joint, the problem arises of locating welds in relatively low-stressed positions or increasing the sectional size in the weld vicinity to give more load-carrying area. In locating the weld joints, it is necessary to account for rigidity; therefore, the complete weldment must be studied as a unit, not as a pinjointed structure. Correct joint preparation is important relative to the particular process that is to be used.

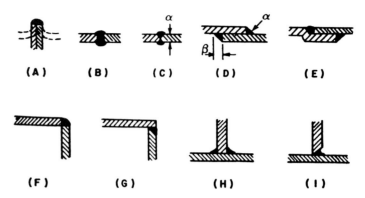

Figure 5-39 Forms of welded joints (weld shown in black)

A variety of types of welded joints is used in fabricating equipment, depending on the service, the thicknesss of the metal, fabrication procedures, and code requirements. There are five basic forms of welded joints: edge, butt, lap, corner, and tee. Figure 5-39 illustrates these forms. Figure 5-39A shows a plain edge joint. The dotted lines shown are referred to as leaf-and-edge joints and are obtained by bending the plates over after welding. Figures 5-39B and C show single- and double-edged butt joints, respectively; D and E are lap joints; F and G are corner joints; H and I are tee joints. With the exception of the joint in A, the other joints can be obtained by applying either a butt weld, where the weld metal is deposited between the plate edges to be joined (Figures 5-39 B, C, and G), or fillet weld, where the weld metal is deposited in the corner formed by two surfaces (Figures 5-39D, E, or H). These welds are classified as: *reinforced* (Figures 5-39B and F), *flush* (Figures 5-39D, E, and G) and *concave* (Figure 5-39I), if we consider them from the standpoint of strength. The dimension α is called the *throat* and β the *leg* of a weld. A butt weld is considered the best welding joint. The selection of the edge form depends on the welding procedure, the

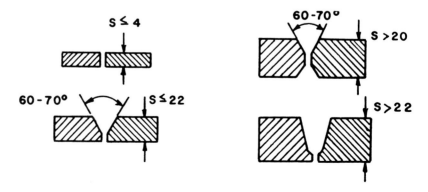

Figure 5-40 Forms of edging of plates for manual welding

Design of Welded Equipment

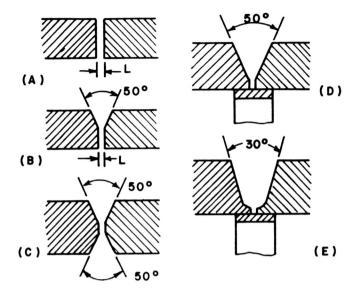

Figure 5-41 Forms of edges of plates for automatic welding

plate thickness, and whether both sides of the joint or only one side can be welded. Figure 5-40 illustrates various forms of edges of the plates for manual welding. These are used for square, butt-welded joints, and single-welded butt joints with and without backing strips. The forms of edges of the plates used for automatic welding are shown in Figure 5-41. The gap 1 between sheets is taken as a function of sheet thickness S as shown in Table 5-4.

For welding vessels, longitudinal joints of shell plates should not be thicker than 6–8 mm. Thicker shell plates require beveling, as shown in Figure 5-41B. Beveling of plates is necessary if it is desirable to increase the role of the weld metal in the joint formation, for example, in welding stainless steel with carbon steel to avoid weld cracks. The double-welded butt joint (Figure 5-41C) is used widely in weld metals in joints, as well as for welding sheets more than 50-mm thick. In the latter, it is desirable to use electro-slag welding without beveling with a gap of 30 mm between the plates. The edges of the plates (Figure 5-41D and E) are used only for transverse joints, single-welded butt joints with backing strip, and one-sided access to the joint. The edge (Figure 5-41E) is used for plates up to 18 mm.

The minimum size of fillet welds, K, (Figure 5-42) used for tee and lap welds depends on the plate thickness to be welded. Sufficient penetration may be achieved by following the recommended sizes of fillets, K, given in Table 5-5.

TABLE 5-4 SHEET THICKNESS AS A FUNCTION OF GAP DISTANCE

S_1, mm	Up to 8	10–12	14–18	20–30	30–40	40–50
l_1, mm	0–1	1–2	2–3	3–4	4–5	7–8

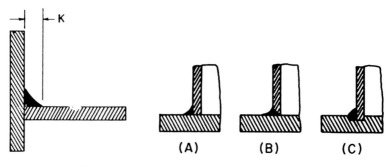

Figure 5-42   Fillet weld

Figure 5-43   Welded joints of shell and tube plate

The maximum fillet size is as follows:

$$K \leq 1.2\, S_{min}$$

where S_{min} is the thickness of the thinnest plate to be welded. The length of the longitudinal fillet weld should not be less than $50K$ because of incomplete fusion at the beginning and end of the weld.

Bevelless automatic hidden-arc welding shells with tube plates (Figure 5-43A) provide better results than beveled shells (Figure 5-43B). There is no need to turn a groove in the tube plate and groove the shell as the plate becomes weakened and welding becomes more complicated (Figure 5-43C). In addition, the cost of machining increases.

The design of a weld joint should provide a sound seam. Wherever possible, avoid overhead welds. Automatic hidden-arc welding is only possible in downhand and horizontal positions. Any configuration of weld joints is possible when using manual welding.

Butt-welded sheets should have a uniform thickness, wherever possible. If their thickness differs more than 20 percent to 25 percent, it is necessary to make the edges of the sheets the same. The slope of the thicker sheet should be 1:5 (Figure 5-44). The weld joint must provide a free shrinkage of deposited weld metal and have a gap between sheets.

In welding processes, access to the weld position with welding equipment is essential. Figure 5-45 illustrates this for manual welding. The example given does not exhaust the complexity of the problem. It depends to a certain extent on the size of the weldment. A very small one is manipulated successfully by hand or by a small fixture to bring welds into workable positions. With very large sizes, welding equipment usually has to be manipulated about the work.

TABLE 5-5   RECOMMENDED FILLET SIZES

S_1, mm	Up to 4	4–8	9–15	16–25	25
K_{min}, mm	3	4	6	8	10

Design of Welded Equipment

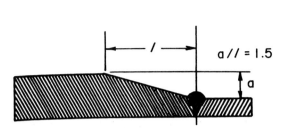

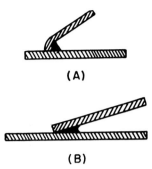

Figure 5-44 Single-welded butt joint for sheets of different thicknesses

Figure 5-45 Joint accessibility for welding

In this case, automatic or even semiautomatic progressive welding processes can be realized economically to long lengths of readily accessible joints.

Certain welding processes require certain types of weld joints, each with their own problems of accessibility and accommodation. For example, automatic and semiautomatic hidden-arc welding processes require protection of the weld metal and flux and must be maintained in one place, as shown in Figure 5-46. The first design version in Figure 5-46 is less successful, but the second one requires a temporary welded strip to keep the flux in place. In some cases where the weld joint is undercut and welding is impossible, for example, in small-diameter tubes and vessels or transverse assembling seams, there is a need to insert a backing strip ring or provide a half-lap joint (Figure 5-47).

Strength factors of a welded joint should be considered in the same way as in any other fabricating system; that is, the weakest links in the system have to be found and their effects evaluated. The mechanical properties of a continuous

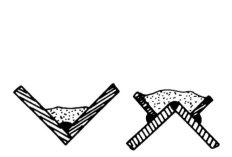

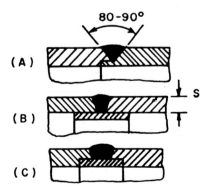

Figure 5-46 Devices that prevent the flux from draining down

Figure 5-47 Weld joints preventing weld metal passage: (A) half-lap joint; (B) butt-joint with backing ring strip; (C) the same for tubes

welded joint are not uniform. The following three zone conditions of a material must be considered:

1. unaffected parent material
2. heat-affected zones of parent material
3. weld deposit or bonding zone of the weld

Because of low welding temperatures, in some cases the second condition listed will be nonexistent; however, the first and third conditions always will be present.

The tensile strength of a welding joint gives a value at which complete rupture would take place under steady-load conditions. It provides a safer indication in welding than some other applications. The mode of fracture of welded material in a tensile test can give a good guide to the material's probable responses to constrained loading coditions. A brittle type of test fracture is a danger sign. The mode of development of a fracture often may be correlated with weldability. A fracture showing marked lipping is a good sign. A transcrystalline cleavage fracture means high static strength but little resistance to notches, and so on. A rough shearing-type fracture indicates a heterogeneous material is liable to crack during welding.

A relatively low yield stress indicates low strength but good shock resistance to welding stresses. For welding applications, it is recommended that the yield stress should be at least 10 percent lower than the fracture stress.

Shearing strength is one of the most important factors in all lap joints. However, in many applications the shearing strength of the weld metal is not likely to be a critical factor.

A low elastic modulus in a material can help absorb strain and shock loading. In cases in which the elastic behavior of the weld material is similar to that of the adjacent parent material, its effect usually is not great. The effect is pronounced if the elasticities of adjacent materials are very different. It is then possible for stress concentrations to build up, which can lead to a failure under shock loading, or even steady load in some cases; or it can lead to premature fatigue failure. A high modulus of elasticity usually implies relatively high residual stress in almost any kind of fusion weld.

For withstanding the welding stresses, as well as normal service, it is desirable that every weld joint provide a reasonable degree of plasticity. Light mechanical shocks to the apparatus may cause severe shocks to the area containing weld joints. If they are not shock resistant, the joints must be placed in the middle of slender parts, rather than toward a corner near a rigid member. A material that has poor shock resistance is also sensitive to the presence of a notch because welds do not arrest a crack once one is started.

Estimation of fatigue in welded applications is very important wherever vibration and other fluctuation loads are applied. Fatigue is not an inherent weakness of weld joint metal and, therefore, is not the main source of trouble. Rather, the effect of transferring stress from one kind of metallurgical structure

Design of Welded Equipment

into another, weld defects, and stress concentration due to weld contour are major problems.

In dissimilar structures existing side by side, stress passes from one to the other and lowers the fatigue resistance due to elastic response. Weld defects due to stress concentrations will affect fatigue strength adversely, the amount of which depends on the type and location of the defects. If a weld with reinforcement is left as welded, there must be a notch effect at the discontinuity that lowers the fatigue strength. This is dependent on the sharpness of the changes of the section and the nature of the weld metal.

A welded structure at elevated temperatures is liable to the same limitation of creep resistance as any other structure used; therefore, it is necessary to know the creep properties of the weld joint metal. The designer should account for the state of the weld with respect to structure and residual stresses. As a result of the interaction of these factors in the case of very large welds, the resistance to creep may be lowered. To increase creep resistance it is advisable to bring the weld structure into as stable a condition as possible by heat treatment before use at elevated temperatures.

Four effects are related to the application of welding in process equipment fabrication:

1. transient stresses
2. residual stresses
3. transient distortion
4. permanent distortion

The severity of each depends on the welding method used, the design configuration of the weldment, and the local circumstances.

It is often necessary to disperse weld joints. Figure 5-48 illustrates the location of longitudinal and transverse weld joints of a vessel shell. The distance between the weld joints a must be no less than 35 or 40 mm. Note that it is not possible to connect several parts with one weld joint (Figure 5-49).

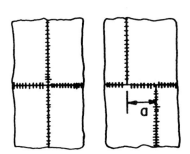

Figure 5-48 Location of weld joints

Figure 5-49 Poor design of a weld connection of three parts

Longitudinal weld joints should not be interrupted by holes and nozzles. There should be regular inspections of weld joints in process equipment. The weld joints should not be overlapped by any kind of supports. If the pickup overlaps the transverse weld joint, it should be designed as shown in Figure 5-50.

Stress concentration can arise within a material in service: (1) if it is deliberately designed so that some parts of the structure have different specific stress intensities from others immediately adjacent to parts, or if certain parts are subjected to combined stresses imposing more strained conditions than on adjacent parts; or (2) accidentally, when it is difficult to consider the general behavior of the materials of construction, because of the effects of the geometry whose significance could not be foreseen, or as a result of damage or defect.

If different stress intensities of one type of stress exist in one member in adjacent zones, the transverse contractions or expansions relative to the stress planes will differ and each zone will impose a transverse restraint on the other, which is proportional in intensity to the difference in strain behavior.

If the zone subject to a basic principal tensile stress has further lateral tensile stress added to the principal stress, then it is most likely a source of trouble.

Residual stresses and distortion must be at a minimum. This can be achieved by good design (that is, by decreasing the volume of weld metal, avoiding local metal accumulation and joint intersection, locating the weld joint symmetrically with respect to the axis). The influence of thermal stresses is especially unfavorable for construction formed from flat sheets, which can result in high warping. Such designs should be avoided wherever possible. A considerable decrease in stress may be achieved by incorporating flexible elements in welded connections, for example, a butt-flanged connection or a nozzle with a

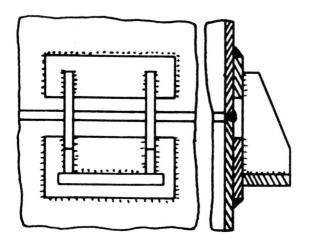

Figure 5-50 Pickup design providing inspection of a transverse butt weld

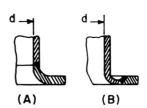

Figure 5-51 Welded joint of a flanged nozzle

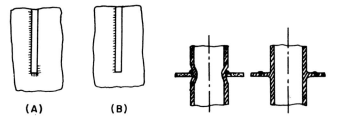

Figure 5-52 Fixing the welded ribs

Figure 5-53 Separation of weld joint for strain decrease

shell (Figure 5-51). There are two versions of this design: (A) flanged opening in the shell, and (B) flanged rim of the nozzle. Beading of a nozzle is more convenient than that of the shell; however, version A is preferable because the walls of a shell almost always are thicker than those of a nozzle. With a shell flanging, the rim of an opening becomes thinner, approaching the nozzle thickness, which is favorable for welding. (See Figure 5-52.)

It is not recommended to weld nozzles, rims, and other parts at the ends of an apparatus. In such cases, the stresses cannot be distributed evenly over the section of the material and can cause weld cracks and rim warping. Welded rim ends should be welded from both sides. Shrinkage during cooling can cause distortion of welded parts but can be avoided by separating the weld joint of the weld in parts (Figure 5-53). Considerable welding strains determine the sequence of the apparatus fabrication. If, for example, the connection is made by rolling out, riveting, caulking, and so on, then these connections must be made after welding, or thermal strains will break the tightness and strength of such connections. Due to an inevitable warping, the machining of welded vessels, checking, and flange turning must be done after welding if there is a need for finished parts. The location of a weld joint then must be separated from a finished surface as far as possible to avoid damaging during welding (Figure 5-54).

Because the weld metal is always more brittle than the base metal, the weld joint must be carried out of the bending stress zone. Construction should be avoided where such stresses exist (Figure 5-55). Any residual stress left in a weld joint may react with applied working stresses as additive stress and bring the material into its plastic flow range or to its fracture stress if it has limited plasticity. In the case of an angle residual tensile stress orientation to the working tensile stress, a restraint on the freedom to deform under the working stress

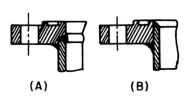

Figure 5-54 Separation of finished surface from the weld joint: (A) good design; (B) poor design

Figure 5-55 Poor weld joint designs where bending stresses arise

will arise, causing the apparent yield stress to increase and thus reducing the ductility of the material. For sensitive materials, ductility may be considerably diminished and brought to embrittlement under shock loading. It is important to account for the effects of any residual stress in the entire range of service temperatures to which materials are to be subjected. Changes in temperature can lead to changes in resistance. Some steels may become more sensitive with a fall in temperature, whereas others become more sensitive with rise in temperature. Lap joints and T-joints are vulnerable to severe stress concentrations and never should be used to withstand either shock or fatigue loading conditions. A weld that has a reinforced shape or extrusion of metal outside the uniformity of the main contour is a potential stress raiser. Hence, the weld profile should be machined smooth in the event that either fatigue or shock loading is critical. If the material is sensitive, then even a gentle reinforcement of a weld (Figure 5-56) can be dangerous to the sharpness of corners. These are likely zones for orientation of fatigue failure.

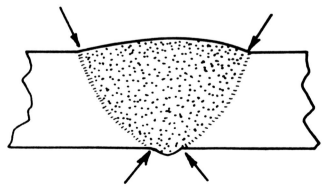

Figure 5-56 Possible stress concentration positions on a normal fusion weld profile

Design of Welded Equipment

There may be dangers also in the interior of a weld joint (Figure 5-57), even if full allowance is made for the external shape. These kinds of joints are sometimes used for relatively unimportant work but are generally not recommended. In the case of bad workmanship full penetration is not achieved, and there is the chance of weld discontinuity, which may bring about notch-induced failure. Even if measures are taken to avoid apparent geometric discontinuities in a weld joint, stress concentrations may arise if adjacent zones of the material do not react in the same manner as to the applied stress.

From economic considerations, it is desirable to minimize the amount of welding by combining bending and stamping with weld joints (Figure 5-58), combining continuous and intermittent weld joints, and using a minimum angle of beveling. Unfortunately, the possibilities of decreasing the metal weld volume are quite limited in fabrication since weld joints not only must be strong, but tight as well. In welding small parts, such as strips and studs for bubble caps in distillation columns, pressure contact welding is used. This will not only decrease the length of weld joints but also increase the accuracy of assembly.

The joint should be arranged to minimize any tendency to pry off the elements to be welded. A rule of thumb in designing a welded joint is to provide strength by length rather than by size, as it is necessary to use more weld metal for developing the required strength when a large short weld is applied rather than a small long weld.

One of the most important principles in the design of welded joints is to arrange the welds to be symmetrical in location, size, and length. For members such as structural steel angles (Figure 5-59), the length of a weld is divided in accordance with the following equation:

$$L_1 X a = L_2 X b$$

where a and b are dimensions from the centroidal axis of the section. Welded joints subjected to tension are divided into three classes according to their internal stresses (Figure 5-60). In class 1, the resistance of the weld is exclusively in

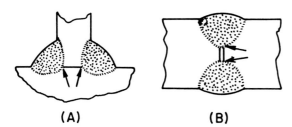

Figure 5-57 Joint conditions likely to give internal stress concentrations where indicated by arrows: (A) normal fillet-welded T-joint; (B) double V-joint with unblended roots

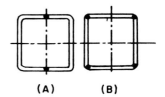

Figure 5-58 Decreasing the weld-joint length by combination bending and welding: (A) good design; (B) poor design

178   Fabrication Methods   Chap. 5

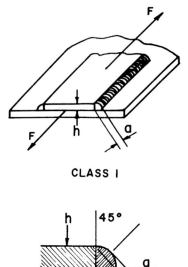

CLASS 1

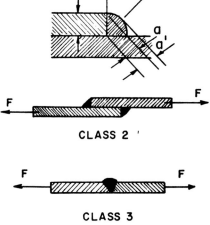

CLASS 2

CLASS 3

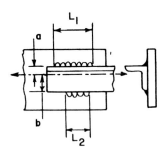

Figure 5-59 Symmetry applied to welding an angle

Figure 5-60 Force action in various classes of welds

TABLE 5-6 RATIO OF STATE STRENGTH OF SHIELDED WELDS TO BASE METAL

Joint classification		Weld type			
Class type	Condition	Leak proof	Concave	Flush	Reinforced
1	Shear	0	0.40	0.90	1.10
2	Shear and tension	0	0.30	0.75	0.95
3	Tension	0	0.20	0.60	0.75

Design of Welded Equipment

TABLE 5-7 STRENGTH OF SHIELDED ARC-FLUSH STEEL WELDS (psi)

		Tension	Compression	Bending	Shear	Shear and tension
Limit Stress	Elastic Limit					
	Base metal	32,000	35,000	35,000	20,000	—
	Deposited metal	40,000	44,000	44,000	24,000	—
	Endurance limit of deposited metal	22,000	—	26,000	—	—
Recommended Design Stress	Static load (safety factor 2)	16,000	18,000	18,000	11,000	11,000
	Load varies from 0 to F (safety factor 2)	14,500	16,000	16,000	10,000	10,000
	Load varies from +F to −F (safety factor 2.75)	8,000	8,000	9,000	5,000	5,000

shear; in class 2, the resistance is in shear and tension; in class 3, the resistance is in pure tension.

The approximate relative strengths of welds in percentage of the base metal strength for various classes of joints and types of weld are presented in Table 5-6. These data should be used only for orientation in selecting the type of joint and weld. For properly designed and skillfully made welds, the strength is almost equal to that of the base metal.

Table 5-7 gives the design stresses for shielded flush welds. The throat dimension a of a reinforced fillet (Figure 5-60) should not be taken more than $1.2a$, and for a concave weld $a' = 0.5a$. The effective length of each weld should be taken 5/8 in. shorter than the total length to allow for imperfections caused by starting and stopping a weld bead.

Welds should be proportioned to be stiffer than the adjacent material, and thus throw the deflection into the latter in all cases when the joints are exposed to bending or reverse stresses. The endurance limit of the weld metal in the case of repeated stresses should be taken no higher than 18,000 psi. A safety factor of 2 or 2.5 may be applied.

Under static loads initial stresses are not dangerous. This is true when the yield point exceeds the maximum stressed point, causing the metal to creep and the load to redistribute more evenly over the cross section. After several local overloadings, the stress distribution in the structure approaches the distribution that would exist without initial stresses. For variable loads the initial stress actually may reduce the safety factor. In the case of impact the presence of initial stresses is extremely undesirable.

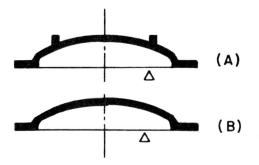 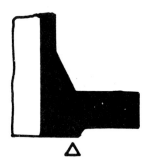

Figure 5-61 Ribs on a shell head for assistance during machining

Figure 5-62 Flange projection for machining

Influence of Machining on Half-Finished Parts

Half-finished parts must have a shape convenient for making alterations in machines. For castings, it is necessary to provide special ribs to permit further work in a chuck (Figure 5-61) on machines. These ribs may be removed after machining. The rough and clean bases must be selected properly for easy marking and fixing. For cast vessels it is convenient to take parting planes as a base. For steel apparatuses, one should take the face edge of shell.

The shape and size of half-finished parts should be considered so that the machined surface is minimal. The surface to be machined should be clearly separated from the surface that is nonmachined. Therefore, all surfaces to be machined must have a projection over nonmachined ones as, for example, in cast flanges (Figure 5-62), or parts machining with another surface roughness.

The thickness of similar lugs should not be small (that is, not less than the thickness provided by the casting tolerance). Otherwise, warpage of the pattern

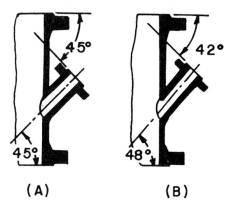 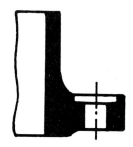

Figure 5-63 Surfaces must be situated at angle convenient for machining: (A) good design; (B) poor design

Figure 5-64 Flange machining by countersink

Design of Welded Equipment

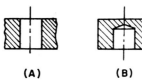

Figure 5-65 The hole drilled through (A) is better than the blind hole (B)

Figure 5-66 A slot for passing out the cutting tool from a nut screw cap

and casting may prevent lug formation. The ribs must not reach the machining surface to prevent a shock load on the cutting tool when turning (Figure 5-60C).

Machining surfaces should be mutually situated either parallel or perpendicular. Only when absolutely necessary (Figure 5-63) is it permissible to make an angle arrangement different from 0, 30, 45, 60, or 90°. It is sometimes profitable to use countersinks instead of total flange machining (Figure 5-64). For machining massive parts by speed methods, sufficient balance should be provided to avoid vibration.

Wherever possible, parts must provide for the passing-through of machining of surfaces and planes (Figure 5-65). In this relation, version A is better than B in Figure 5-65.

One also may provide a slot for passing out a cutting tool (Figure 5-66). Three versions of boring a vessel for centering a head extension are given in Figure 5-67. Versions A and B are equally correct, but version C is not recommended.

As already noted, a part must be accessible for machining. For example, it is impossible to drill a hole for a screw in a pulley hub (Figure 5-68) since the pulley rim prevents this. Therefore, it is necessary to drill an angular hole, or to drill the pulley and then a hub (Figure 5-68).

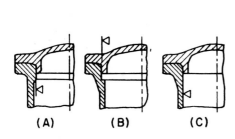

Figure 5-67 Versions of boring of a vessel for centering the head: (A, B) good designs; (C) poor design

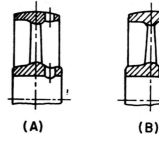

Figure 5-68 Drilling on the pulley hub: (A) good design; (B) poor design

The shape of a part must not make the task of machining difficult. A drill should be started wherever possible on a surface perpendicular to the axis of the hole to be drilled. The surface at the drill exit should be perpendicular to prevent the tool from breaking (Figure 5-69).

Holes should be drilled through. Where blind holes are used, sufficient thickness should be left at the bottom of the hole on the opposite side of the wall. When using ordinary drills and drilling methods, a depth equal to six to eight times the diameter of the drill is permissible. With the use of special drills, the depth of drilling can be increased. It is recommended that half-finished parts have a shape providing minimum drilling depth (Figure 5-70).

Internal threads for steel bolts and screws subject to a tension load should provide an engagement equal to the thread diameter (1.5 times the thread diameter for cast iron or brass, and two times the thread diameter for aluminum or zinc). In threads tapped in blind holes (Figure 5-71), the depth of the hole should be deeper than the depth of the thread by a distance of at least five times the pitch of the thread. A thread relief should be provided in the hole with a width that is at least three times the thread pitch, and a diameter that exceeds the maximum major diameter. The thread engagement for standard studs should be at least 1.25 times the diameter of the thread for steel, 1.5 times the diameter for cast iron, and as much as 2.5 times the diameter for other mate-

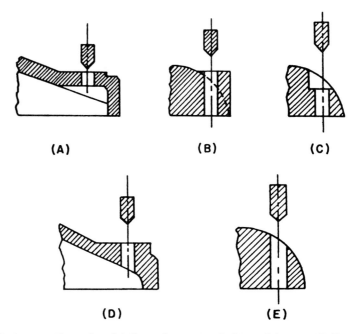

Figure 5-69 Examples of drilling schemes: (A, B, D) good designs; (C, E) poor designs

Design of Welded Equipment

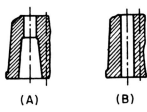

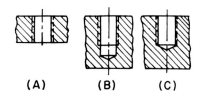

Figure 5-70 Decreasing the drilling depth: (A) good design; (B) poor design

Figure 5-71 Hole tapped for a stud: (A, B) good designs; (C) poor design

rials. It is not recommended to align the parts by studs because the accuracy of stud fabrication is lower than that of smooth finished parts, and the alignment must be done by a special cylindrical driving band (Figure 5-72).

Precise holes made with a reamer should be done all the way through and be provided with an entering chamfer. Shaped surfaces, which must be machined with a master form or pattern gauge, should be provided only in extreme need. It is better to use conventional methods of machining (for example, milling and turning).

A part to be produced by milling must be rigid enough to withstand the cutting force without deflection. The part should be designed so that the maximum number of surfaces can be machined in one pass. Removal of the part and placing it in another position will take more time and allow opportunity for error.

All internal and fillet radii should be the same size on parts produced with an end mill. The fillet radius produced by the radius on the corner of the cutter should not be too large as it can interfere with the cutting action. Parts must be so designed that the portion produced by the bottom of the end mill is at a 90° angle to that produced by the side, as shown in Figure 5-73A.

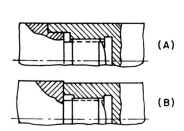

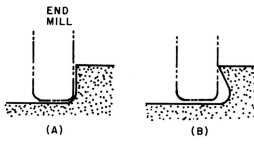

Figure 5-72 Example of threads that do not provide alignment: (A) alignment with a driving band, good design; (B) alignment by a thread, poor design

Figure 5-73 (A) Preferred configuration when using an end mill; (B) unsatisfactory configuration

Slots should be made with a regular milling cutter (Figure 5-74A) rather than with an end mill (Figure 5-74B) because such slots are more accurate and cheaper.

Turning is a widespread method for parts fabrication. It is versatile with respect to both the operations that can be performed and the size of the parts that can be produced. Parts that have a length in excess of 15 times their diameter require special precautions because of the deflection caused by the force of the cutting. The concentricity of the various machined diameters can be held closer when they are all machined in a single setup. All fillet radii that can be made by the same cutting tool should have the same dimension. Corners should have a chamfer rather than a radius.

Shafts and other parts bounded by surfaces of revolution should be made smooth, without sharp changes in diameters. In the finished shafts, the adjacent diameters should not differ by more than 3–5 mm. Square ends should be avoided on shafts (preferring in this case the use of keys). If there is a need for coaxial collars of large diameters, removable thrust rings (Figure 5-75) are recommended as opposed to turning them from a thick round blank.

The grooves and slots should be designed so that they can be machined with the use of an engine or vertical lathe, as well as with a milling machine. As far as is practicable, the depths of grooves and slots should be located in the same plane. For fulfillment, for example, of the last condition, the transverse slot depth in tube plates should match the depth of the rum bore for a gasket (Figure 5-76).

In designing equipment, we are concerned not only with determining dimensions to assure proper assembly and function, but with providing for tolerances great enough to permit economical manufacture. To lower the cost of fabrication, select maximum tolerances (the total variations that are permitted) and a minimum degree of surface finish. Wherever possible, the surface should be left unfinished. It should never be machined to present a better appearance because the increase in class of accuracy and degree of surface finish are related to the cost of equipment. Surfaces of a part may be divided in two groups: functional and nonfunctional. The nonfunctional surfaces are those that usually

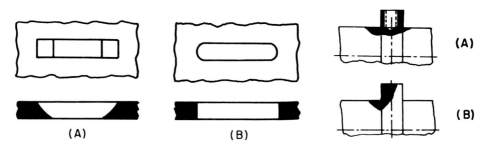

Figure 5-74 (A) Preferred slot; (B) unsatisfactory slot

Figure 5-75 (A) Solid and (B) removable collars

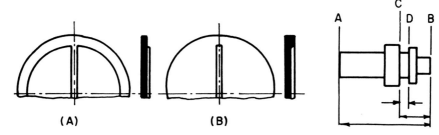

Figure 5-76 Slot boring for a gasket in a tube plate: (A) good design; (B) poor design

Figure 5-77 Relation of dimensions

do not come in contact with other surfaces in the assembly or operation of the machine, for example, the rough surfaces of castings and forgings.

Functional surfaces may be either primary or secondary: the first ones control the position of motion of parts, such as holes, which have locations in two directions. Secondary surfaces do not control position as a primary surface, such as a spotface, for example. Surfaces have a direct or first-degree relationship if there is only one dimension between them; for example, a second degree, as A and C, and third degree, as A and D in Figure 5-77.

A relationship between surfaces is required when they are functional surfaces. One should try to utilize the relationship between surfaces that results from fabricating operations for primary functional surfaces that require an accurate relationship.

Whenever a close relationship must be maintained between two surfaces, they should be directly related. If they have an indirect relationship, several dimensions must be held much more accurately than would otherwise be necessary. For example, in Figure 5-78 dimension C should be used rather than A and B if the size of the slot must be strictly maintained. If the width of slot C is critical and must be accurate, dimensions A, B, or D do not necessarily have to be closely held.

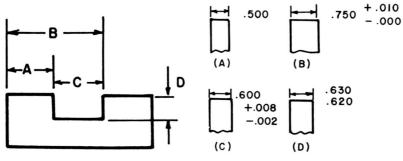

Figure 5-78 Effect of dimensioning on tolerance required

Figure 5-79 Tolerance systems

Parts fabrication to exact dimensions is not economically feasible; in most instances it is necessary to keep only a few dimensions strictly held. The tolerance must have an acceptable magnitude of variation from the exact size. Because tolerance has a great effect on cost, one must be certain that a large tolerance will not cause difficulty in assembly or operation.

Tolerances may be specified in several ways (Figure 5-79). If the tolerance was not necessary, it would be referred to as the design size (Figure 5-79A). The unilateral system (tolerance all in one direction) is illustrated in Figure 5-79B; the bilateral system (tolerance allowed in both directions) in Figure 5-79C; and in Figure 5-79D the tolerance is indicated by specifying the maximum and minimum sizes that are acceptable. The method shown in Figure 5-79C may be used to indicate the desired size.

The interpretation of the dimension (Figure 5-80A) allows the conditions in Figure 5-80B and C as variation in size, but not the lack of parallelism (Figure 5-80B) or the lack of flatness (Figure 5-80C).

Geometric tolerances applied regardless of feature size are used to avoid extremely close tolerances. In such a system a note may specify that a surface is to be "flat within 0.002 total." The interpretation is shown in Figure 5-81A. The entire surface must lie between two parallel planes 0.002 in. apart. To avoid this condition (Figure 5-80B), a tolerance on parallelism (Figure 5-81B) would be used, the interpretation of which is shown in Figure 5-80C. The surface should be located between two parallel planes 0.003 in. apart and parallel to the datum plane, A, established by the high points of the surface, as defined in Figure 5-82. When a system of dimensions and tolerances (Figure 5-83A) is used for locating several holes, it is possible to obtain the situation shown in Figure

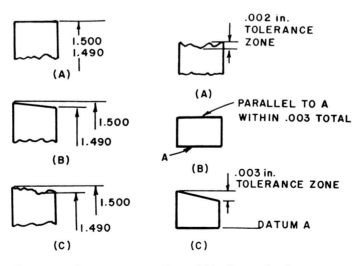

Figure 5-80 Interpretation of a dimension

Figure 5-81 Geometric tolerance

Design of Welded Equipment

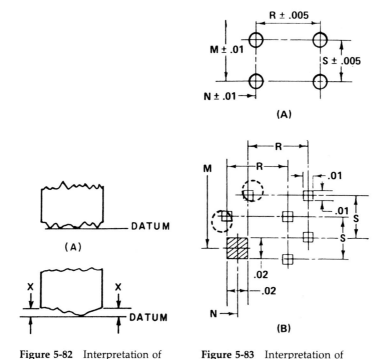

Figure 5-82 Interpretation of datum plane

Figure 5-83 Interpretation of location dimensions

5-83B, where two extreme (tolerance is exaggerated) positions of the upper left-hand hole are presented by the two dotted circles. Following this figure, one must make allowance for the size of the holes required, which will be rather involved. Consequently, the hole will have to be oversized by a considerable amount if reasonable tolerances are to be permitted.

The design may permit considerable tolerance in the initial location of a part attached by bolts in the holes. Once located, however, it has to be limited to only a small variation from that position. In such a situation the method shown is obviously undersizable, and it can be avoided by using true-position dimensioning when the dimensions locating the holes have no tolerance. The holes then are permitted to vary from this exact location by a specified amount, as shown in Figure 5-84. In this figure the shaded circles denote a tolerance zone of 0.010 in. diameter, within which the center of the hole has to be located and the dotted circles show the two extreme positions of the upper left-hand hole. Dealing with a true position when the holes are smallest, assembly would be possible with larger holes when the distance between them exceeds the allowed tolerance. Generally, this additional tolerance is acceptable, and the true position is considered to apply to the maximum material condition (MMC). But if this additional tolerance is not to be permitted, the notation RFS is used, which means that the true position is to be held regardless of a feature size. Locating

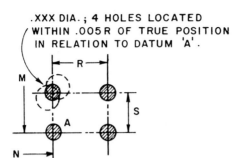

Figure 5-84 True-position dimensioning

holes in parts that must be assembled also may be done by placing a note on each set of holes stating that one must match the other. With this method one can achieve the desired result with less effort on the part of fabricating. Critical portions of the tooling of both parts can be made simultaneously by assuring proper matching without having to work to very close tolerances.

A layout should be carefully drawn to scale showing the mean dimensions and locations in the final stages of a design. The layout illustrates a clearance where required, and if the parts were made to the mean dimension the actual clearance would be close to that shown. The allowed tolerances may combine to create an interference. Tolerance also can combine to create an excessive clearance. Both these unsatisfactory conditions are referred to as tolerance stacks. Calculations should be made wherever an interference or excessive clearance appears possible to determine whether a tolerance stack exists.

Sheet Materials

Many articles of process equipment are made from sheet metal. Such articles are usually lighter in weight and often less expensive than castings or forgings. Such important parts as shells and heads for vessels, pipes, elbows, hoppers, and so on may be made from sheet iron and steel, galvanized iron, copper, aluminum, or brass.

The shape to which the sheet material is cut must be simple. The parts should be cut by power-squaring shears and the amount of scrap should be minimized. Inner corners must be avoided where possible as shown in Figure 5-85. A small projection is shown in Figure 5-86.

In general, it is easier to cut a rectangular sheet and weld it to two strips. This simplifies cutting and decreases the amount of scrap. It is possible to obtain parts of a more complicated shape with a cutting torch. It is easier to manufacture circular pipes than square ones.

In sheet bending, elastic and plastic deformation can arise. The minimum bend radius R under cold conditions depends on the sheet thickness S and is determined as follows:

$$R = K'S$$

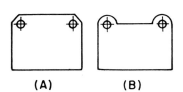

Figure 5-85 Shape of a part cut from sheet: (A) good design; (B) poor design

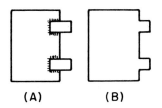

Figure 5-86 Welded strip simplifies the cut: (A) good design; (B) poor design

where K is a coefficient, depending on the material and the direction of plate rolling. Table 5-8 shows typical values of K'.

If the direction of sheet rolling is unknown, the larger K value should be selected. Cold bending of sheets with a small radius is accompanied by cold hardening. It is desirable that bent sheets be heat treated to improve their structure. To avoid cracking, bends should be made across the grain of the metal. The grain will be parallel to the long edge of the stock. The minimum bend radius in which there is no cold hardening is $R = 20S$ inches. Hence, the minimum diameter of a cold-rolled shell should be $D > 40S$.

The minimum bend radius for pipes is determined by their diameter, wall thickness, and the method of bending. Pipe bending should be in the direction of the elongation of the outer fibers of the pipe, which tends to decrease the wall thickness. This also produces wrinkle formation in the compressed zone of the pipe. Denoting the pipe wall thickness as $S/D = K'$ and assuming that $R_{min} = 20S$, we will obtain a minimum bend radius of a pipe (Figure 5-87A):

$$R = D(20K' + 0.5)$$

The wall thickness S usually is taken in the limits from 0.04 to 0.1D. To avoid defective bends, the following minimum bend radius for seamless pipes when fabricating with mandrel can be used:

At $D < 51$ mm and $S > 1.2$ mm, $R_{min} = 2D$

At D from 51 to 73 mm and $S > 1.6$ mm, $R_{min} = 3D$

At D from 37 to 103 mm and $S > 2$ mm, $R_{min} = 4D$

TABLE 5-8 K' COEFFICIENT VALUES

Material	Across fibers	Along fibers
Steel	0.5	1.2
Brass (Hard)	0.5	1.2
Mild Brass and Aluminum	0.3	0.5
Copper	0.25	0.4

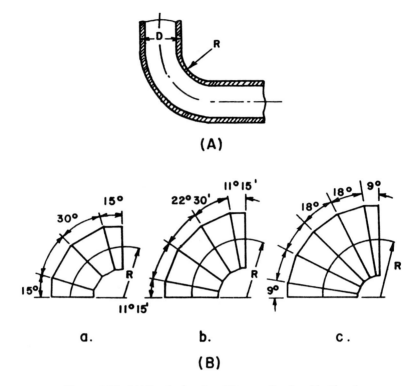

Figure 5-87 (A) Bend of a pipe; (B) normalized welded bends

For bending of a welded pipe with a seam situated along a neutral line:

$$R_{min} = 6D$$

and for a seam situated along the stretched side:

$$R_{min} = 9D$$

The large radius bendings cause an increase in the apparatus' gabarids. To decrease them, use special-shaped and stamped parts with beveled edges for welding. This improves the design considerably and lowers its price.

Bending large pipes is very difficult. Their sizes become excessively large. Therefore, the bends for large pipes usually are welded (Figure 5-87B).

Assembly/Disassembly Considerations

When designing, consider the influence of assembly and disassembly for the process equipment. There are three reasons for considering assembly: reliability, the cost of initially assembling the parts, and the cost of assembly when maintenance is performed. The design should provide safeguards against incorrect assembly.

Design of Welded Equipment

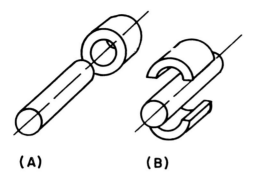

Figure 5-88 Methods of assembly: (A) axial method; (B) radial method

The assembly of process equipment may be done by the radial and axial methods. In the axial assembly method, one part is slipped on the other and moved along the axis. For example, a ball bearing is slipped on the shaft. In the radial assembly method, an exterior embracing part is provided by a radial parting plane. Because of this there is no need to push the interior part through the exterior one. An example providing radial assembly is a split-sliding bearing. Both these methods of assembly are shown in Figure 5-88.

The axial method is convenient because it provides good coaxial alignment of parts and simplifies the design and fabrication. However, the assembly itself to some extent becomes more complicated. The radial assembly method is used when the interior part is very long, heavy, and awkward, or when there are some parts that prevent disassembly. All parts assembled by the axial method must be slipped on the covered part from one side.

It is always desirable to design equipment so that it would be possible to preassemble and adjust subassemblies and sets (for example, a bubble-cup tray) and then couple them together. In so doing, it is important to keep the number of "type-size" parts used in the apparatus as small as possible. This is convenient both for assembling and for decreasing the amount of spare parts needed.

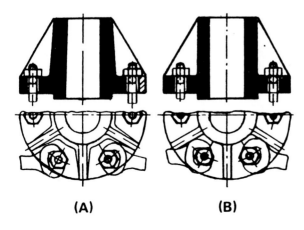

Figure 5-89 Relative position of bolts: (A) good design; (B) poor design because it does not consider the casting draft

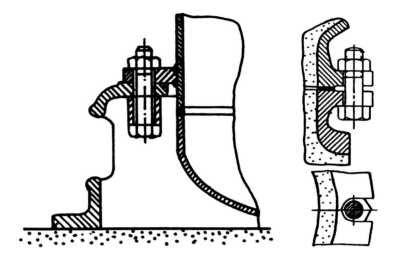

Figure 5-90 Opening in a frame for bolt insertion and support

Figure 5-91 Inserted bolts

Standard parts, such as bolts, nuts, and screws should be used wherever possible because drawing, tooling, and manufacturing paperwork are saved, as well as assembly and maintenance operation.

It should be noted that hands and tools are required to accomplish assembly and hence sufficient access to components must be provided. The distance between adjacent nuts, or nuts and walls, for example, must be sufficient to allow the nut to be put on and to apply a wrench to it (Figure 5-89).

Figure 5-92 Example of a design with nonparallel axes of fasteners

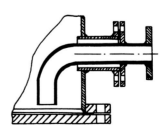

Figure 5-93 A drain nozzle

Design of Welded Equipment

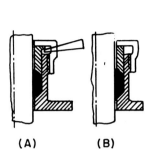
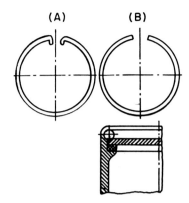

Figure 5-94 The design of small stuffing boxes: (A) good design; (B) poor design because the gland withdrawal is difficult

Figure 5-95 Spring catch lock for a head: (A) good design; (B) poor design

If bolts are used it is necessary to provide a support for a bolt head. For example, it is important to insert and support the bolt in the opening in a frame (Figure 5-90); the high washer makes it easier to reach the bolt head.

Free access to all holders and fasteners as well as to the parts that must be aligned during operation (for example, stuffing boxes) should be provided. The closeness of adjacent parts or the configuration of apparatus parts may prevent insertion and withdrawal of bolts. In such cases, slots are provided and bolts are inserted in the slots from the side (Figure 5-91).

Bolt withdrawal from such a slot need not require complete screwing off. It may be sufficient to provide one to two turns and to move the bolt in a radial direction. This withdrawal often is used in cases in which rapid assembly and disassembly are desired. However, the slots for bolt withdrawal weaken flanges, and for conservation of their strength they must be reinforced.

The design should be foolproof so that parts cannot be assembled incorrectly. For parts hooked up in two ways, the bolts could be made of different diameters to permit only the proper way. Bolts and screws should not require a closely limited length to provide the required clearance, especially where the end cannot be checked. Designs should be avoided where the axes of fasteners are not parallel to each other (Figure 5-92).

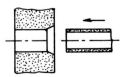

Figure 5-96 Chamfer on the tube plate to simplify the assembly

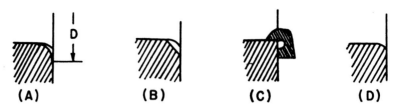

Figure 5-97A Mate parts: (A, B, C) good designs; (D) poor design

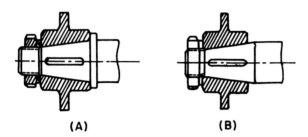

Figure 5-97B The fit of a boss on a conical shaft: (A) poor design; (B) good design

The design of assemblies and subassemblies must provide a simple and convenient assembly. This possibility has to be checked with great care. The drain nozzle of an apparatus (Figure 5-93) is made removable to allow for apparatus disassembly. However, the dimensions chosen do not permit the job to be done because of the bent part. To do it, one needs to make either the nozzle diameter larger, or to increase the bend radius of the pipe to withdraw it through the nozzle. The parts (such as shafts, sleeves, glands of stuffing boxes, sealing rings, and so on) that are withdrawn must be so designed to be conveniently clamped when disassembled. Threaded holes for screws and studs should be provided that will simplify the withdrawal process. Pins must be designed to be easily detached by a screw or to be knocked out.

Proper (Figure 5-94A) and poor designs (Figure 5-94B) of stuffing boxes are shown. The gap formed by conical machining of the sleeve provides for the insertion of a screwdriver, allowing easy detachment of the gland. The proper example (Figure 5-95) is a spring catch lock of a small apparatus. The spring catch lock is made as a split-snap elastic ring. It can be inserted easily into a groove, but withdrawal is very difficult. Simply bending the ring ends will provide easy withdrawal. For reasons of disassembly, it is necessary to insert a ball bearing with an interference fit. It is sufficient to use a push fit. In the latter case the ball-bearing withdrawal is much easier.

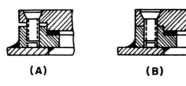

Figure 5-98 Assembly of a blank: (A) good design; (B) poor design

Design of Welded Equipment

The shape of adjacent parts must simplify the assembly: the mate part must be beveled to simplify the assembly (Figure 5-96). The sizes and shape of chambers and fillets must be so designed that the parts cannot be pinched. For low-stressed parts, it is possible to use undercuts (Figure 5-97(D)). The design must provide a good alignment when assembled. The alignment of bolted parts are made either by a centering projection or by register pins. It is important to provide parts that are mated only in one surface, that is, statically determinate. Extra mate surfaces complicate the fitting and the design. Figure 5-97(A) gives an example in which it is difficult to machine the shaft end of the collar and cone, which are supported simultaneously with the force against the boss. A proper design is shown in Figure 5-97(B).

Another example of a statically determinate design is shown in Figure 5-98A. In the proper design the full gasket strain should be left as a gap providing the possibility of further tightening of the connection. In version B, (Figure 5-98B), further gasket tightening is impossible. The statically indeterminate designs are sensitive to temperature variations.

The lengthened parts must be fixed only on one end to provide for the free expansion of the other one. To simplify an axial assembly, it is desirable that the extensions on the interior part enter the exterior one, not simultaneously but stepwise. In designing, one must keep in mind that disassembling an apparatus, especially in a functioning unit, is much more difficult than assembling on the construction site, where special devices and tools are required.

NOMENCLATURE

A	cross-sectional area (in.^{2})
a, b	dimensions (in.)
D	diameter (in.)
d	distance or dimension (in.)
E	modulus of elasticity (psi)
F	force (lb$_f$)
h	height (in.)
I	moment of inertia (in.^{4})
K	maximum fillet size (in.)
K'	coefficient
L	length (in.)
M	bending moment (lb$_1$-in.)
MMC	maximum material condition
N	casting size (m)
n	safety factor
R	bend radius (in.)
r	junction radius (in.)
S, S'	thickness (in.)

T	temperature (°F)
W	weight (lb_m or kg)
Z	section modulus (in.^{3})
α	coefficient of thermal expansion (in./C)
δ	interference (in.)
τ_t	tensile stress (psi)

6

Design and Material Properties

DESIGN FROM HIGH-ALLOY STEELS

Equipment of austenitic steels and alloys is widely employed in the chemical and allied industries. Prolonged overheating of austenitic steels, even if they contain titanium or niobium, results in a burning out of alloy components and a reduction in chemical resistance.

Welded joints should be designed so that the parts to be connected are heated simultaneously to the melting point. This condition is achieved automatically when parts of the same thickness are butt-welded. For welding parts of different thicknesses, the thicker part should be made beveled to match the thinner part with a slope of 1/4 to 1/5. Butt-welding of a tubular shaft end, for example, is considerably better than that of a conventional method. This is illustrated in Figure 6-1.

Similarly, the welding of a blade to a hollow shaft is more favorable than to a solid one, as shown in Figure 6-2. The same considerations are valid for connecting any different-walled parts, for example, welding a shell to a tube plate, flanges to thick-walled pipes, and so on. The conventional method of welding the shell to the tube and plate (Figure 6-3) used in fabricating carbon steel apparatuses cannot be recommended for austenitic steels. The plate thickness decrease is achieved by turning on two grooves (Figure 6-3A). The other method sometimes used (Figure 6-3B) is more expensive, as machining produces

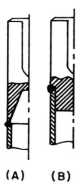 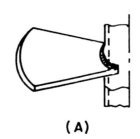 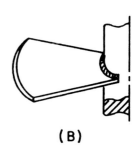

Figure 6-1 Welding of a tubular shaft end: (A) good design; (B) poor design

Figure 6-2 Blade welding to a shaft: (A) good design; (B) poor design

more waste from trimmings. For connecting a thin-walled shell to a plate, the design shown in Figure 6-4 is recommended.

A circular slot (3–4 mm wide) is turned onto the plate and then the thin rim of the plate is butt-welded to the shell. In welding parts of different thicknesses, an intermediate ring (Figure 6-5) or connection of a thin-walled shell with a massive tube plate (Figure 6-6) is employed. A deep circular slot is turned onto the plate with the radius no less than 1.5 times the shell wall thickness. The upper end of the plate is rounded. Owing to its flexibility, such a connection decreases the boundary stresses as well.

All the thick-walled parts are connected with thin-walled ones in the same manner (Figures 6-7 and 6-8). If it is not possible to make a butt joint, the piece should be turned on a slot as a massive part to decrease its thickness (Figure 6-9). Such a design decreases the risk of burning out the alloy components. However, border stresses may develop as a result of shrinkage after welding.

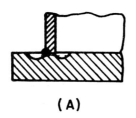 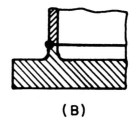

Figure 6-3 Welding a shell to a plate: (A) good design; (B) poor design

Design from High-Alloy Steels

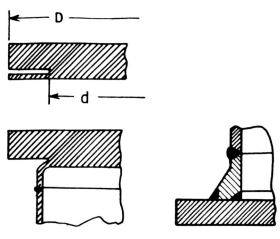

Figure 6-4 Welding a shell to a flanged plate

Figure 6-5 Welding parts of different thicknesses with the use of an intermediate ring

Lap joints are unsatisfactory for two reasons: First, the thickness of a lap joint is equal to the sum of the thicknesses of two sheets, which is equivalent to the metal accumulation. This is dangerous because there is a possibility of burning through a thin sheet. Second, in such joints the strip of metal between welds would be under tension, and every tension in austenitic steels decreases their strength because of a recombination of the crystalline lattice as a result of deformations. Proper and poor designs are shown in Figure 6-9.

It is important to maintain the composition of steel and, therefore, not to overheat many times in the same place. The weld joints of equipment of stainless steel should be separated from each other. In connecting shells, the distance between longitudinal seams should not be less than five times the width of a weld and not less than 50 mm; therefore, piling up welds is not permissible (Figure 6-10B). The version shown in Figure 6-10A is preferable.

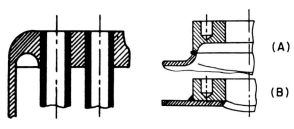

Figure 6-6 Welding a shell to a tube plate

Figure 6-7 Welding of a boss to a shell: (A) good design; (B) poor design

Figure 6-8 Welding a flange to a shell: (A) good design; (B) poor design

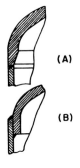

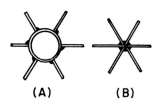

Figure 6-9 Welded assembly of head to shell: (A) good design; (B) poor design

Figure 6-10 Rib welding: (A) good design; (B) poor design

Welding of austenitic with nonaustenitic steels is difficult due to the danger of diluting the seam with nonaustenitic metal and the formation of fragile transition layers between the austenitic seam and nonaustenitic steel. The influence of the difference between these two types of steels on the quality of equipment will become clear if we compare some physical properties of carbon and stainless steel, for example, the 18-8 type.

The coefficient of thermal expansion of 18-8 steel is $\alpha = 0.0000173$, or 1.5 times larger than that of carbon steel ($\alpha = 0.0000112$), and the coefficient of thermal conductivity of chrome-nickel steel is equal to 12–16 kcal/m °C hr, which is three to four times less than for carbon steels ($\lambda = 40$–50 kcal/m °C hr). This results in lower heating along the width of the seam during welding, which, together with a larger coefficient of thermal expansion, may cause excessive stresses that break the apparatus and deteriorate the corrosion resistance of the steel.

To prevent the formation of excessive stresses, flexible elements are used that play the role of compensators. Because of their deformation, the stresses

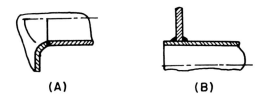

Figure 6-11 Welding a nozzle to a shell: (A) good design; (B) poor design

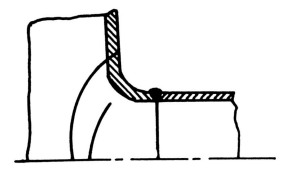

Figure 6-12 Welding in a reinforcing ring

are decreased in the zones around the seams. The typical design solution to this problem is shown in Figure 6-11A. As shown, the nozzle is butt-welded to a flanged rim of a shell, which serves as a compensator. The usual design for an apparatus of carbon steel given in Figure 6-11B is not valid for stainless steel because of crack formation. The same principle is used in connections in which a reinforced ring is inserted (Figure 6-12).

The use of flexible connections near weld seams is quite reasonable because they help to maintain the corrosion resistance of an apparatus.

Relatively low heat conductivity of austenitic steels should be considered when designing heat exchangers. If one heat-transfer coefficient is small compared to the other, then the thermal resistance of a wall, S/λ, has almost no influence on the value of the heat-transfer coefficient, k. For example, let $\alpha 1 = 10$ and $\alpha 2 = 10,000$ kcal/m² °C hr, and wall thickness $S = 0.5$ cm. For a wall of carbon steel, the heat-transfer coefficient will be

$$k = \frac{1}{\frac{1}{10} + \frac{1}{10,000} + \frac{0.005}{12}} \approx 10 \text{ kcal/m}^2 \text{ °C hr}$$

For a wall of chrome-nickel steel, we will obtain approximately the same value:

$$k = \frac{1}{\frac{1}{10} + \frac{1}{1000} + \frac{0.005}{12}} \approx 10 \text{ kcal/m}^2 \text{ °C hr}$$

The situation changes greatly when designing high-efficiency equipment, where change of phases takes place on both sides of the heat-transfer surfaces and both heat-transfer surfaces and their heat-transfer coefficients, $\alpha 1$ and $\alpha 2$, are comparable to each other, for example, 10,000 kcal/m² °C hr. Then, for a wall of carbon steel, we obtain

$$k = \frac{1}{\frac{1}{10,000} + \frac{1}{10,000} + \frac{0.005}{50}} \approx 3,300 \text{ kcal/m}^2 \text{ °C hr}$$

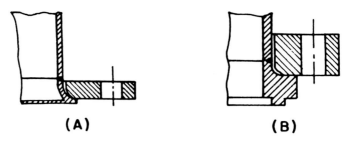

Figure 6-13 Flanges for apparatuses of austenitic steel: (A) on beading; (B) on bead ring

and for a wall of austenitic steel,

$$k = \frac{1}{\frac{1}{10{,}000} + \frac{1}{10{,}000} + \frac{0.005}{12}} \approx 1{,}600 \text{ kcal/m}^2 \text{ °C hr}$$

Thus, for the heat-transfer surface made from austenitic steel, the heat-transfer coefficient decreases by about 2. Consequently, the low heat conductivity of austenitic steels may prevent the design of highly efficient heat-exchanger equipment. The high cost of austenitic steels (mostly because of the high content of chrome and nickel) demands that the designer be economy conscious.

Only parts contacting aggressive substances should be fabricated from chrome-nickel austenitic steels. All other parts beyond the working space of the apparatus, for example, flanges, lugs, and so on should be manufactured from conventional steel. Flanges are preferred to be loose, using beading or bead rings (Figure 6-13).

Welding austenitic steels with carbon steels is possible because the melting points of both types are similar. However, such welding is difficult due to the

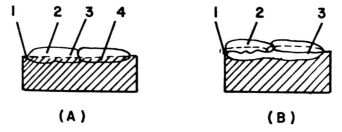

Figure 6-14 Beading of a layer of austenitic steel on the surface of carbon steel: (A) without intermediate layer (breaks along the martensile layer); (B) with an intermediate layer of noncarbon soft iron; 1-carbon steel; austenitic beading; martensite layer; 4-breaks; intermediate layer (soft Armco iron)

Design from High-Alloy Steels

danger of diluting the seam with nonaustenitic metal and the potential formation of fragile transition layers between the austenitic seam and carbon steel. To prevent dilution of the seam, electrodes and wire are used with an increased content of nickel, chromium, and other alloying elements, and an attempt is made to provide minimum melting of the carbon steels. During manual arc welding of parts of heterogeneous steels with electrodes of ordinary nonalloyed steels, cracks generally do not appear in the metal of the seam. However, this does not eliminate the danger of local destruction of the zone near the seam.

The appearance of fragile layers between the austenitic seam and the nonaustenitic steel can occur from the following conditions: as a result of diffusion of carbon in the steel into the metal of the seam, and/or insufficient mixing of the metal of the seam with the nonaustenitic steel.

Diffusion of carbon into the seam is accompanied by decarbonizing of the nonaustenitic steel in the zone near the seam, causing its weakening. However, destruction can occur not only in this zone, but also on the austenitic carbide layer in the austenitic seam because of carbon enrichment. To prevent carbon diffusion, it is recommended to weld combined joints with electrodes using an increased nickel content.

If there is an insufficient amount of alloying components in the filler metal, breaks may develop on the fragile layer, not only during the welding of heterogeneous steels but also during seam welding of a layer of austenitic metal on the surface of carbon steels. This is illustrated in Figure 6-14A.

One way to prevent breaks is by preliminary seam welding of the intermediate layer of noncarbon soft iron (Armco iron) on the surface of steel being subjected to seam welding with austenitic metal. This is shown in Figure 6-14B. In welding heterogeneous steels, there is always the danger of a decrease in alloying substances in the welding zone (shown in Figure 6-15 by the dotted

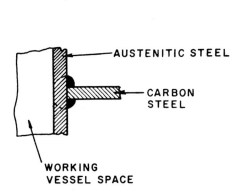

Figure 6-15 Metal zone deteriorated by diffusion of alloy elements (shown by dotted line)

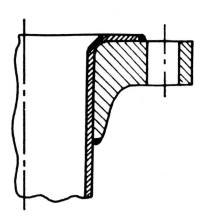

Figure 6-16 Poor flange design

line). This will occur only when the steel alloy's sheet thickness is sufficient that corrosion can be avoided in the place opposite the weld. The deterioration of corrosion resistance in such connections may arise as a result of residual welding stresses.

For thin austenitic metal sheets (3–5 mm thick), a reduction in corrosion resistance almost always occurs in places opposite the weld joints with parts of carbon steel. Hence, such assemblies as that shown in Figure 6-16, in which a carbon steel is welded on the outside of stainless pipe, is completely unsatisfactory because the metal of the pipe opposite the weld will corrode at a faster rate than the pipe itself, eventually breaking the connection.

To avoid deterioration of the base metal, it is sufficient that the seam between the austenitic and carbon steel be moved as far as possible from the corrosion site (generally a distance larger than the depth of diffusion). This is realized by inserting between austenitic and carbon steel parts an additional piece of austenitic steel. Thus, the seam between the carbon and austenitic steels would be located far away from the region subjected to the aggressive media, and the strength of the equipment would not be violated. Figure 6-17 shows examples of such a design welding a support shell to the column of austenitic steel (Figure 6-17A); welding a jacket (Figure 6-17B) and a shell of the heat exchanger to the tube plate (Figure 6-17C).

The same purpose is served by a pad of austenitic steel between a vessel made from stainless steel and a lug from carbon steel. This enables the load to distribute over the larger surface of the shell, as shown in Figure 6-17D.

After welding parts of austenitic steels, the seams should be cleaned carefully, and the apparatus itself should be polished and pickled. This increases the corrosion resistance by a factor of 3 to 5.

In the overwhelming majority of cases, the application of welding austenitic steels can be managed without heat treatment. However, when it is necessary to remove welding stresses to avoid corrosional cracking of weld joints or to increase durability properties of the steel and seam, heat treatment is necessary. Elimination of the inclination of austenitic seams to intercrystalline corrosion is attained by hardening the welded structures.

If it is possible to subject only the small parts to hardening without difficulties, then the entire structure does not require treatment. It has been determined that a considerable increase in corrosional stability can be attained without hardening from high temperatures (1050–1150°C). It is sufficient to heat the weld joint to a temperature of 850–900°C, to hold it at this temperature for two to three hours, and cool it in the air to give the metal even greater stability against corrosion. This is obtained as a result of hardening. Such heat treatment is called stabilizing or diffusion annealing.

Welding evokes the appearance of residual stresses, which may cause one of the most dangerous forms of destruction of welded structures of austenitic steels in certain aggressive media (corrosion under stress or corrosional cracking). To avoid this, it is necessary to remove the stress in the structure by heat treatment. Heating to a temperature of 650°C is sufficient to remove stresses in

Design from High-Alloy Steels

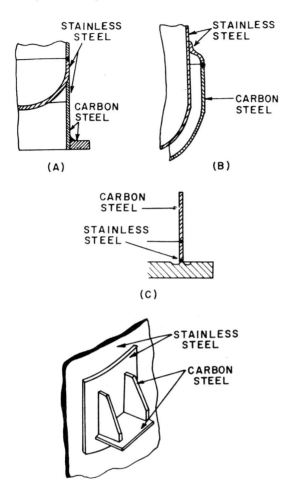

Figure 6-17 Examples of the use of intermediate parts when welding austenitic steels with carbon steel

parts of ordinary carbon steel. Austenitic steels possess higher indices of yield; therefore, with heating to 650°C, internal stresses caused by welding are maintained. Furthermore, at such a temperature the formation of chromium carbides results on the boundary of grains causing the appearance of the inclination of austenitic seams (resulting in intercrystalline and knife corrosion). Therefore, to remove stresses from a part made from austenitic steels, heat treatment at higher temperatures (800–850°C) is used. Consequently, stabilizing annealing increases the stability of the basic metal and the weld seam, not only against intercrystalline corrosion, but also against corrosion under stress. Considering this, parts of particular importance are subjected to stabilizing annealing. If it is not possible to subject the entire part to annealing, one must resort to local heat treatment of weld seams. Thus, for instance, pipelines of austenitic pipes are treated this way during assembling of chemical and other process plants.

DESIGN OF EQUIPMENT USING COPPER

Pieces of equipment made from copper are fabricated from rolled copper sheets or tubes. Copper has excellent ductility and, hence, can be readily formed into complex shapes. Since it is relatively inexpensive, it has enjoyed wide use as half-finished parts, as well as in the form of large apparatuses and their connections. The shells of apparatuses are easily rolled, with heads stamped or beaten out. To increase the rigidity of thin-walled shells, they are bead formed by a creasing machine (Figure 6-18). Bead forming also is used for fixing parts inside the equipment.

Permanent connections of copper parts are made by soft and hard soldering, welding, and riveting. Soft soldering is always made with the use of lap joints. The butt joints generally are not strong enough because soft-soldering strength is low. For example, butt-soldering will be 2,200/320 = 7 times weaker than the copper sheet. From considerations of full strength and taking an equal safety factor for sheet and solder, one obtains the proper amount of overlapping. It is approximtely equal to seven times the sheet thickness, $b = 7S$ (Figure 6-19).

Soldered joints made from soft solder can be used up to +120°C. At this temperature the strength of the connection decreases considerably. Taking this into account, as well as a possible seam inequality, it is recommended that the allowable stress for soldered seams be no more than 25 kg/cm². The soldered joints are excellent at low temperatures up to cryogenics. Of the three examples given in Figure 6-19, example A is more favorable because of its symmetry and the absence of additional bending moments. However, its adjustment is difficult for large-diameter vessels. The seam in B is used for connections with limited parting, and it assists parts alignment in assembling. The assembling

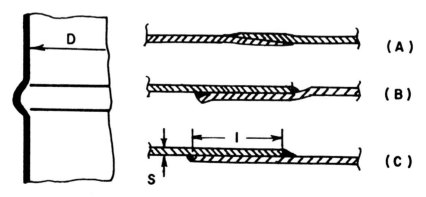

Figure 6-18 Bead forming of copper shell

Figure 6-19 Examples of soldered joints

Design of Equipment Using Copper

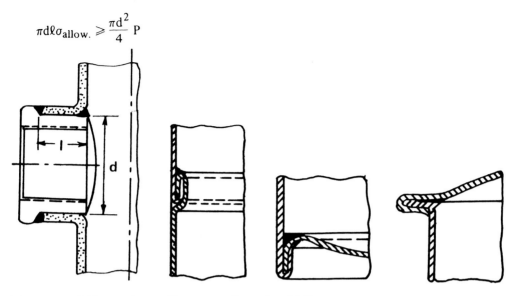

Figure 6-20 Soldering of a boss

Figure 6-21 Examples of folders

force creates a bending moment, which gives rise to additional stresses in the seam; therefore, such connections should be avoided.

In designing any loaded-soldered connection, the connection area must be determined according to the condition that the seam should withstand all acting forces. For example, in the design of the seam between a shell and a boss (Figure 6-20), the following criteria should be fulfilled.

$$\pi d l \sigma_{allow} \geq \frac{\pi d^2}{4} P$$

where σ_{allow} = allowable stress for the solder. The value to be determined is l.

The increase of strength and seam unloading are achieved by the use of locks and folders, which play the role of rigid elements at the same time. Figure 6-21 shows three examples of folders, and Table 6-1 gives the widths of folders for different sheet thicknesses.

Soft soldering is the basic method of making tube connections in heat-exchanger plates. For installations working under pressure (in cryogenics units, pressures can be up to 200 atm), tube diameters must be small (from 5 × 1 to 10 × 1.5) because of copper's low strength. It is difficult to roll out such small tubes, and generally a tight fit in tube sheets is not achieved because of the metal's softness. A recommended design for soldered connections of tubes is given in Figure 6-22.

The tube sheet shown is provided with grooves. Tin-coated tubes are drawn through the sheet so that they extend 5–10 mm over the sheet. Then the

TABLE 6-1 TYPICAL SHEET THICKNESS AND FOLDER WIDTHS

Sheet Thickness (mm)	0.75	1.25	1.5	1.75	2.00	2.50
Folder Width (mm)	7	13	14		16	19

sheet with tubes is heated and poured by solder. To increase strength, solder pouring sometimes is done on both sides of the sheet. Soft soldering is used in many cases, including in the tightening of riveted connections.

The greatest advantage of soft soldering is its low melting temperature, which minimizes the formation of thermal stresses that can cause part warping. Soft-soldering devices that do not permit load increases on the seam above the design rating should be used (for example, stresses arising from uneven heating during starting or assembling of the equipment, and so on). This is helpful in the use of lens or bellows compensators.

Hard solders and brass are used for connecting parts that are heavily loaded statically or dynamically. Brass melts at 910°C, whereas most solders have a melting temperature of 840°C. Silver solders containing 10 percent to 70 percent silver are used for vital soldering, where not only tightness and strength are important, but surface finish of seams as well, and where there is also a danger of burning through thin parts. The tensile strength for mild brass is higher than that of copper at the same temperatures. The relative elongation for all three solders is quite sufficient (20 percent to 40 percent); therefore, sheets can be butt-welded.

In some cases hard soldering is needed. One example is in connecting tubes for heat exchangers for oxygen units (Figure 6-23). The connecting tubes are soldered in the upper head of the condenser-vaporizers of copper-oxygen columns. The hard solders are diffused into the material of connecting parts; therefore, the connections are very strong. In hard soldering, the possibility of warping connecting parts exists because of residual thermal stresses in the seam. It is important that parts that are to be connected be well heated and not be moved out of their positions. For soldering thin sheets, a connection with set teeth is used, as illustrated in Figure 6-24.

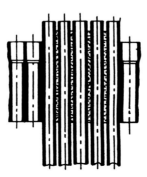

Figure 6-22 Soldering copper tubes to a tube sheet

Design of Equipment Using Copper

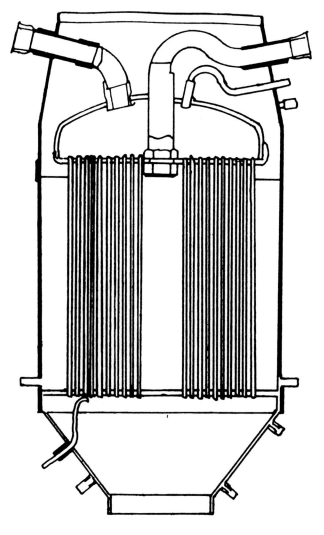

Figure 6-23 Tube soldering by silver solder

Figure 6-24 Design of a longitudinal seam of a shell

The stretched end of one sheet is cut at a length equal to five times the sheet thickness. The teeth are set and the other sheet also stretched, but not until it is inserted among the teeth. After hammering, the seam is hard soldered, then cleaned, rolled out, and smoothed. The shells for oxygen columns are rolled once more to restore them to their correct shape.

For important seams, such as longitudinal seams on vessels, hard soldering or welding is recommended. Soldering permits parts to be connected to different metals; for example, copper and brass, copper and steel, steel and brass, and so on, provided that the surfaces of the connecting parts are moistened.

Welding successfully replaces the other methods of connecting parts of copper equipment but demands more delicacy in fabrication because of difficulties in avoiding warping. Welded joints of copper equipment should be butt-welded for the same reasons that apply to steel. The safety factor for copper-welded joints is equal to $\psi = 0.8$.

Riveted connections (no longer used in manufacturing steel equipment) are still used in the fabrication of copper apparatuses. Riveting eliminates warping and produces a strong connection. Riveting with copper rivets is made under a cold condition. There is no need to heat rivets because copper is sufficiently malleable. The minimum sheet thickness for riveting is 3 mm. In the case of sheet connections, with sheet thicknesses at 7 mm or more, the tightness

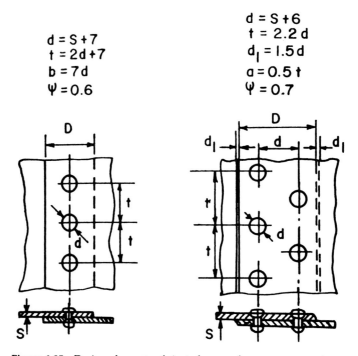

Figure 6-25 Design elements of riveted seams for copper apparatuses

Design of Equipment Using Copper

TABLE 6-2 ALLOWABLE SHEAR STRESS FOR COPPER RIVETS

Wall Temperature (°C)	120	121–140	141–160	161–180	181–200	201–220	221–240	241–250
Allowable Shear Stress (kg/cm²)	360	350	340	320	300	280	260	240

of the seam can be achieved by caulking. If the sheet thickness ranges from 3 to 7 mm, caulking cannot be used and the tightness must be achieved by soft soldering. For strength, soldering is not taken into account in the design. It is assumed that all loads are taken by the rivets. The design elements of riveted joints are shown in Figure 6-25.

If sheets or parts of different thicknesses are connected, all design sizes of a seam should be determined on the basis of the thinner sheet. Parts whose total thickness is more than four times the diameter of a rivet should not be connected because in this case the stem of the rivet may be curved and the quality of the connection would be spoiled. The strength factor of a row seam is $\psi = 0.6$; for a two-row seam it is $\psi = 0.7$.

The allowable shear stresses for copper rivets are given in Table 6-2. The allowable bearing stresses are equal to double the shear stress at the same temperature. The detachable connections of copper apparatuses are specific because copper is too mild and expensive to subject flanges to large loads. Therefore, flanges for these systems usually are made from regular carbon steels as lap-joint flanges fastened on a lap-joint stub or bead ring, which, in turn, are soldered to the shell with hard solder. The inner flange edge should be rounded so the beading or the bead ring will not be spoiled. To protect the flange from damage during disassembling, it should be riveted to the beading as shown in Figure 6-26A, or provided with bead forming (Figure 6-26B).

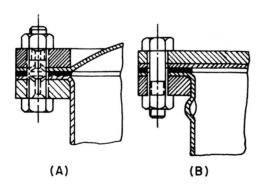

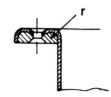

Figure 6-26 Fastening of a flange to the shell: (A) by riveting; (B) by bead forming

Figure 6-27 Building in the head of a rivet

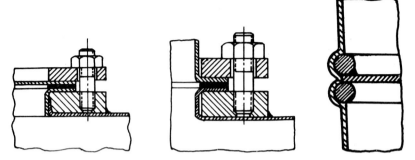

Figure 6-28 Design of a boss for copper equipment

Figure 6-29 Fastening a tray by the use of rings

For providing a smooth sealing surface, countersunk riveting should be performed. If the beading thickness is not enough to countersink the head of the rivet, counterboring should be done to build it in as part of the sheet (Figure 6-27). The connection bosses of bronze or brass are bended round by copper, then soldered as shown in Figure 6-28.

The contact area of a sheet and a boss should be sufficient from the standpoint of equality of acting forces of pressure and elasticity in the solder. Nozzles are either soldered or welded. Sometimes they are connected by a socket joint. Bosses and other parts covered by copper are fabricated from copper alloys. Figure 6-29 shows an example of fastening a plate in a small-diameter distillation column. The shell is bead formed with the split ring inserted in the lower part of the two beads and then soldered to the shell with a soft solder. The tray is put on this ring and then inserted into the other ring above, and the entire assembly

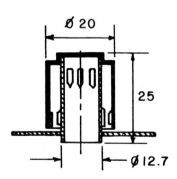

Figure 6-30 Bubble cap soldered to a tray

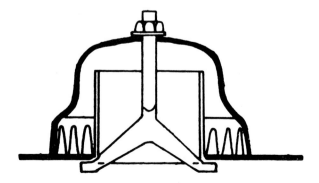

Figure 6-31 Removable bubble cap for a copper column

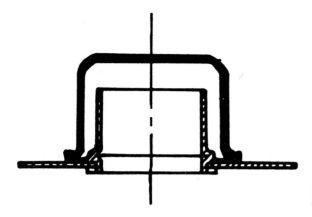

Figure 6-32 Fixing a bubble cap with a tab washer

is soldered together. Bead soldering is possible only for shells up to 3.0-mm thick.

Beads formed for shells that work under pressure (for example, shells of lower oxygen columns; $P = 6$ atm) should be unloaded. This shell is inserted inside a more thick-walled shell (6–8 mm thick) and the same pressure is created in the gap between two shells. Typical examples of bubble caps are shown in Figures 6-30, 6-31, and 6-32.

Figure 6-31 presents a bubble cap of a cryogenic column fixed to the tray by soldering. An example (Figure 6-31) is a bubble cap that could be disassembled by unscrewing only one nut.

DESIGN OF EQUIPMENT USING ALUMINUM

Design characteristics of equipment fabricated from aluminum are infuenced by the metal's fluidity, its instantaneous oxidizability, and its low mechanical strength. Aluminum's instant oxidizability makes soldering practically impossible. Hence, the main method of aluminum parts connection is welding.

Because of its low melting temperature (657°C) and the low viscosity of the melted metal, it is important to melt the two parts to be connected simultaneously. Therefore, all connections should be butt-welded. The parts to be connected must have equal thickness at linkup; consequently, the welded joints are designed in the same manner as for vessels of austenitic steels. For example, all nozzles are welded to a shell beading. This is done not because we wish to compensate for thermal stresses, but to provide butt-welding. Flanges usually are fabricated from carbon steel. Lap-joint flanges and lap-joint stubs are used for aluminum equipment. Because aluminum is easily flanged, there is no need for beading rings. Steel parts, for example, support lugs, cannot be welded or soldered to aluminum apparatuses. Hence, these components also are fabricated from aluminum. However, it is preferential to fabricate them from steel and then weld them to a detachable steel ring, which is fastened with bolts

214 Design and Material Properties Chap. 6

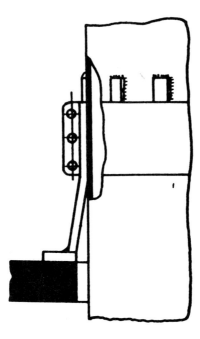

Figure 6-33 Lugs for aluminum apparatuses

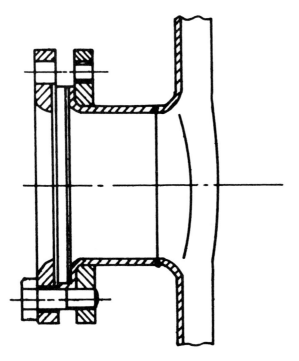

Figure 6-34 Sight glass for an aluminum apparatus

tightly pressing against the vessel's shell. This is shown in Figure 6-33. Sight glasses generally are fastened as ordinary blind flanges onto the nozzle beading, as shown in Figure 6-34.

DESIGN OF CERAMIC-LINED EQUIPMENT

Enameling is a very delicate process. The quality of enameling depends on many factors but is largely established by the properties of the base metal.

Cast iron used for fabrication of enameled equipment must be strong and tight. It should have little inclination to grain growth, and its coefficient of thermal expansion should be close to that of the enamel. The following composition of cast iron is recommended for enameling: C = 3.5–3.5 percent; Si = 1.4–1.8 percent; Mn = 0.55–0.65 percent; P = 0.3–0.38 percent; S = 0.10–0.12 percent; and the cross-breaking strengths should be 37–40 kg/mm^2.

Steel used for enameled equipment should have a carbon content of no more than 0.12 percent and should be of high purity because the gases released from slag insertion cause bubble formation in the enamel. Steel used in enameling should have the following composition: C = 0.005–0.09 percent; Mn = 0.35–0.4 percent; P = 0.10–0.02 percent; S = 0.03–0.04 percent; Si traces.

For obtaining a high-grade enamel coat the equipment should have a special shape, the necessity of which is determined by the following conditions: preparation of equipment for enameling; conservation of even thickness of coating during roasting; simultaneous and uniform heating and roasting of enamel; and elimination of bubble formation during roasting as well as dangerous deformations (especially tensile strain due to residual thermal and other stresses, which could break the enamel coating). For obtaining a high-grade enamel coating, the shape of the equipment should be simple and smooth without sharp rims, corners, and cavities. If there is insufficient rounding, the enamel will drain down the corners or crack. All parts of the equipment should be prepared carefully before enameling. The seams should be polished and the corners rounded; the greater the radius of rounding, the better the quality of the coating. Consequently, the equipment should be designed such that its internal surfaces have rounded shapes. This provides easy access for mechanical work and inspection. Local cavities are unwanted in apparatuses subjected to enameling. Some enamel accumulates after cooling, which can cause cracks due to thermal stresses.

Larger parts enclosed by flat sheets are very unfavorable. They tend to undergo warping during roasting, as well as dent formation where accumulated enamel would be broken away after cooling. Planes should be reinforced by ribs. It is better to avoid abrupt planes where possible by employing rounded surfaces. High-grade and uniform enamel roasting, as well as coating conservation during cooling, is provided by a uniform rate of heating and cooling of all the equipment parts. This condition is easily fulfilled when the walls of an

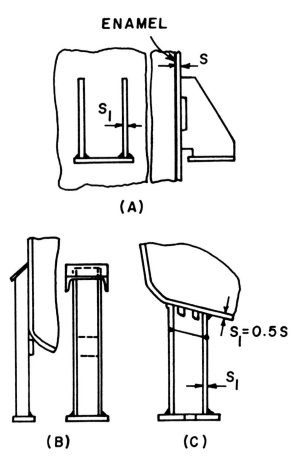

Figure 6-35 Examples of lug welding to an enameled apparatus

apparatus have the same thikness. Therefore, in designing, care should be taken to avoid using massive parts and other large metal accumulations. Difficulties arise in the design of flanges, lugs, supports, and so on. Lugs and similar parts are welded on steel vessels subjected to enameling by an intermittent weld (Figure 6-35) without strengthening plates.

Apparatuses with tubular lug supports should be welded to short tubes (150–200-mm diameter) before enameling. The lugs are then welded to these tubes afterward. The main cause of bubble formation in enamel is gas release from seams during roasting, and especially dangerous is the root of a seam where slag is always accumulated, providing a source of air holes. In Figure 6-36 an example is shown where bubble formation develops against the root of the weld. Welds should be cut out carefully and welded onto the surface from the inside, and the seam itself should be polished. A proper design for enameling is shown in Figure 6-37. An example of a poor design is shown in Figure 6-38B.

The bubbles were formed as a result of air penetration from the gap between the tube and flange through microcracks and pores in the seam. The

Design of Ceramic-Lined Equipment

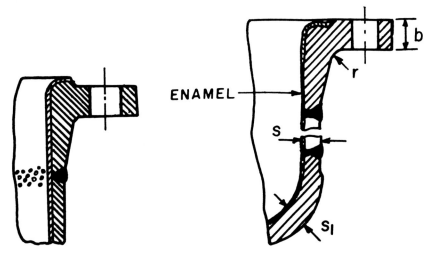

Figure 6-36 Defective enameling as a result of gas formation during roasting

Figure 6-37 Proper design of apparatus for enameling

change of the lower continuous seam in the intermittent seam provided a free air exit from the gap outside and eliminated spillage completely (Figure 6-38A).

Elimination of residual stresses in enamel vessels is achieved by the following measures: equal wall thickness, reasonable seams location, and butt-welding of nozzles to beaded shells. Nozzles up to 200 mm in diameter often are fabricated in a conical shape, as shown in Figure 6-39. The beading is important to coat the place where the enamel is connected.

Designs that could develop local edge stresses during cooling should be avoided (Figure 6-40) because they are very dangerous for brittle enamel. Also to be avoided is the use of ribs because they prevent free vessel shrinkage during cooling and may generate local stresses. The nozzles on the head should be located symmetrically. Care should be exercised during assembly and disassembly of enameled vessels to avoid local overstressing and shocks.

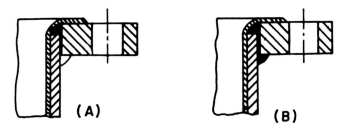

Figure 6-38 Welding of a flat flange: (A) good design; (B) poor design

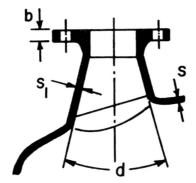

Figure 6-39 Conical nozzle of a vessel to be enameled

The flange design for enameled vessels is different compared to non-enameled equipment. For a weight decrease of cast and welded flanges, complex shapes must be used to approach equal wall thicknesses (Figure 6-41A).

The flanges of main parting are especially complicated in fabrication (Figure 6-41B); however, such a complication is justified because it reduces the amount of defective equipment that is manufactured. The other feature of such connections is the use of a greater number of thin bolts or studs (12–16 mm in diameter). Small-diameter bolts decrease the distance between the gasket and bolt circle and, consequently, decrease the bending moments applied to the head flange. Being more elastic than thick bolts or studs, they provide more

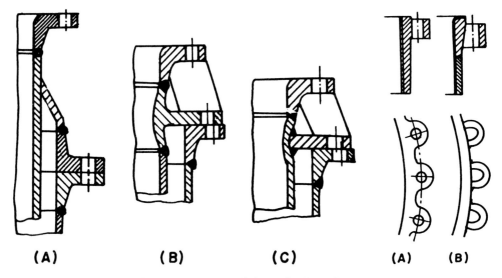

Figure 6-40 Connection of a jacket to an enameled vessel: (A) good design; (B, C) poor designs

Figure 6-41 Main parting flanges for an enameled vessel: (A) cast; (B) welded

uniform tightening in the connection. The gasket thickness is taken as three to four times more than for the same equipment without enamel. This results in a decrease in the seating load, bolting force, and bending moments. To tighten enameled equipment it is desirable to use soft gasket materials (polyethylene, fluoroplastics, rubber, and so on). Loose flanges are quite suitable for enameled vessels. After roasting, enameled vessels may undergo warping. The danger of warping should be considered in designing, especially with mixers. Consequently, a sufficient gap should be provided between the impeller and the walls of the vessel, as well as the shaft and guiding sleeve. To decrease the frequency of defective parts, all small parts and assemblies, stuffing boxes, thermometer sockets, devices for sampling, and so on should be detachable.

The difficulty of enameling and the possibility of covering only streamlined and smooth surfaces considerably limit design possibilities, forcing the designer to fulfill heat-transfer surfaces either as jackets or field tubes, and columns fabricated either from packed or stuffed grid plates. The high corrosion resistance of enamel vessels has justified their use in the manufacturing of a variety of high-purity products.

DESIGN USING MOLDING MATERIALS

Equipment fabrication from prefabricated sheets of plastic materials generally is not difficult. For thermosetting materials the process is reduced to the formation of items from mild sheet and their hardening by heat treatment at 100–160°C.

Thermoplastics (polyvinyl chloride plastic, acrylics, and so on) are stamped from sheets heated to a softening temperature, then welded or glued. After stamping, the parts usually are machined. The fabrication of vessels from sheets is simple and does not present special limitations.

The fabrication of parts by stamping of molding material is specific and demands careful design considerations. In parts designing it is necessary to provide shapes that produce the following features:

- high-quality fabrication and a minimum of defective items
- minimum consumption of material
- ease of parts removal from lower and top stamping dies
- minimum final machining of a part

The defective products result in incomplete material filling of stamping dies, inadequate pressing, cracks formation and warping due to poor selection of part shape, and nonuniform hardening.

Smooth shape and equal wall thickness are very important to obtain high-grade parts. Therefore, the first criterion for obtaining quality parts is good design. The wall thickness should not be excessive. Excessive wall thickness results in additional consumption of material, which increases the time a part is

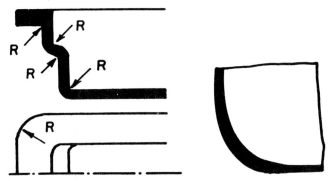

Figure 6-42 Radii of curvature for a plastic part

Figure 6-43 Reinforcement of an angle

punched in the machine. The minimum wall thickness that provides filling the mold is 2–4 mm. The minimum wall thickness for medium parts is approximately equal to 1 mm. The maximum wall thickness must not exceed 10–12 mm. With thick walls in regions of material accumulation, incomplete hardening occurs even if the part is kept in the mold for a long time. This tends to decrease the strength of a part. It is best to maintain the small wall thickness over the entire piece. If different wall thicknesses are inevitable, then the transition from one wall thickness to another should be as smooth as possible. The walls of plastic parts must be coupled smoothly.

Sharp corners on walls result in crack formation and stress concentration; therefore, they should be avoided. The following radii of curvature are recommended (Figure 6-42) in vertical projections and internal radii in horizontal projections:

$$R \geq 0.5\text{–}15 \text{ mm}$$

and for outside horizontal projections:

$$R = 5\text{–}10 \text{ mm}$$

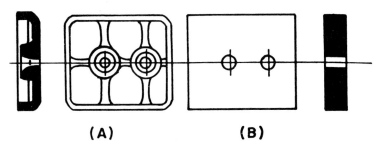

(A)   **(B)**

Figure 6-44 Change of a massive part by a ribbed one

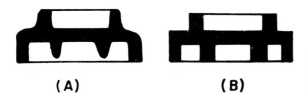

Figure 6-45 Decreasing the rib height for ease of machining: (A) good design; (B) poor design

For loaded parts, maximum possible radii of curvatures should be taken. To decrease bending stresses in the corners, it is permissible to make a smooth thickening by bringing it to the thickness of the wall (Figure 6-43).

Changing massive thick-walled parts to thin walls with ribs results in a savings of molding material. This increases quality and avoids warping of a part. An example of changing a massive part with ribs is given in Figure 6-44.

The rib thickness should be equal to the wall thickness and the edges of walls and the coupling walls and ribs are rounded. The ribs should be located in the direction of the top die so the part can be removed from the mold. The ribs should not prevent machining, and thus should not be brought to the machining surface (Figure 6-45A).

The wall draft located in the direction of pressing is used to permit easy withdrawal of a part. The draft is made no less than 1:200; the internal draft is recommended to assume more than the external one. Examples of proper and poor designs are given in Figure 6-46.

Increased rigidity and minimal warping can be achieved by changing flat surfaces into spherical or shaped surfaces (Figure 6-47). For example, parts with high walls should be fabricated with bent-in bottoms (Figure 6-46). Sometimes both ribbing and shaping are used to increase rigidity.

The shape of a part's edges should be considered carefully. Too thin an edge can be broken and would violate the principle of equal thicknesses of walls (Figure 6-48C, D).

If the part must be elastic, it is best to make it both without a bead and without edge thickening (Figure 6-48A), or to make a bead of minimum thickness (Figure 6-48BB). The bead should extend around the perimeter of the part (Figure 6-49A). Bead interruption is not desirable because it may result in crack formation (Figure 6-49B).

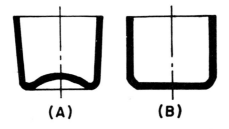

Figure 6-46 Draft made for easy part withdrawal: (A) good design; (B) poor design

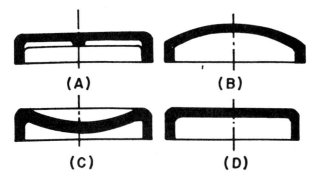

Figure 6-47 Flat surfaces such as A and D are poor designs. B and C represent acceptable designs that minimize warping

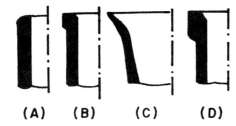

Figure 6-48 Examples of part edges: (A, B) good designs; (C, D) poor designs

Figure 6-49 Design of an edge: (A) good design; (B) poor design

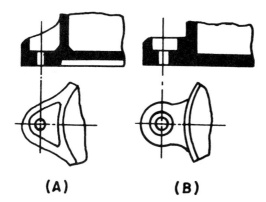

Figure 6-50 Supporting projections: (A) good design; (B) poor design

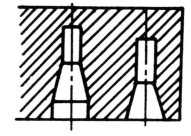

Figure 6-51 The shape of holes

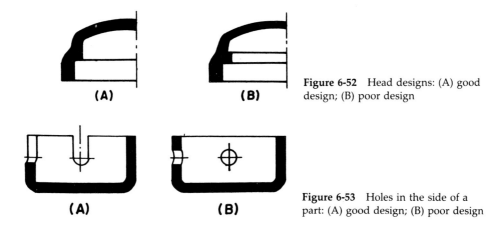

Figure 6-52 Head designs: (A) good design; (B) poor design

Figure 6-53 Holes in the side of a part: (A) good design; (B) poor design

It is not recommended that the part be supported around the entire base. Because of warping, the supported surface always becomes somewhat curved. Tightening with bolts may warp parts and induce bending stresses. Curved surfaces will hinder the support stability in this case. The supporting surface is better handled with separate lugs or projections, which should have a shape sufficient to withstand the stresses resulting from fastening. Stress concentration should be avoided, and the design should account for the maximum material volume (Figure 6-50A, B).

Holes for bolts and screws should be drilled with a clearance. This will provide easy fit and assembly even if some part warping occurs after stamping. Holes in parts of molding materials may be fabricated by two methods: molding a through-hole by pressing, or molding a blind hole with subsequent drilling after pressing.

Vertical-through concentric holes may be pressed up to 10 diameters long. Vertical blind holes may be pressed up to 7.5 diameters long. Vertical blind

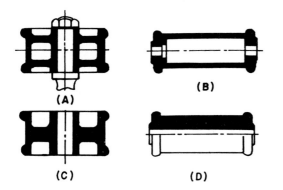

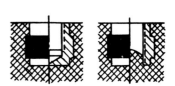

Figure 6-54 Avoiding undercutting by the use of section parts: (A, B), good designs; (C, D) poor designs

Figure 6-55 Pressing of threaded sleeves: (A) good design; (B) poor design

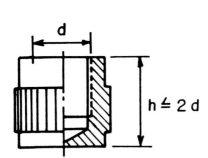

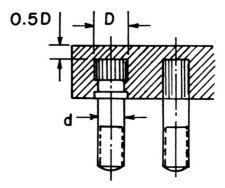

Figure 6-56 The sizes of threaded sleeves

Figure 6-57 Pressing of studs

holes with eccentric hole geometries should be done up to 2 diameters long. If a larger length is required, the stud base should be thickened as shown in Figure 6-51. The studs should have a slope of 1:50.

Undercutting should be avoided because it complicates the design of press molds and decreases the efficiency of pressing. For example, the head design according to Figure 6-52A is considerably simpler than that of Figure 6-52B; however, they work equally well in equipment. For the same reason holes should be avoided on the sides of the part. Such holes complicate the press mold fabrication, which, in this case, should be made detachable. Bringing the hole up to the upper edge of the part simplifies the formation considerably (Figure 6-53).

Often it is possible to avoid undercutting by fabricating a part in two sections, as shown in Figure 6-54(A, B). The designs in Figure 6-54(C, D) are not recommended. In changing the shape of the part, one also can avoid undesirable undercutting. The thread in plastic parts at diameters greater than 5 mm is formed by pressing. The threaded connections may be given any profile. To avoid breaking the connection, it should be started at some distance from the edge of the part. If a metal screw must be screwed into the threaded hole, it should be pressed in the plastic part as a threaded metal sleeve.

Reinforcement is required to increase the strength of a plastic part. Reinforcement should not be located very close to the edge of the part; otherwise cracks, chipping, and swelling of the plastic may occur. The section of metal reinforcement should be smaller than that of the plastic to avoid cracking. Reinforcement should have knurling, or grooves, to eliminate turning over and withdrawal. If the threaded sleeves are pressed, then holes should be blind to prevent the plastics from being penetrated (Figure 6-55).

The sizes of threaded sleeves are given in Figure 6-56. The heads of pressed studs should be larger than the stud diameter. The head should overlap the breech in the press mold to fit the stem with no less than 0.5 mm on each side; that is, $D \geq d + 1$ mm; otherwise, the plastic will pour the thread (Figure

6-57). The length of threaded studs should not be less than 1.5 times the diameter or the stud will stand crooked. Plastic flash should be accessible for elimination by grinding or machining.

NOMENCLATURE

b	overlap length (in.)
d	diameter (in. or mm)
F	effective thread length (m)
h	height (mm)
K	overall heat-transfer coefficient (kcal/m^2 °C hr)
l	length (in.)
P	pressure (kg/cm^2)
S	wall thickness (cm)
α	coefficient of thermal expansion (in./(in. °F))
λ	thermal conductivity (kcal/m °C hr)
allow	allowable stress (N/m^2)
ψ	safety factor or strength factor

7

Pipes, Compensators, and Valves

PIPES

The pipes and conduits are the veins and arteries of any plant handling fluids. Piping and its components provide the means for transporting materials between and from various pieces of processing equipment. In many cases, they are an integral part of the equipment itself as, for example, in tubular heat exchangers. In addition, many internal and external assemblies and parts of equipment are constructed from piping.

Among the major types of piping are piling pipe, line pipe, mechanical tubing, pressure tubing, and standard pipe. Piling pipe is a round-welded or seamless pipe, employed as foundation pipe. The pipe cylinder acts as a permanent load-carrying unit and often is filled with concrete to form cast-in-place concrete piles. It is employed by the construction industry in buried installations for foundation work. Piling pipe is designed according to American Society for Testing and Materials (ASTM) specifications.

Line pipe is used for transporting gas, oil, and water. Pipe diameters range from 1/8 in. to 42 in. This pipe is fabricated to American Petroleum Institute (API) and American Water Works Association (AWWA) specifications.

Mechanical tubing includes welded or seamless tubing fabricated in sizes ranging from 3/16 in. to 10 3/4 in. outside diameter (o.d.) inclusive for carbon and alloys. Specifications are not tight, and only exact outside diameter and decimal wall thickness must be met.

Pressure tubing is used for conveying fluids at elevated temperatures and/or pressures. It is suitable for head applications and is fabricated to meet exact o.d. and decimal wall thicknesses. Sizes range from 0.5 in. to 0.6 in. o.d. inclusive, at ASTM specifications. Examples of pressure tubing are air-heater tubes, boiler tubes, header tubes, heat-exchanger and condenser tubes, oilstill tubes, superheater tubes, and pressure-tubing couplings and coupling stock.

Standard pipe is used in low-pressure applications, for example, in transporting air, steam, gas, water, oil, and so on. Applications are in machinery, building, sprinkling and irrigation systems, and water wells. This type of pipe can transport fluids at elevated temperatures and pressures that are not subjected to external heat. Diameters range from 1/8 in. to 42 in. o.d. and are designed according to ASTM specifications. Examples of applications are machine piping, pressure piping, water and gas service pipe, nipple pipe, and pipe for plating or enameling.

Structural pipe and tubing includes welded or seamless pipe and tubing that is often used in structural or above-ground load-bearing operations. The construction industry is the largest user of this type of pipe. It is manufactured in nominal wall thicknesses and sizes according to ASTM specifications in round, square, rectangular, and odd cross sections.

Fabricating assemblies and parts from pipes often requires that the piping be bent. Bending pipes fabricated from plastic materials with outside diameters less than 50 mm is most often accomplished without heating. For diameters greater than 50 mm, heat must be applied to prevent cracking or breakage. The average bending radius, R_{av}, is recommended to be greater than, or equal to, three times the pipe's nominal diameter. Pipes constructed from nonferrous metals and their alloys may be bent at $R_{av} = 2D$.

For steep bending of carbon steel pipes, stamped elbows with a small bend radius ($R = D$) are recommended. If there is a need to provide a pipe bend with the radius $R < 2D$ (at $P < 5$ Mn/m^2) as well as for bends of large diameters, welded sectoral bends should be employed. The basic design for 90° bends is shown in Figure 7-1.

Bends with either larger or smaller radii may be fabricated in a similar fashion. The design of a return bend for $P < 5$ Mn/m^2 with an angle of 180° and small distance between its parallel branches is shown in Figure 7-2.

Examples of nondetachable pipe connections between the pipes themselves are shown in Figure 7-3. The length of the socket, l, in connections with soft soldering according to types F and G, is determined from the following formula:

$$l = \frac{p^D}{4\tau_{sh}}$$

Examples of nondetachable pipe connections to different parts of apparatuses without reinforcement of their walls are presented in Figure 7-4.

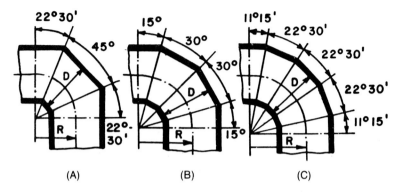

Figure 7-1 Examples of welded sectoral bends with a 90° angle: (A) $R > 0.75D$; (B) $R > 0.9D$; (C) $R > D$

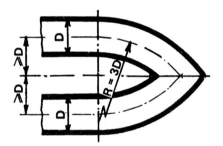

Figure 7-2 Example of welded return bend ($\alpha = 180°$) with a small distance between its parallel branches

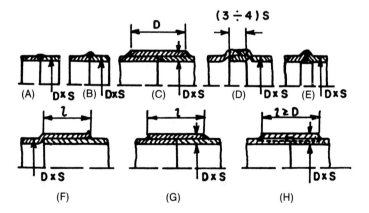

Figure 7-3 Examples of nondetectable pipe connections between the pipes themselves: (A–E) welded connections [A, B $p < 20$ MN/m^2; C $p < 10$ MN/m^2; D, E $p < 20$ MN/m^2]; (F, G) soldered connections, $p < 5$ MN/m^2; (H) threaded connections with soldering or welding, $p > 5$ MN/m^2

Pipes

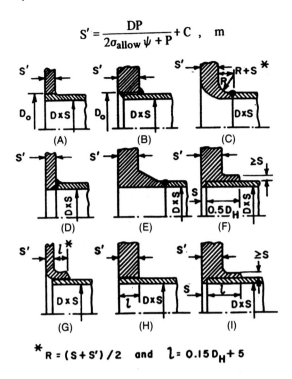

$$S' = \frac{DP}{2\sigma_{allow}\psi + P} + C, \text{ m}$$

*R = (S+S')/2 and l = 0.15D_H + 5

Figure 7-4 Examples of nondetachable pipe connections to the different parts of apparatuses. (A–E) welded connections (A), $S \geq 0.6$, S_1 to walls for $p \leq 5$ MN/m² and to flanges at $p \leq 2.5$ MN/m²; (C) to walls for $p \leq 10$ MN/m²; (D) at $S \geq 0.4$ S_1 to walls and flanges for $p \leq 5$ MN/m²; (E) at $S < 0.6$ S_1 to flanges for $p > 2.5$ MN/m²; (F) threaded connection on tinning to walls and flanges for $p \geq 5$ MN/m²; (G–I) soldered connections (G) to walls for $p \leq 1.6$ MN/m²; (H) to walls and flanges for $p \leq 5$ MN/m²; (I) to walls and flanges for $p \leq 1.6$ MN/m².

Design of Pipes Operating Under Internal Pressure

The design thickness of the wall of the pipe S', straight or bent (with $R \geq 2D$), and operating under internal pressure p is determined from the following formulas: At $P/\sigma_{allow} < 0.4$ from the following formulas,

$$S' = \frac{DP}{2\sigma_{allow}\psi + P} + C, \text{ m}$$

At $P/\sigma_{allow} > 0.4$,

$$S' = 0.5D \frac{10^{(P/2.3\sigma_{allow})} - 1}{10^{(P/2.3\sigma_{allow})}} + C, \text{ m}$$

Allowance $C = C_1 + C_2 + C_3$, where

$$C_3 = S_o K > 0.5 \times 10^{-3} \text{ m}$$

where K is the coefficient as taken from Table 7-1.

TABLE 7-1 COEFFICIENT K VALUES FOR EQUATION

Tolerance on Wall Thickness (%)	−15	−12.5	−10	−5
K	0.2	0.18	0.15	0.1

The assumed thickness of the pipe S should be checked for stress at the hydrostatic test from the following formula:

$$\sigma = \frac{d + (S - C)P'}{2(S - C)\psi} \leq \frac{\sigma_y}{1.2}, \text{ N/m}^2$$

if the design was based on the preceding formula; otherwise, use the following formula:

$$\sigma = \frac{P'}{\ln\beta} \leq \sigma_y/1.2, \text{ N/m}^2$$

where

$$\beta = \frac{d + 2S}{d + 2(C_1 + C_2)}$$

Design of Pipes Operating Under External Pressure

Pipes operating under external pressure should be constructed to almost exact cylindrical shapes. The permissible out-of-roundness in any cross section of such pipe should not be more than 1 percent of the pipe diameter. The design thickness of the pipe S', straight or bent (with $R > 2D$), operating under external pressure is calculated from formulas found in Figure 7-5.

Figure 7-5 (from the 1956 ASME code) gives the allowable design pressure for pipes subjected to external pressure as a function of the allowable stress of the material and the ratio of t/d_o (pipe thickness, in./pipe outside diameter, in.). In case of corrosion or erosion, additional metal must be supplied. If the pipe is threaded, additional metal equal to $(0.8/n)$ inches, where n is the number of threads per inch, must be provided.

Reinforced Plastic Piping

Steel, cast iron, and various other conventional piping materials have been replaced in recent years by special-grade plastics and fiberglass-reinforced plastics (FRP). Plastic piping offers the advantages of light weight, durability, and strength. In addition, FRP is especially resistant to corrosion and chemical attack, provided the proper bonding resin is employed. Corrosion is a key consideration since the steel industry alone devotes roughly 40 percent of its production toward the replacement of corroded materials. If FRP pipe and components are prepared properly, significantly fewer replacements are needed. This, in turn, results in significant savings in energy, and manufacturing and maintenance costs. Specific mechanical properties depend on the plastic's and resin's resistance to heat and chemicals being handled.

Plastic and FRP pipe are available in a wide range of diameters and lengths. In the United States, inside diameters range from less than 1 in. to more than 12 ft. FRP pipe diameters of 60 ft to more than 100 ft have been

Pipes

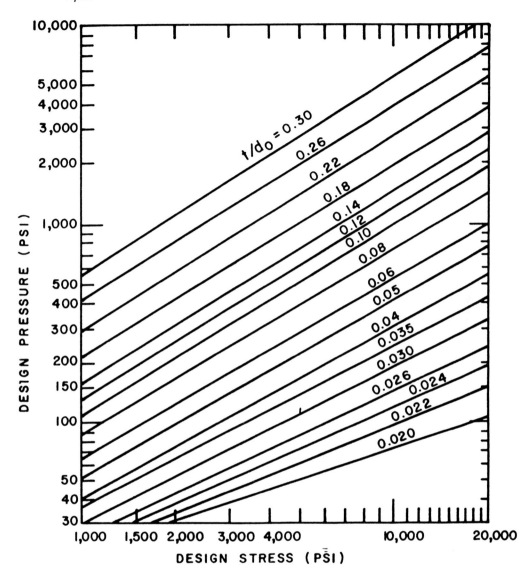

Figure 7-5 Chart for determination of wall thickness of tubes under external pressure

proposed for ocean thermal applications in which enormous volumes of water must be circulated between the surface and ocean floor.

The most widely used FRP pipes are fabricated from polyesters and vinyl esters. Maximum operating temperatures are as high as 300°F, depending on the resin. Operating pressures vary with pipe diameter, wall thickness, and the method of fabrication.

TABLE 7-2 NOMINAL PROPERTIES OF VINYL ESTER, FILAMENT-WOUND PIPE

Temperature (°F)	Hoop stress on reinforced thickness (psi)	Axial stress (psi)	Axial tensile (psi)	Coefficient of elasticity	Beam bending modulus of elasticity (psi × 10^6)	Beam bending (psi)	Pipe-fitting layup joint shear (psi)
77	50,000	25,000	9,000	2.0	1.0	9,000	600–1,000
150	50,000	25,000	7,000	1.5	0.4	7,000	600–1,000
180	50,000	25,000	5,000	1.2	0.15	5,000	600–1,000
200	40,000	20,000	4,000	0.8	0.04	2,000	600–1,000

Pipe diameter = 14–72 in.; specific gravity = 1.86; coefficient thermal expansion = 10.5×10^{-6} in./in./°F

Filament-wound pipe and ductwork can be employed in a wide number of high-pressure applications and are used extensively in chemical processing plants, pulp and paper mills, auto manufacturing plants, food processing facilities, bulk storage and loading complexes, and various conventional and nuclear power generating plants. Table 7-2 lists properties of large-diameter, filament-wound pipe.

FRP is well suited for many fluid flow applications, with a weight that is roughly one sixth that of steel and one twelfth that of concrete. It has the strength of steel with additional properties of inertness. FRP pipe will not decay, corrode, or support combustion, is dimensionally stable, and has some flexibility.

FRP piping can be made corrosion resistant to many kinds of chemicals or compounds, which explains why it is used widely in process industry and pollution control applications.

Pipe and fittings can absorb more than 40 percent diametrical deflection without undergoing structural change, and components will return to their original configuration once the overload has been removed.

Filament-wound fiberglass-reinforced pipe has excellent flow properties. The interior surfaces of piping and components are generally smooth and glass-like. This smooth finish greatly reduces and, in many cases, eliminates, material buildup. This, in turn, significantly reduces pumping costs. Figure 7-6 is a flowchart that can be used for rough estimates for large-diameter pipes.

There are several types of wall construction used with FRP filament-wound pipe. Standard pipe generally has no circumferential reinforcement ribs. These pipes are designed in this manner to meet the pressure/bending stress criteria of various applications and are suitable for buried pipe installations or for above-ground or hanger-suspended pipe systems. Standard wall designs also are often used as tunnel liners.

Rib-wall construction is used on large-diameter pipes to provide additional strength for buried and subaqueous installations. Usually, soil compaction is difficult to achieve in these situations. Prestressed ribs are circumferentially wound over cured pipe after the specified overall wall thickness has been reached. The rib reinforcement must be positioned based on closely calculated spacing requirements. This addition greatly increases the pipe wall stiffness. The design reduces the nominal pipe wall thickness while maintaining the same strength-to-weight ratio. Two basic types of rib reinforcement are the half-elliptical and the trapezoidal designs.

In the former design, a solid filament-wound, half-elliptical rib is provided to give additional pipe wall stiffness (Figure 7-7). These are employed in sub-standard burial conditions or excessive burial depths. The half-elliptical rib is constructed from continuous strands of tensioned high-strength glass reinforcement, which are helically wrapped at high helix angles over the completed outer wall. The rib generally is wider than it is high.

The trapezoidal rib design (Figure 7-7) is employed under conditions in which the ultimate degree of pipe stiffness is required. The rib also is con-

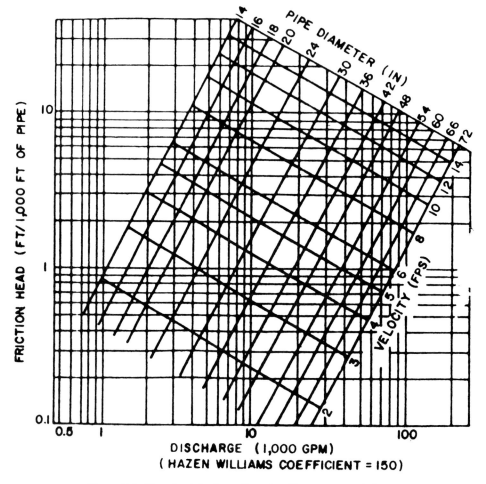

Figure 7-6 Flowchart for large-diameter, filament-wound pipe

structed from highly tensioned continuous strands of impregnated fiberglass. A helix angle is employed that gives a much thicker rib profile. This type of rib configuration enables the pipe to withstand excessive burial loads and extremely high impact strengths. As such, it provides the most rugged construction, capable of withstanding shock that normally would crumble or damage concrete or steel piping.

There are a variety of procedures for connecting FRP filament-wound pipe, in addition to the numerous structural components often encountered in large, complex construction jobs. If the pipe is to pass through a wall, for example, a filament-wound thrust ring is recommended. This is wound onto the pipe during fabrication. The thrust ring ensures anchoring and serves as a medium to transfer hydrodynamic and thermal stresses to the wall. When subsidence is expected, the pipe can be adapted to accommodate the anticipated settlement.

Pipes

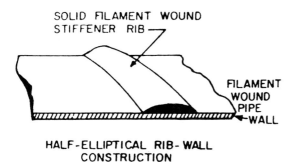

HALF-ELLIPTICAL RIB-WALL CONSTRUCTION

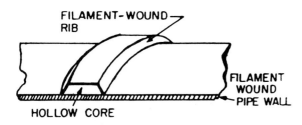

TRAPEZOIDAL RIB-WALL CONSTRUCTION

Figure 7-7 Two types of rib reinforcements used in filament-wound pipe

Usually, a neoprene rubber compression wrap can be employed at the pipe-to-concrete termination encasement to dampen or minimize shear possibilities at this termination point.

When expansion joints are used, a flex coupling can be employed. Expansion joints usually are used less frequently in filament-wound FRP piping networks than with such systems as steel. They are important, however, when severe expansion thrusts are anticipated at connection points, for adjusting misalignment at equipment connections, or where provisions for value replacements are necessary. This is illustrated in Figure 7-8.

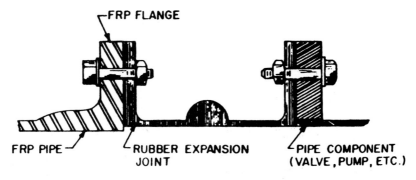

Figure 7-8 Typical expansion joint tie-in

Because of its great versatility, FRP pipe can be readily incorporated into existing systems constructed from other materials, provided proper connections are used. Two methods are employed for making connections to concrete pipe. One encloses the mated sections of pipe in a concrete encasement; the other approach employs a specially designed spigot joint to fit a concrete bell.

Mating FRP pipe with steel pipe can be done in a number of different ways. One way is to use a concrete-encased adhesive joint connection, as shown in Figure 7-9A. Here, the steel pipe is sealed adhesively within the bell end of the plastic pipe and the entire joint encased in concrete. A second method, shown in Figure 7-9B, employs flange ends on both pieces of pipe. An appropriate-size flange must be fabricated onto the end of the FRP pipe and drilled to prespecified orientation. FRP pipe also may be joined to steel by means of a flex coupling, similar to that shown in Figure 7-8.

Pipe can be supported on hangers or saddles, which may be constructed from either plastic or steel. FRP saddles can be cut from actual portions of the pipe to be supported.

Recommended saddle lengths should be at least one half the pipe diameter. Note that the pipe and saddle will not have exactly the same radius; there-

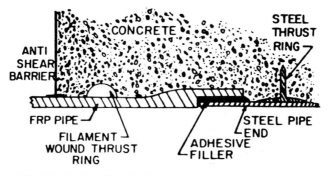

(A) FRP PIPE TO STEEL PIPE CONNECTION.

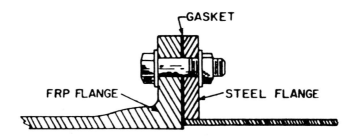

(B) FRP FLANGE TO STEEL FLANGE.

Figure 7-9 Two methods for connecting FRP pipe to steel pipe: (A) connection of FRP pipe to steel pipe; (B) FRP flange to steel flange

Pipes

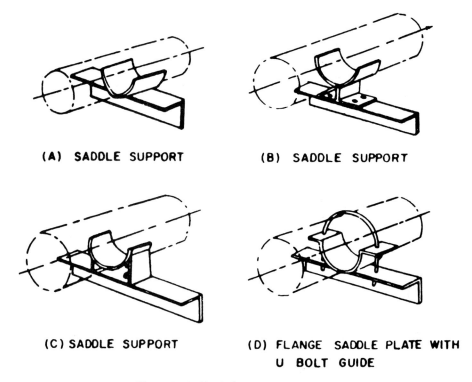

Figure 7-10 Typical support arrangements

fore, the gap must be filled with a suitable filler material to provide full bearing support. Clamps and U-bolts should be used to provide snug, but not excess, clamp pressure on the pipe. Some typical support designs are shown in Figures 7-10 and 7-11.

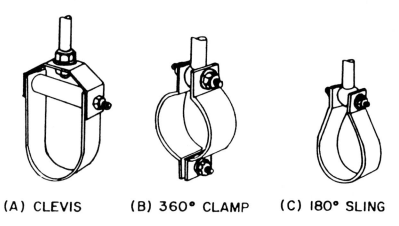

Figure 7-11 Typical hanger support arrangements

The recommended saddle angle should be 180° for any one of the following conditions:

1. when pipe diameters exceed 18 inches.
2. when the pipe wall is rated at less than 50 psig.
3. when the pipe is designed for vacuum.

Valves, flow meters, or other flow equipment that cause weight concentrations to occur in the line should be supported rigidly. This additional support should be independent of the remainder of the piping system.

Support spacings should provide for maximum tolerable deflection as well as maximum stress at the points of maximum deflection and maximum stress at the hangers. Table 7-3 gives recommended support spacings. (Note that when complete line drainage requirements exist, lines should be sloped at a minimum of 0.5 in. for every 10 ft of pipe length. The system should be provided with periodic low spots with flanged drain connections for positive drainage during a shutdown or cleanout.)

The most effective way to restrain pipe movement by applied forces is through the use of an anchor. Figure 7-12 illustrates anchor arrangements. The design can be used to restrain thermal expansion; as such, it must meet maximum anticipated end forces. It also must prevent the pipe from sliding within the arrangement.

In situations in which a full anchor is not necessary but movement must be restricted, a tie-down can be used. Tie-downs are employed when long lengths of pipe are left to expand. It is generally recommended that the pipe be tied down in the middle to ensure equal expansion in both directions. The expansion may be directed to one end by placing the tie-down away from the center.

TABLE 7-3 RECOMMENDED SUPPORT SPACING SPAN (FT) FOR FLUIDS WITH SPECIFIC GRAVITY 1.2

Inside diameter (in.)	Internal pressure rating (psig)					
	25	50	75	100	125	150
2	8.0	8.0	8.0	8.0	8.0	8.0
4	10.0	10.0	10.0	10.5	10.5	10.5
6	10.5	10.5	11.5	11.5	12.0	12.5
8	11.5	12.5	13.0	13.0	13.5	14.0
10	12.0	13.0	14.0	14.5	15.0	15.5
12	12.5	14.0	15.0	15.5	16.0	17.0
14	14.5	15.0	15.5	16.5	17.5	18.5
16	15.0	15.5	16.5	17.5	18.5	
18	15.5	17.0	18.0	19.0	20.0	
20	15.5	17.5	18.5	20.0		
24	12.5	19.0	20.5	22.0		

Pipes

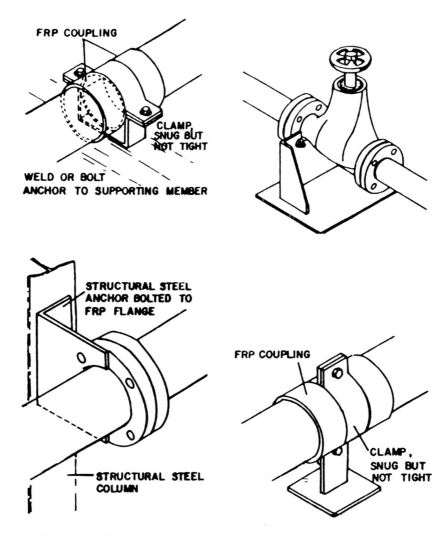

Figure 7-12 Various anchor arrangements that restrain pipe movements in all directions

Another situation in which tie-downs can be used is when the velocity and surge forces promote pipe movement.

When expansion must be directed away from equipment to prevent the expanding line from overstressing a tank shell or vessel nozzle, tie-downs also should be used. Tie-downs basically consist of a snug U-bolt and a 6-in.-wide strap joint, which is placed over the U-bolt to maintain the pipe in a fixed condition. The support must be fastened securely to a rigid foundation. This arrangement should not be used for supporting vertical runs.

Guides should be used to minimize snaking when the pipe has been restrained from expanding. A guide is essentially any type of device or support that prevents the pipe from transverse movement (but not axial movement). Under conditions in which axial movement is excessive, sliding guides should be employed. These will protect the pipe from abrasion. Figure 7-13 illustrates this type of guide.

Most FRP piping systems can be designed without expansion joints because although plastic piping systems have roughly twice the rate of thermal expansion of steel, they have a comparatively low modulus of elasticity. Usually, a slip-type joint or a bellows-type expansion joint is recommended for added protection. Anchors and supporting structures must be designed properly for the end force from operating line pressures when expansion joints are employed.

Between any two anchors only one expansion joint should be installed. The pipe should be guided close to the expansion joint to prevent shear and bending loads and to eliminate any misalignment during axial movement. Guides should be positioned on both sides of the expansion joint at 4 and 12 pipe diameters away from the joint.

FRP pipe is commercially available in a wide range of diameters, wall thicknesses, and pressure ratings. Wall thickness specifications are based on an allowable strain, which can yield structural safety factors, in some cases in excess of 20:1. The strain level is defined as the maximum strain that can be tolerated to ensure long-term integrity of the inner liner. Note that when considering FRP pipe versus concrete pipe, corrugated metal, or other types for buried installations, the choice should be based on the safety factor. Table 7-4 gives wall thicknesses of filament-wound FRP pipe at various pressures.

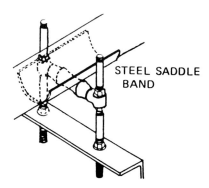
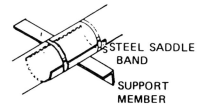

(A) ADJUSTABLE PIPE SUPPORT (PIPE CAN ROLL OR MOVE AXIALLY)

(B) PIPE SUPPORT (PIPE CAN MOVE SIDEWAYS OR AXIALLY)

Figure 7-13 Typical sliding guide arrangements: (A) adjustable pipe support (pipe can roll or move axially); (B) pipe support (pipe can move sideways or axially)

TABLE 7-4 RECOMMENDED WALL THICKNESS FOR FILAMENT-WOUND FRP PIPE AT VARIOUS PRESSURES

Inside diameter (in.)	Internal pressure rating, psi									
	25		50		75		100		125		150	
	Wall (nominal) (in.)	wt/ft (lb/ft)	Wall (nominal) (in.)	wt/ft (lb/ft)	Wall (nominal) (in.)	wt/ft (lb/ft)	Wall (nominal) (in.)	wt/ft (lb/ft)	Wall (nominal) (in.)	wt/ft (lb/ft)	Wall (nominal) (in.)	wt/ft (lb/ft)
2	0.188	0.945	0.188	0.945	0.188	0.945	0.188	0.945	0.188	0.945	0.188	0.945
2.5	0.188	1.18	0.188	1.18	0.188	1.18	0.188	1.18	0.188	1.18	0.188	1.18
3	0.188	1.42	0.188	1.42	0.188	1.42	0.188	1.42	0.188	1.42	0.188	1.42
4	0.188	1.89	0.188	1.89	0.188	1.89	0.188	1.89	0.188	1.89	0.188	1.89
5	0.188	2.36	0.188	2.36	0.188	2.36	0.188	2.36	0.188	2.36	0.188	2.36
6	0.188	2.63	0.188	2.83	0.188	2.83	0.188	2.83	0.188	2.83	0.188	2.83
8	0.188	3.78	0.188	3.78	0.188	3.78	0.188	3.78	0.188	3.78	0.188	3.78
10	0.188	4.72	0.188	4.72	0.188	4.72	0.188	4.72	0.188	4.72	0.188	4.72
12	0.188	5.67	0.188	5.67	0.188	5.67	0.188	5.67	0.188	5.67	0.214	6.45
14	0.188	6.61	0.188	6.61	0.188	6.61	0.188	6.61	0.208	7.32	0.250	8.80
16	0.188	7.56	0.188	7.56	0.188	7.56	0.190	7.64	0.238	9.57	0.286	11.5
18	0.188	8.50	0.188	8.50	0.188	8.50	0.214	9.68	0.268	12.1	0.321	14.5
20	0.188	9.45	0.188	9.45	0.188	9.45	0.238	12.0	0.298	15.0	0.357	17.9
24	0.188	11.3	0.188	11.3	0.214	12.9	0.286	17.3	0.357	21.5	0.429	25.9
30	0.188	14.2	0.188	14.2	0.268	20.2	0.357	27.0	0.446	33.6	0.536	40.4
36	0.188	17.0	0.214	19.4	0.321	29.0	0.429	38.8	0.536	48.5	0.643	58.2
42	0.188	19.8	0.250	26.4	0.375	39.6	0.500	52.8	0.625	66.0	0.750	79.2
48	0.188	22.7	0.286	34.5	0.429	51.8	0.571	68.9	0.714	86.1	0.857	103.4
54	0.188	25.5	0.321	43.6	0.482	65.4	0.643	87.3	0.804	109.1	0.964	130.8
60	0.188	28.3	0.357	53.8	0.536	80.8	0.714	107.7	0.893	134.7	1.07	161.4
72	0.214	38.7	0.429	77.6	0.643	116.4	0.857	155.1	1.07	193.6	1.29	233.4
84	0.250	52.8	0.500	108.1	0.750	158.3	1.00	211.1	1.25	263.9	1.50	316.7
96	0.286	69.0	0.571	137.8	0.857	206.8	1.14	275.1	1.43	345.0	1.71	412.6
108	0.321	87.1	0.642	174.3	0.964	261.7	1.29	350.0	1.61	437.0	1.93	524.0
120	0.357	107.7	0.714	215.3	1.07	322.7	1.43	431.0	1.79	540.0	2.14	645.0
144	0.429	155.3	0.857	310.2	1.29	467.0	1.71	619.0	2.14	775.0	2.57	930.0

In addition to having high-pressure ratings, filament-wound pipe can be fabricated with the proper resins to be highly resistant to corrosion and chemical attack. It is used by chlorine manufacturers in transporting hot chlorinated acid brine; in the paper industry when handling corrosive wastes at high temperatures; in transporting caustic, chlorine, and sodium hypochlorite from bleach plants to waste treatment facilities; and in the food processing industry when handling both acetic and citrus acids in addition to process treatment chemicals.

Unlike most metals that undergo electrochemical attack, plastic pipe generally will be affected by chemical reaction or by solvation, which is the penetration of the plastic by a corrosive element. This causes softening, swelling in the walls and, eventually, failure. Because of the fundamental differences between electrochemical and solvation reactions, conventional corrosion rates should not be used in evaluating or comparing the chemical resistance of plastic pipe to other materials. Chemical resistance for FRP usually is determined by immersing the laminates in a bath of the corrosive medium and simulating operating conditions. Tests of strength retention, surface hardness, and visual inspection for chemical attack provide detailed qualitative information on plastic and FRP pipe characteristics. Table 7-5 gives the chemical resistance of filament-wound pipe to various fluids and the range of maximum operating temperatures for which FRP pipe is commercially available.

TABLE 7-5 CHEMICAL RESISTANCE OF FILAMENT-WOUND FRP PIPE (POLYESTER RESIN SYSTEMS)

Chemical	Strength (%)	Maximum operating temperature (°F)
Acetic Acid	0–10	100–150
	10–50	100
Acetic Acid, glacial	—	100
Aceto	0–5	150
Acrylic Acid	—	75
Adipic Acid, air	Solution	200–250
	—	300
Alcohol, ethyl	0–10	150
Alcohol, isopropyl	10	150
Alcohol, methyl	10	150
Alcohol, methyl isobutyl	10	150
Alcohol, secondary butyl	10	150
Allyl Chloride	—	100
Aluminum Chloride	—	100–300
Aluminum Fluoride	—	150
Aluminum Hydroxide	—	150
Aluminum Nitrate	—	150–250
Aluminum Sulfate	—	200–300
Alums	—	200–300
Ammonia Gas, dry	—	150

TABLE 7-5 *Continued*

Chemical	Strength (%)	Maximum operating temperature (°F)
Ammonia, wet	—	100
Ammonium Carbonate	—	100–200
Ammonium Chloride	—	200
Ammonium Fluoride	25	150–200
Ammonium Hydroxide	0–10	150–200
	10–20	150
	20–30	100
Ammonium Nitrate	—	200–250
Ammonium Persulfate	—	200
Ammonium Phosphate	—	150
Ammonium Sulfate	—	200–250
Amyl Acetate	—	75
Amyl Chloride	—	75
Aniline	—	75
Antimony Trichloride	—	150–200
Barium Carbonate	—	200–250
Barium Chloride	—	200–250
Barium Hydroxide	0–10	150–200
Barium Sulfate	—	200–250
Barium Sulfide	—	150–300
Beer	—	200
Benzene	—	100
Benzene Sulfonic Acid	—	200
Benzoic Acid	—	200
Black Liquor	—	200
Borax	—	200–250
Boric Acid	—	200
Bromic Acid	—	150
Bromine, liquid	—	NR[a]
Bromine Water	—	100
Butadiene	—	100
Butane	—	100
Butyl Acetate	—	75
Butyl Cellusolve	—	150
Butyric Acid	0–15	150–200
	25–50	150
Calcium Bisulfide	—	200
Calcium Carbonate	—	150–300
Calcium Chlorate	—	200
Calcium Chloride	—	200–300
Calcium Hydroxide	0–50	150–200
Calcium Hypochlorite	0–20	200
Calcium Nitrate	—	200–250
Calcium Sulfate	—	200–250
Carbon Dioxide	—	200–250
Carbon Disulfide	—	NR
Carbon Tetrachloride	—	100

TABLE 7-5 Continued

Chemical	Strength (%)	Maximum operating temperature (°F)
Carbonic Acid	—	150
Chloroacetic Acid	0–25	100–150
Chlorine, dry	—	200
Chlorine, wet	—	200
Chlorine Dioxide	15	150
Chlorine Water	—	200
Chlorobenzene	100	100
Chloroform	100	100
Chromic Acid	5	150
	10	150
	20	75–150
	100	NR
Chromic Fluoride	—	75
Citric Acid	—	200–250
Copper Chloride	—	200–250
Copper Fluoride	—	200–250
Copper Nitrate	—	200–250
Copper Sulfate	—	200
Crude Oil, sour	—	200–300
Crude Oil, sweet	—	200–300
Deionized Water	—	200–300
Diacetone Alcohol	—	150
Dichlorobenzene	100	150
Dichloroethylene	100	75
Diethylene Triamine	100	NR
Dimenthylamine	—	NR
Ethyl Acetate	—	150
Ethyl Ether	—	75
Ethylene Chlorohydrin	—	NR
Ethylene Diamine	—	NR
Ethylene Glycol	—	200
Ethylene Oxide	—	NR
Fatty Acids	—	200
Ferric Chloride	—	200–300
Ferric Nitrate	—	200–250
Ferric Sulfate	—	200
Ferrous Chloride	—	200–250
Ferrous Sulfate	—	200
Fluorine Gas, wet	—	75
Fluorosilicic Acid	10	200
Fluoroboric Acid	—	200
Formaldehyde	40	150–200
Formic Acid	25	100
Freon®	—	150
Gas, natural	—	200
Gasoline, sour	—	200–300
Gasoline, 108 octane	—	150

TABLE 7-5 *Continued*

Chemical	Strength (%)	Maximum operating temperature (°F)
Glucose	—	200–300
Glycerine	—	150–300
Glycol, ethylene	—	200
Glycol, propylene	—	200–250
Heptane	—	150
Hexane	—	75
Hydraulic Fluid	—	150–200
Hydrobromic Acid	50	150–200
Hydrochloric Acid	0–37	200
Hydrocyanic Acid	10	150
Hydrofluoric Acid	10	150
Hydrogen	—	150
Hydrogen Peroxide	10	150
	20–30	75–150
Hydrogen Sulfide		
Aqueous	—	200–250
Dry	—	200–250
Hypochlorous Acid	—	200
Kerosene	—	200–250
Lactic Acid	—	200
Lauric Acid	—	200
Lead Acetate	—	200–250
Levulinic Acid	25	200–250
Magnesium Carbonate	—	200–250
Magnesium Chloride	—	200–300
Magnesium Hydroxide	—	150–250
Magnesium Nitrate	—	200–250
Magnesium Sulfate	—	200–250
Maleic Acid	100	150–200
Mercury	—	200–300
Mineral Oils	—	200–300
Naphtha	—	200
Naphthalene	—	100–150
Nickel Chloride	—	200–300
Nickel Nitrate	—	200
Nitric Acid	0–15	100
	20	75
Oleic Acid	—	200
Perchloric Acid	70	75
Phenol	1	150
Phosphoric Acid	0–75	200–250
	75–85	200
	85–110	200
Phosphorus Pentoxide	0–54	200
Picric Acid	—	75
Potassium Bicarbonate	—	150–250

(continued)

TABLE 7-5 *Continued*

Chemical	Strength (%)	Maximum operating temperature (°F)
Potassium Bromide	—	200
Potassium Carbonate	—	150–250
Potassium Chloride	—	200–300
Potassium Dichromate	—	200–250
Potassium Hydroxide	—	150–200
Potassium Permanganate	10–25	150
Potassium Sulfate	—	150–200
Propane	—	100
Silver Nitrate	—	200–250
Soaps	—	200–250
Sodium Acetate	—	150
Sodium Bicarbonate	—	150–250
Sodium Bromide	—	200
Sodium Carbonate	—	100–250
Sodium Chlorate	—	150
Sodium Chloride	—	200–300
Sodium Cyanide	—	200–250
Sodium Fluoride	—	200–250
Sodium Hydroxide	0–5	150–200
	>5	150–200
Sodium Hypochlorite	0–10	75–100
Sodium Methoxide	40	150
Sodium Nitrate	—	200–300
Sodium Peroxide	—	75
Sodium Sulfate	—	200–300
Sodium Sulfite	—	200
Sodium Thiosulfate	—	150–200
Stearic Acid	—	150
Sulfamic Acid	25	200
Sulfite Liquors	—	200
Sulfur Dioxide, wet and dry	—	200
Sulfuric Acid	10	200–250
	25	150–200
	50	100–200
	60	75–200
	70	75
Sulfurous Acid	0–7	200
Tartaric Acid	—	200–250
Tetraethyl Lead	—	100
Toluol	100	150
Trichloroacetic Acid	—	200
Trichloroethylene	—	150
Triethylamine	—	75
Tung Oil	—	200
Turpentine	—	75
Urea	—	150
Vinegar	—	150

TABLE 7-5 *Continued*

Chemical	Strength (%)	Maximum operating temperature (°F)
Vinyl Acetate	—	150
Water, deionized	—	200–250
Water, distilled	—	200–300
Water, fresh	—	200–300
Water, salt	—	200–300
Xylene	—	150
Zinc Chloride	—	200–250
Zinc Sulfate	—	200–250

[a] NR = Not recommended.

The specific corrosion-resistant service for a reinforced plastic pipe is determined from the resin-to-glass ratio used in the laminate construction. The nature of the resin plays the dominant role in corrosion resistance to specific chemicals. The interior layer of the pipe generally has a low glass content (on the order of 20 percent to 30 percent glass content by weight). This ratio provides high corrosion resistance. To meet desired wall thicknesses and strengths, a higher glass content is used in fabricating laminates for the remaining thickness. A typical overall ratio is roughly 70 percent resin to 30 percent glass reinforcement by weight. The outer wall is often covered with a glass surfacing mat that is resin rich. Such a surface provides excellent resistance to weathering (sunlight, heat, cold, and other weathering conditions), fumes, and spillage.

Contact-molded or hand layup piping often is used where maximum chemical resistance is required throughout a laminate and where pressure requirements are not severe. For many applications, however, filament-wound FRP pipe is preferred over contact-molded piping because of the former's higher structural properties. These properties in turn possess economic advantages since thinner laminate thicknesses can be employed to meet physical requirements. Table 7-6 gives recommended wall thicknesses of contact-molded pipe at various pressures.

Contact molding is a process in which layers of resin-impregnated fabrics are built up, one layer at a time, on a mold-forming pipe. Little or no pressure is applied in forming the resultant laminate during curing. The structural wall of the pipe usually consists of chopped glass-strand mat and/or woven roving. Normally, a 60 percent resin, 40 percent glass content by weight is used. An inner wall with a very high resin content is made for protection against chemical attack and corrosion. The resin content at the inner wall is on the order of 80 percent to 90 percent, and the reinforcement consists of chopped strand mat (roughly 0.10-in.-thick mat). An intermediate layer usually is formed, with resin contents in the range of 70 percent to 80 percent.

TABLE 7-6 RECOMMENDED WALL THICKNESS FOR CONTACT-MOLDED FRP PIPE AT VARIOUS PRESSURES

	Internal pressure rating, psi											
	25		50		75		100		125		150	
Inside diameter (in.)	Wall (nominal) (in.)	wt/ft (lb/ft)	Wall (nominal) (in.)	wt/ft (lb/ft)	Wall (nominal) (in.)	wt/ft (lb/ft)	Wall (nominal) (in.)	wt/ft (lb/ft)	Wall (nominal) (in.)	wt/ft (lb/ft)	Wall (nominal) (in.)	wt/ft (lb/ft)
2	0.187	0.863	0.187	0.863	0.187	0.863	0.187	0.863	0.187	0.863	0.187	0.863
2.5	0.187	1.08	0.187	1.08	0.187	1.08	0.187	1.08	0.187	1.08	0.250	1.44
3	0.187	1.30	0.187	1.30	0.187	1.30	0.187	1.29	0.250	1.73	0.250	1.73
4	0.187	1.73	0.187	1.73	0.187	1.73	0.250	2.30	0.250	2.30	0.250	2.30
5	0.187	2.17	0.187	2.17	0.250	2.89	0.250	2.89	0.312	3.61	0.375	4.33
6	0.187	2.59	0.187	2.59	0.250	3.45	0.250	3.45	0.312	4.32	0.375	5.18
8	0.187	3.45	0.250	4.60	0.250	4.60	0.312	5.76	0.375	6.91	0.437	8.06
10	0.187	4.32	0.250	5.76	0.312	7.20	0.375	8.64	0.437	10.07	0.500	11.9
12	0.187	5.18	0.250	6.91	0.375	10.4	0.437	12.1	0.500	13.82	0.625	17.3
14	0.250	8.05	0.312	10.7	0.375	12.1	0.500	16.1	0.625	20.15	0.750	24.2
16	0.250	9.21	0.312	11.5	0.437	16.1	0.562	20.7	0.687	25.3	0.812	30.0
18	0.250	10.36	0.375	15.5	0.500	20.7	0.625	25.9	0.750	31.1	0.937	38.9
20	0.250	11.51	0.375	17.3	0.500	23.0	0.687	31.7	0.875	40.3	1.00	46.1
24	0.250	13.82	0.437	24.2	0.625	34.5	0.812	44.9	1.00	55.3	1.12	62.2
30	0.312	21.59	0.500	34.5	0.750	51.8	1.00	69.1	1.25	86.4	1.50	103.6
36	0.375	31.08	0.625	51.8	0.937	77.7	1.25	103.6	1.50	124.4	1.81	150.3
42	0.375	36.3	0.750	72.5	1.06	102.8	1.44	139.0	1.75	169.3	2.12	205.9
48	0.437	48.4	0.812	89.8	1.25	138.2	1.62	179.6	2.00	221.1	2.44	269.4
54	0.500	62.2	0.937	116.6	1.37	171.0	1.81	225.4	2.25	279.8	2.75	342.6
60	0.500	69.1	10.00	138.2	1.50	207.3	2.00	276.3	2.50	345.4	3.00	415.0
72	0.625	103.6	1.25	207.3	1.81	300.5	2.44	404.0	3.00	497.0	3.62	601.8
84	0.750	145.1	1.44	278.1	2.12	411.0	2.81	545.0	3.50	677.0	4.25	820.8
96	0.812	179.6	1.62	359.2	2.44	540.0	3.25	720.0	4.00	885.0	4.81	1,065.0

Pipes

TABLE 7-7 PHYSICAL PROPERTIES OF CONTACT-MOLDED FRP PIPE

Wall thickness (in.)	Ultimate tensile strength (psi)	Flexural strength (psi)	Flexural modulus of elasticity (psi)	Compressive strength (psi)
0.187	9,000	16,000	700,000	18,000
0.250	12,000	19,000	800,000	20,000
0.312	13,500	20,000	900,000	21,000
0.375	15,000	22,000	1,000,000	22,000

Properties of contact-molded FRP pipe of various wall thicknesses are given in Tables 7-7 and 7-8.

FRP pipe also can be manufactured by centrifugal casting and pressure molding. Wall thicknesses vary, and the use of fillers in both methods is optional. The manufacturing processes produce pipe that is greatly different in construction and physical properties; however, the general benefits each has to offer are approximately the same. Table 7-9 compares various physical properties of the three principal manufacturing techniques.

Service Considerations for Plastic Pipe

Physical properties and strengths vary with the resin system (for FRP) and fabrication method. The coefficient of thermal expansion is approximately 10.5×10^{-6} in./in./°F for centrifugally cast FRP pipe and 13×10^{-6} in./in./°F for filament-wound FRP pipe. Figure 7-14 gives typical data on the total expansion per 100 ft of pipe as a function of temperature change. In general, FRP pipe acts as a thermal insulator (that is, it conducts heat more slowly than metal, which expands and contracts more rapidly). This must be considered carefully to prevent any excess stress on FRP pipe or components when connections are made to metal piping systems. Tanks, steel piping, and other structures can transmit significant stresses to plastic pipe; hence, these deflections created by the nonplastic system must be compensated for by independent means. In a piping

TABLE 7-8 GENERAL PROPERTIES OF CONTACT-MOLDED FRP PIPE

Specific Gravity	1.3–1.7
Impact Strength (ft-lb)	30–40
Barcol Hardness	35–45
Thermal Conductivity @ 212°F (Btu/ft²-hr-°F/in.)	1.5
Linear Coefficient of Thermal Expansion (32 − 212°F) (in./in.-°F)	15×10^{-6}
Heat Distortion Temperature at 264 psi (°F)	250–350
Electrical Resistivity (ohm-cm)	10^{14}
Dielectric Constant	3.16
Machinability	Good
Light Transmission	Translucent

TABLE 7-9 AVERAGE PROPERTIES OF VARIOUS TYPES OF FRP PIPE

	Filament-wound, epoxy, or polyester resins	Centrifugally cast epoxy or polyester resin	Contact-molded polyester resin
Modulus of Elasticity in Axial Tension @ 77°F (psi)	1.0–2.7×10^6	1.3–1.5×10^6	0.8–1.8×10^6
Ultimate Axial Tensile Strength @ 77°F (psi)	8,000–10,000	25,000	9,000–18,000
Ultimate Hoop Tensile Strength @ 77°F (psi)	24,000–50,000	35,000	9,000–10,000
Modulus of Elasticity in Beam Flexure @ 77°F (psi)	1–2×10^6	1.3–1.5×10^6	1.0–1.2×10
Coefficient of Thermal Expansion (in./in./°F)	8.5–12.7×10^{-6}	13×10^{-6}	15×10^{-6}
Heat Deflection Temperature @ 264 psi (°F)	200–300	200–300	200–250
Thermal Conductivity (Btu/ft^2-hr-°F/in.)	1.3–2.0	0.9	1.5
Specific Gravity	1.8–1.9	1.58	1.3–1.7
Corrosive Resistance	E[a]	E	NR[b]

[a] E = Excellent, will resist most corrosive chemicals.
[b] NR = Not recommended for highly alkaline or solvent applications.

system, nonplastic lines should be anchored immediately before the plastic connection to ensure that any thermal strain is not transmitted to the plastic system.

Usually, directional changes in piping systems can compensate for small expansions due to temperature changes (the same is often true for contractions due to a drop in temperatures). Expansion joints or expansion loops should be designed into the system wherever possible to prevent overstressing.

When expansion joints are used, they should be preset to properly allow for both expansion and contraction. The amount of precompression necessary for an expansion joint can be computed from the following formula:

$$\tilde{C}_p = \frac{R_M(T_i - T_m)}{\Delta T}$$

where $\tilde{C}_p$ = amount of precompression (in.)

R_M = expansion joint rated movement (in.)

T_i = temperature at which system was installed (°F)

T_m = minimum operational temperature anticipated (°F)

ΔT = operational temperature difference anticipated (°F)

If expansion joints are not used, then the piping arrangement must be designed to absorb the expansion or contraction. End loads are caused when

Pipes

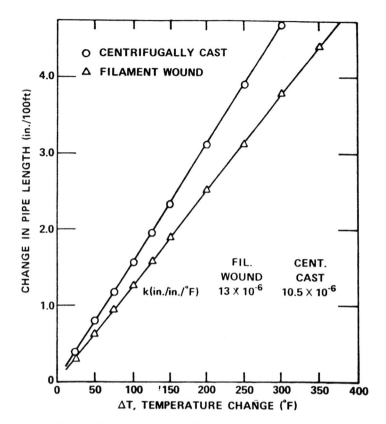

Figure 7-14 Expansion as a function of temperature change

the pipe is restrained and no proper means of expansion is provided. Such situations require guiding to keep the pipe straight. Guides direct forces in the axial direction and prevent buckling.

When the piping system is allowed to move freely with temperature change, care must be taken to ensure that all fixed pints, anchors, laterals, elbows, and so on are within the recommended safe stress level. High-stress levels can be minimized by employing one or more 90° bends with adequate pipe leg between anchors or rigid mounting stations.

The amount of expansion that can be handled by right-angle pipe legs can be estimated from the plot shown in Figure 7-15, which is a plot of the allowable deflection versus pipe length with pipe diameter as a parameter. For pure thermal stress, the amount of expansion at the point of leg attachment is expressed by

$$\Delta x = \alpha L \Delta T$$

where Δx = expansion (in.)

α = coefficient of thermal expansion (in./in./°F)

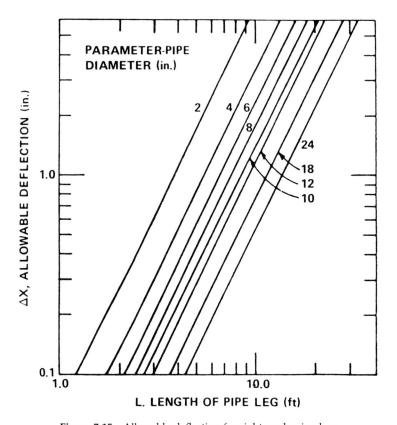

Figure 7-15 Allowable deflection for right-angle pipe legs

L = pipe length (in.)

ΔT = temperature change (°F)

The necessary pipe-leg length can be obtained to take up the anticipated expansion.

Another approach to this problem is to anchor the pipe at some intermediate point and direct the expansion to another portion of the system. To eliminate expansion completely, long, straight runs can be anchored at both ends. If the latter arrangement is selected, anchors must be designed to meet the end force, which is the total force resulting from both the internal operating pressure and the thermal expansion stress. The end force resulting from the internal operating pressure of the system can be estimated from the plot shown in Figure 7-16, which is a plot of end force F versus operating pressure P with pipe diameter as a parameter. The amount of end force exerted on anchors due to thermal stress from expansion and contraction can be estimated from the plots shown in Figures 7-17 and 7-18, respectively.

Note that when U-bolts are used as anchoring devices, the pipe should be

Pipes

protected from excessive pressure. In general, supports that make only a point contact or have narrow supporting areas should not be used unless additional protection is supplied.

Pressure and Vacuum Service Considerations

The wall thickness for FRP pipe that is to be fabricated for pressure service can be computed from Barlow's formula:

$$\delta = \frac{PD}{2\sigma}$$

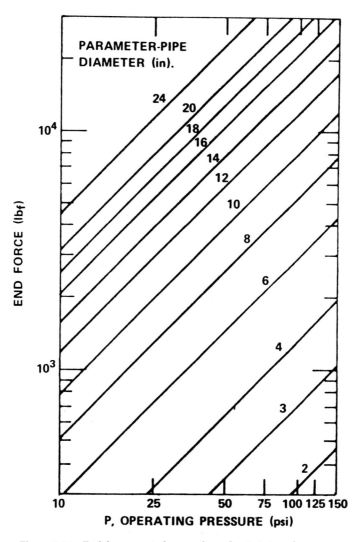

Figure 7-16 End force exerted on anchors due to internal pressure

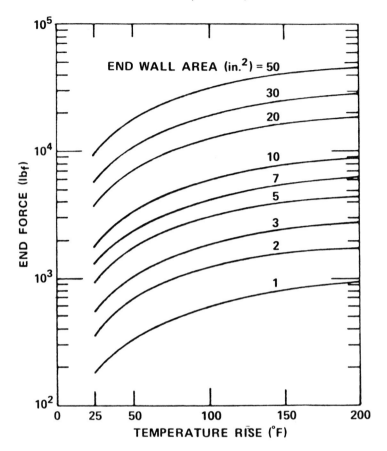

Figure 7-17 End force exerted on anchors due to expansion

where δ = laminate wall thickness required (in.)

P = operating pressure (psi)

D = inside diameter of pipe (in.)

σ = design stress (psi)

After the system has been installed, it should be tested for leaks by filling the lines with water and pressurizing to 1.5 times the rated pressure of the pipe. Testing with compressed air or gases is dangerous and is not recommended.

Consideration should be given to protection against the occurrence of high vacuums during operation, as reinforced plastics have relatively low moduli of elasticity. Adequate vacuum protection can be accomplished by employing vacuum-release valves or stand pipes. Vacuum lines should be supported by 180° saddles and should not be exposed to any external loads. Whenever flow veloc-

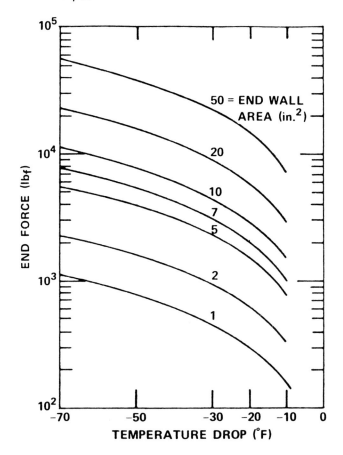

Figure 7-18 End force exerted on anchors due to contraction

ity is disrupted suddenly (for example, pump shutoff or valve closure), vacuum conditions may arise.

Disturbances or changes in a flow line can cause excessive or instantaneous rises in line pressure. This is caused most often by closing a valve or by sharp expansions or contractions in the line or other constrictions, such as sharp bends, tees, and so on. Water hammer can occur in any size pipeline, and it is important to note that the normal working pressure does not play a role in the intensity of the hammer. The impact of the flow stoppage is proportional to the mass of liquid holdup being stopped and the square of the flow velocity. The potential for these disturbances should be accounted for in the design.

Air chambers are recommended for cushioning the shock; however, frequent maintenance and spot checking is required as they often become inoperable due to waterlogging. Water hammer does not always create a noise problem. It is most often detected by violent vibration of the pipeline. Furthermore, the point or region where water hammer is detected may not be the source. Shock waves can travel over a considerable distance through the line and be the cause of damage or noise problems far from the wave source. Under

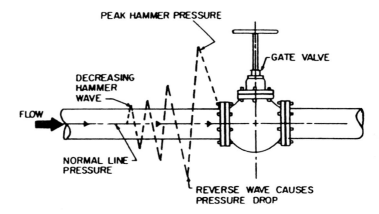

Figure 7-19 Buildup of shock waves by quickly closing a valve at the end of a line

certain conditions, shock forces caused by water hammer can reach levels of sufficient magnitude to rupture the piping system. The buildup of shock waves is illustrated in Figure 7-19.

Proper design practices can eliminate, or at least minimize, water hammer. Check valves on pumps should be designed to close as quickly as possible to maintain at a minimum the liquid velocity flowing back through the check valve. Shock waves from quick-closing valves are most difficult to eliminate in short lines. Reliable estimates should be made of the impact of water hammer on pressure rise to ensure that the system will function properly.

The velocity of the shock wave through the line can be estimated from the following expression:

$$\omega = \frac{12}{\left[\frac{\rho}{g_c}\left(\frac{1}{\gamma} + \frac{D}{\varepsilon \chi}\right)\right]^{1/2}}$$

where ω = shock wave velocity (fps)

γ = bulk modulus of compressibility of the fluid (psi)

ρ = fluid density (lb/ft^3)

D = pipe inside diameter (ft)

ε = modulus of elasticity of the pipe (psi)

χ = reinforced wall thickness of the pipe (ft)

For an estimate of the instantaneous rise in pressure that would be experienced from a fast valve closure at the end of a line, the following expression can be used:

$$\Delta P' = \frac{\rho \omega \overline{U}}{144 \, g_c}$$

Pipes

where $\Delta P'$ is the pressure rise above normal (psi). The preceding equation is subject to the following constraint:

$$\dot{\theta} \leq \frac{2L}{\omega}$$

where $\dot{\theta}$ = elapsed time for closing the valve completely (sec)
L = length of straight pipe into the valve (ft)

For a situation in which the valve is closed slowly, the following set of expressions are recommended for use along with Figure 7-20:

$$\hat{K} = \frac{\omega \overline{U}}{2gh_0}$$

$$N = \frac{\omega t}{2L}$$

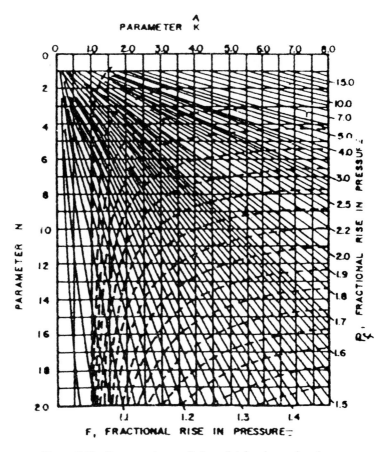

Figure 7-20 Pressure rise prediction plot for slow valve closure

where g = acceleration due to gravity (32.2 ft/sec^2)

$\hat{K}$ = top axis of plot in Figure 7-20

N = left axis of plot in Figure 7-20

h_0 = normal head for steady flow (ft)

Equations are subject to the following condition:

$$t > \frac{2L}{\omega}$$

Once parameters K and N are known, Figure 7-20 can be used to estimate the fractional rise in pressure, P_f, defined by

$$P_f = \frac{h_0 + \Delta P'}{h_0}$$

Buried Pipe Installations

The depth, grade, and general layout or contour of a trench will depend on the project specifications. However, there are some general guidelines.

The width of the trench floor should be roughly 1–2 ft wider than the diameter of the pipe to be installed. This will allow sufficient clearance for working crews and is optimum for compacting the backfill. If unstable soil conditions exist, the trench width should be widened to remove unwanted matter adjacent to the pipe. Stable backfill material should be used to replace the additional removal. As a standard safety practice, trenches should be shored and drained.

Bell and spigot joints allow the pipe to be deflected at each point for alignment adjustments and for curves, as shown in Figure 7-21. Table 7-10 gives rough estimates of the amount of joint deflection possible and the resulting radius of curve for two different lengths of pressure and nonpressure pipe. The table should be used in conjunction with Figure 7-21.

If the design calls for curves of shorter radius than can be obtained with deflected joints, then fittings will be required. Extra trench space should then be required for working.

The trench bedding should be uniform and continuous. The floor should be free of rocks greater than 1 in., lumps, or debris. It is often recommended that excavation be at least 1 in. below grade so that the bottom can be built up with rock-free granular soil. For rougher terrain, where the soil is saturated with stones or the trench has a rock base, overexcavation is recommended to at least 6 in. below the grade. This can then be replaced with granular soil or crushed stone.

High spots in the trench floor promote uneven bearing on the pipe and may cause damage due to stress from backfill. They may also cause excessive wear, particularly when pulsation occurs in lines due to pumping.

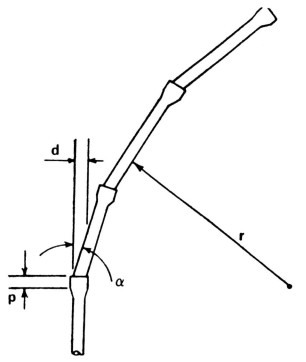

Figure 7-21 Important dimensions in bending lines in a trench: α = maximum angular deflection per joint; p = maximum pull per joint; d = maximum linear deflection per joint; and r = minimum curve radius

Bells should be provided with depressions on the trench bottom to provide clearance and provide the proper support along the pipe length. For assembly, pipe should be strung along the top of the trench, with the bells facing the direction in which they are to be installed. For small-diameter pipes, the following procedure is recommended:

1. If the trench wall is straight, having firm edges, lower the pipe by hand.
2. If the trench is not too deep, a crew can be stationed on the bottom to reach the pipe as it is lowered.

For larger-diameter pipes, or for relatively deep trenches that do not fit the criteria in steps 1 and 2 just given, the following is recommended:

1. Wooden skids should be placed along the side of the trench. (This prevents dirt and rocks from entering the bell and protects against any damage from rocks extending out of the trench wall.)
2. Using rope or cable, slowly lower the pipe by hand. Each crew member should stand firmly on the other rope end.
3. Stop-knots should be used on the ropes under the workers' feet to prevent heavy pipe lengths from slipping. For very large pipe, cranes or other types of heavy-duty hoists may be necessary.

TABLE 7-10 PERMISSIBLE CURVE RADIUS AND JOINT DEFLECTION FOR PRESSURE AND NONPRESSURE

Inside diameter (in.)	α, Maximum angular deflection per joint (°)		p, Maximum pull per joint (in.)	Length = 10 ft		Length = 20 ft	
				d, Maximum linear deflection per joint (in.)	r, Minimum curve radius (ft)	d, Maximum linear deflection per joint (in.)	r, Minimum curve radius (ft)
Nonpressure Pipe							
8	5	0	3/4	10½	115	21	230
10	5	0	1	10½	115	21	230
12	5	0	1¼	10½	115	21	230
14	5	0	1⅜	10½	115	21	230
15	5	0	1⅜	10½	115	21	230
16	4	0	1¼	8⅜	145	16¾	290
18	4	0	1⅜	8⅜	145	16¾	290
20	4	0	1½	8⅜	145	16¾	290
21	4	0	1⅝	8⅜	145	16¾	290
24	4	0	2	8⅜	145	16¾	290
27	4	0	2	8⅜	145	16¾	290
30	3	50	2⅛	7 5/16	177	14⅝	354
33	2	45	1¾	5 3/16	250	11½	500
36	2	35	1¾	5 7/16	266	10⅞	532
39	2	35	1¾	5	286	10	572
42	2	15	1¾	4 11/16	308	9⅜	616

45	2	0	1¾	4¼	344	8½	688
48	2	0	1¾	4¼	344	8	688
54	2	0	1⅞	4¼	290	8½	580
60	2	0	2	4¼	290	8½	580
Pressure Pipe							
8	3	0	½	6¼	190	12½	380
10	3	0	⅝	6¼	190	12½	380
12	3	0	¾	6¼	190	12½	380
14	3	0	¾	6¼	190	12½	380
15	3	0	⅞	6¼	190	12½	380
16	3	0	¾	6¼	190	12½	380
18	3	0	1	6¼	190	12½	380
20	3	0	1⅛	6¼	190	12½	380
21	3	0	1¼	6¼	190	12½	380
24	3	0	1⅜	6¼	190	12½	380
27	3	0	1½	6¼	190	12½	380
30	3	0	1⅝	6¼	190	12½	380
33	2	0	1¼	4³⁄₁₆	285	8⅜	570
36	1	50	1¼	3¾	310	7½	620
39	1	40	1¼	3½	345	7	690
42	1	36	1¼	3¹¹⁄₃₂	356	6¹¹⁄₁₆	712
45	1	30	1¼	3⅛	380	6¼	760
48	1	24	1¼	2¹⁵⁄₁₆	410	5⅞	820
54	2	0	1⅞	4¼	290	8½	580
60	2	0	2	4¼	290	8½	580

Once the pipe has been positioned on the trench floor, the joints can be cleaned. The bell's inside surface, spigot groove, and any rubber gaskets should be cleaned thoroughly and solvents given sufficient time to evaporate. The bell-sealing surface and the gasket should be lubricated. The gasket should be stretched over the spigot and snapped into the groove, making sure that it is stretched evenly around the groove and not twisted. Care should be taken not to recontaminate cleaned surfaces.

There are a number of techniques used for making the final connections. For pipe diameters up to 18 in., one or more leverage bars and a block of wood can be used as shown in Figure 7-22A. The wood block protects the pipe from point loads. The block should be positioned horizontally against the bell, and the bar should be embedded firmly in the trench floor behind the block. The spigot should be pushed firmly into the mouth of the bell until it reaches the shoulder.

Another technique is shown in Figure 7-22B, in which a choker cable can be attached to a crane boom. This requires an adequate downward load onto the bell ends of the pipe. Care must be taken to ensure that the cable angle is low to prevent buckling or misalignment. Choker cables and come-along jacks also may be used.

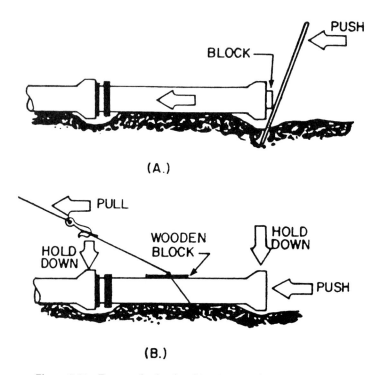

Figure 7-22 Two methods of making joints in buried installations

If fittings are used to make changes in direction, thrust blocking must be provided against solid trench walls. These fitting areas should not be machine dug as the excavator could easily dig too deeply and may damage the bearing surface of the trench wall. The primary function of thrust blocking is to prevent movement of the piping component; that is, it serves as an anchor. A major part of the system design is to determine where thrust blocks are needed and how large they should be. Figure 7-23 shows typical cases in which concrete thrust blocks can be employed. As a general rule when pouring concrete, the widest practical support should be provided without covering piping components.

Closures are also a significant part of the system design. Closures are the arrangement at the last length of pipe that closes the line, and the assembly usually requires a special length of pipe with a coupling. The most common way to do this is to use a long bell-shaped pipe. Another approach is to use a flexible coupling that is to be installed over the pipe. The latter method requires a pipe with either a resin outer wall, smooth outside diameter, or some specially prepared standard pipe from the supplier. A third type of closure makes use of a plain end section of pipe where a full-circle repair clamp can be employed as the coupling.

If pipe is to be laid under road crossings, it is advisable to use conduits through which the pipe can pass. This will protect pipe from damage due to excessive loads or burial depths. Care should be taken to ensure that the pipe is properly bedded at the entrance and exit points of the conduit so that stress or

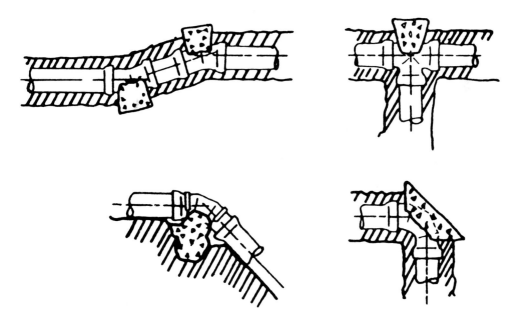

Figure 7-23 Typical examples of the use of thrust blocks

wear does not occur to the pipe. Conduits also are useful in preventing pipe movement caused by pressure surges or flotation.

When installing pipe through casings or conduit, skids must be used. The use of skids prevents the pipe from snagging on the inside wall of the casing during installation. It also prevents the pipe from resting on the bells. Skids should be 2–2 1/2 in. long or continuous and should be sufficiently high to provide clearance between the pipe bells and the casing. The leading ends of the skids should be rounded so that the pipe may be readily pushed through the casing. Skids must be strapped tightly to the pipe by metal strapping.

Care should be taken to install bottom skids parallel to the center line of the pipe, as this will prevent the pipe from rolling in the casing. Normally, six skids are recommended, mounted at 60° intervals about the periphery of the pipe and at selected axial intervals.

Pressure grouting also may be used to restrict pipe movement in the conduit; however, it may be necessary to reposition skids to accommodate the grouting hose. The recommended grout mix ratio is four parts sand to one part cement. Sufficient water should be added to obtain a pea-soup consistency. The grouting machine should be equipped with a pressure gauge at its discharge. Once the delivery pressure has built up, the grout will discharge from the hose and fill the void between the pipe and conduit. Delivery pressure should be monitored carefully as just an excess of 2–3 psi over the initial required delivery pressure might be enough to collapse some pipe.

Once the pipe has been laid out and final connections are made, the installation should be backfilled with sufficient fill to hold the line in place. All fittings and joints should be left open for inspection during the test period. Primary backfill serves the function of a continuous and firm support for the line. Primary backfill, sometimes called the pipe zone, must be placed and compacted so that it provides sufficient and even support about the periphery of the pipe.

Damp, stone-free, nonlumpy soil should be used for primary backfill. The soil or filler material should be relatively clean of vegetation, debris, and stones over 1 in. in size. Filler materials that can be used in place of soil include sand, graded gravel, or crushed rock.

Compaction may be carried out by shoveling the backfill thoroughly under the pipe. Care must be taken not to shovel excessive amounts as this will create voids under the pipe. Backfill should be placed in lifts of 6–8 in. uniformly on both sides of the pipe. When backfilling with gravel or crushed stone, shovel the material under the haunches of the pipe.

Water jetting or flooding can be used as a method for compaction provided there is adequate drainage. If the pipe is laid close to or below the ground-water table, water compaction must be used cautiously to avoid floating the pipe.

When installations are placed under streets, the backfill compaction must match the compaction of the roadbed. A relative compaction of 85 percent to 90 percent is required. Mechanical compactors should not be used directly over the pipe if there is less than 6 in. of cover. For heavy-duty compaction equipment, a minimum of 1 ft of coverage is required.

Pipes

In general, when installing underground installations, only a portion of the line should be covered so that checks can be made for deflections against the 5 percent limit. A good installation should average around 2 percent, with maximum vertical deflections peaking at 3 percent.

Once primary backfilling is completed, the entire trench may be filled in with secondary backfill material. Secondary backfill is most often the native soil. It should be graded adequately so that vegetation and large stones are removed. In situations with live loads, the secondary backfill should be placed and compacted in short lifts. Figure 7-24 illustrates recommended trench configurations and composition. Note that bed depth and recommended compaction densities will vary with the specific pipe used.

Line Testing

During installation, it is often advisable to pressure test segments of the assembly before it is completed. Portions to be tested should be backfilled and com-

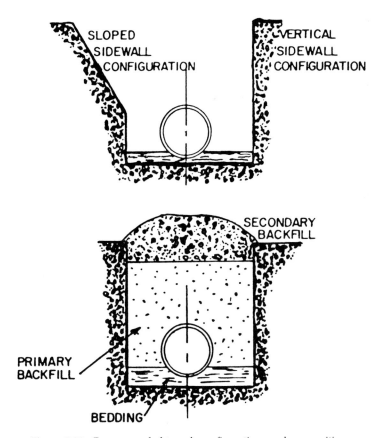

Figure 7-24 Recommended trench configurations and compositions

pacted to prevent movement under pressure. The following factors should be considered for testing:

1. Sufficient backfill must be provided to restrict all movement under test pressures.
2. Thrust blocks at fittings must be permanent and capable of withstanding test pressures.
3. Test ends must be capped and anchored to withstand thrusts generated under test pressures.

Lines can be tested with either air or water, although the latter is preferred. When the line is being filled with water, air becomes entrapped in the water. The air will begin to escape and becomes entrapped in the line once the flowing has stopped and the water settles down. An air vent or air-relief valve should be supplied at the highest point in the line so that entrapped air can be forced out of the pipe; that is, water should displace the air so that the entire length of pipe undergoing testing is filled. This must be done before pressure is applied to the system.

It is generally recommended to fill the line 24 hours in advance at a nominal pressure prior to initiating tests. Once the line air has been expelled, the water can be pumped at the specified test pressure with a pressure gauge located at a low point in the line, where pressures are usually highest. Possible sources of leakage are in joints at saddles, transition fittings, valves, and adaptors. Pipe joints normally are not trouble areas.

The project engineer should establish allowable leakage standards. As a general rule, a pipeline is acceptable if losses are less than 100 gallons per inch of diameter per mile of pipe per day for tests conducted on nonpressure lines with up to 8.7 psi of head. Usually pressure lines are not tested in excess of 1.5 times the pipe's rated pressure. Pressure lines are considered acceptable if the water needed to maintain pressure is less than 30 gallons per inch of diameter per mile per day. Pressure tests at 1.5 times the maximum operating pressure should be conducted for a minimum of two hours. As a rough guide, Table 7-11 can be used for allowable leakage criteria at different test pressures.

TABLE 7-11 ALLOWABLE LEAKAGE GUIDE (gal/1,000 ft/hr)

Pipe inside diameter (in.)	Test pressure, psi				
	50	100	150	200	250
4	0.38	0.54	0.66	0.77	0.85
6	0.57	0.81	0.99	1.15	1.28
8	0.76	1.08	1.32	1.53	1.71
10	0.96	1.35	1.66	1.91	2.14
12	1.15	1.62	1.99	2.30	2.56

For critical operations, a more severe cyclic test is recommended. This involves subjecting the system to five to ten pressurization cycles at 1.5 times the maximum rating of the lowest-rated component in the arrangement. Sudden pressure surges or water hammer should be avoided during testing. Water hammer can cause pressures several times higher than the system's rating.

COMPENSATORS

During operation, different parts of process equipment reach different temperatures, which results in additional thermal stresses to rigid structures. The rigid construction is not applicable if the total stresses (from both the pressure of the processing fluid and thermal stresses) become greater than is permissible. Compensators should be provided for apparatuses operating under such conditions. Compensators are widely used for a multitude of chemical and process equipment. There are two major types:

1. Flexible compensators (lens, bellows, membranes) set between parts of equipment having different temperatures.
2. Stuffing-box compensators, which permit free movement of one part of equipment against another.

The interaction force due to thermal stresses induced between rigidly connected parts of an apparatus, such as a shell and tubes in a heat exchanger, is determined by the following formula:

$$P = \frac{|\alpha_s^t t_s - \alpha_p^t t_p|}{\dfrac{1}{E_s^t F_s} + \dfrac{1}{E_p^t F_p}}$$

where t_s and t_p = the design temperatures of the shell and tubes proceeding from the maximum difference between them during operation, start-up and shutdown of equipment (C°)

α_s^t and α_p^t = the coefficients of linear expansion for the shell and tube material at t_s and t_p (1/°C)

E_s^t and E_p^t = the modulus of elongation for materials of the shell and tube at corresponding temperatures t_s and t_p (N/m²)

F_s and F_p = the cross-sectional areas of the shell and tubes (m²), respectively

Apart from the force, P^t, there may be a pressure force, P, acting on rigidly connected parts of an apparatus and induced by the pressure of the fluid, for example, in single-pass heat exchangers. This pressure force from inside and outside the tubes causes elongation in the axial direction. It can be determined from the following formula:

$$P = 0.785[(D^2 - d_e^2 Z)P_e + d^2 ZP], \quad N$$

where D = the design diameter of tube sheet (m)

d_e and d = the outside and inside diameters of tubes, correspondingly (N/m²)

P_e and P = the outside and inside pressures of the tubes, correspondingly (N/m²)

Z = the number of tubes

The rigid connection created between a shell with tubes in a heat exchanger is allowable if the following conditions are met:

1. At $t_s > t_p$ the shell is compressed in the axial direction, and the tubes are elongated:

$$\sigma_{max}^s = \frac{P^t}{F_s} - \frac{PE_s^t}{E_s^t F_s + E_p^t F_p} \leq \frac{\sigma_{y,s}^t}{1.2}, \quad N/m^2$$

$$\sigma_{max}^p = \frac{P^t}{F_p} + \frac{PE_p^t}{E_s^t F_s + E_p^t F_p} \leq \frac{\sigma_{y,p}^t}{1.2}, \quad N/m^2$$

2. At $t_s < t_p$ the shell is elongated in the axial direction, and the tubes are compressed:

$$\sigma_{max}^s = \frac{P^t}{F_s} + \frac{PE_s^t}{E_s^t F_s + E_p^t F_p} \leq \frac{\sigma_{y,s}^t}{1.2}, \quad N/m^2$$

$$\sigma_{max}^p = \frac{P^t}{F_p} - \frac{PE_p^t}{E_s^t F_s + E_p^t F_p} \leq \frac{\sigma_{y,p}^t}{1.2}, \quad N/m^2$$

The design axial force, P_p, acting in the place of tube fixing in the sheet under operating conditions is determined from the following formula

$$P_p = 0.785(d_e^2 - d^2)\sigma_{max}^P, \quad N$$

If conditions are not fulfilled, it becomes necessary to insert a compensator in the equipment.

Lens Compensators

Lens compensators with one or more lenses are used widely in process equipment (such as heat exchangers, piping, and so on).

Examples of circular metal lens compensators are given in Figure 7-25. Lens compensators may be used in either vertical or horizontal apparatuses, as well as in piping systems for operation under pressures up to 0.6 MN/m² and sometimes up to 1.6 MN/m².

Compensators

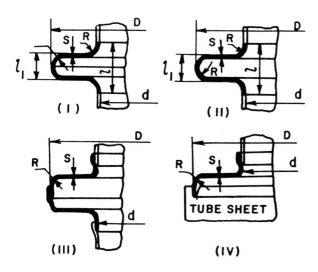

Figure 7-25 Examples of circular metal lens compensators used in process equipment: (A) welded from two stamped semilenses; (B) welded from several sectorial parts connected along radial planes; (C) from copper and brass of two semilenses connected with soft solder; (D) from copper and brass of one semilens inserted between the shell and tube sheet with soft soldering. In all cases, $R \geq S$.

The compensating capability of a compensator is approximately proportional to the number of lenses. However, the use of more than four lenses in one compensator is not recommended. To decrease the resistance to the motion of the fluid, a sleeve is provided inside the shell or tube, which is welded from one side (in vertical apparatuses and piping from above; in horizontal ones from the inside of the liquid motion).

To increase the compensating capacity, lens compensators should be compressed before insertion if they are intended to supplement elongation. If they are intended to supplement compression, they should be stretched. Compressing and extending a compensator is accomplished by its full deformation to double its compensating capability.

The design diameter of a lens compensator for a steel apparatus operating under pressure in the intertubular space $P_{in} > 0.6$ MN/m² as well as for an apparatus of nonferrous metals and their alloys (independent of pressure) is determined from the following formula:

$$D' \geq \frac{0.5d}{0.5 - \sqrt{\frac{P_{in}}{\sigma_{allow}}}}$$

The deformation (tension or compression) of one lens is:

$$\Delta_e = \frac{\alpha_1}{2} \frac{(1 - \beta)P_c d^2}{\pi E S^3}$$

where d = inside diameter of a lens (cm)

P_c = force to induce the deformation of one lens of compensator of $\Delta l (kg)$

S = lens thickness (cm)

The coefficients α_1 and β are

$$\alpha_1 = 6.9 \left(\frac{1-\beta^2}{\beta^2} - \frac{4 ln^2 \beta}{1-\beta^2} \right) \frac{1}{1-\beta}$$

$$\beta = d/D$$

where D = outside diameter of a lens (cm)

d = inside diameter of a lens (cm)

The values of α_1 may be determined from Figure 7-26.

Because of a known rigidity of the system, the tubes become compressed (or elongated) by the value

$$\Delta_p = \frac{P_p L}{E F_p}$$

where P_p is the compression (elongation) force, N, of the tubes. The shell and compensator become elongated (compressed) by the value

$$\Delta_s = \frac{P_s L}{E F_s} + \frac{\alpha_1 (1-\beta) P_s d^2}{2\pi E S^3} = \frac{P_s}{E} \left[\frac{L}{F_s} + \frac{\alpha_1 (1-\beta) d^2}{2\pi S^3} \right]$$

Taking into account that $P_p = P_s$, as well as $\Delta^s = \Delta - \Delta^p$, where $\Delta = \alpha \Delta t L$, we obtain from preceding equations the compensator reaction, P_c, along its axis at given wall thickness, S, and ratio $\beta = d/D$:

$$P_c = \frac{L |\alpha_c^t t_c - \alpha_p^t t_p|}{\frac{L}{E_c^t F_c} + \frac{L}{E_p^t F_p} + \frac{0.159 \alpha_1 (1-\beta) d^2}{E_c^t S^3}}$$

where α_1 = coefficient determined from Figure 7-26

L = length of the shell and tubes (m)

For the compensator in Figure 7-25D (of one semilens), the deformation l computed should be decreased by a factor of 2. The design number of lenses in

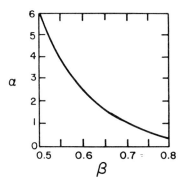 Figure 7-26 Diagram for design of lens compensators

the compensator at a given length of tubes, L, is determined from the following formula:

$$Z'_e = \frac{L}{\Delta_e}\left[|\alpha^t_c t_c - \alpha^t_p t_p| - P_c\left(\frac{1}{E^t_c F_c} + \frac{1}{E^t_p F_p}\right)\right]$$

Stuffing-Box Compensators

Along with lens compensators, stuffing-box compensators are used widely in the fabrication of process equipment, especially for round parts of apparatuses with inside diameters of less than 150 mm.

The important advantage of stuffing-box compensators over lens designs is their ability to be applied under high pressures. At the same time, stuffing-box compensators have a major disadvantage; they may leak and, therefore, there is a need to tighten them periodically. Examples of stuffing-box compensator designs with soft packing are given in Figure 7-27. The packing material is selected according to the fluid and its design temperature. At high fluid temperatures, water cooling of the packing through a jacket is required. In designing stuffing-box compensators with soft packing, the design thickness of the soft packing, S'_c, is determined from the following formula:

$$S'_c = 0.044 \sqrt{D_c}, \quad m$$

The value of S'_c should be $3 < S'_c < 25$ mm. Values for the height of the packing are given in Table 7-12.

The design compression force of the stuffing box, P_c, is determined approximately from the following formula:

$$P'_c = \pi(D_c + S_c)S_c q, \quad N$$

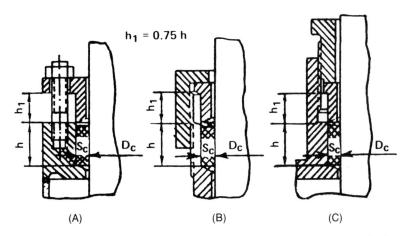

Figure 7-27 Examples of stuffing-box compensators: (A) $D_c > 25$ mm; (B, C) $D_c < 25$ mm

TABLE 7-12 HEIGHT OF THE SOFT PACKING FOR STUFFING-BOX COMPENSATORS

p, MN/m^2	≤ 0.6	>0.6 to 1.6	>1.6 to 2.5	>2.5
h_1, mm	$3S_c$	$4S_c$	$5S_c$	$6S_c$

TABLE 7-13 SPECIFIC LOAD ON THE STUFFING-BOX SLEEVE FOR SOFT PACKING

p, MN/m^2	0.6	1.0	1.6	2.5	4.0	6.5	10.0	>10
q, MN/m^2	1.8	2.5	3.0	5.0	7.5	10.0	12.5	P

where q is the specific load on the stuffing-box sleeve. Values are given in Table 7-13.

The structural design of the basic elements of stuffing-box compensators, such as studs, thread, flange, sleeve, and so on, are made according to the design compression force, P_c, from the corresponding formulas given.

VALVES AND THEIR SELECTION

Valves provide the means to adjust inputs of process control systems. Proper valve selection is essential for a well-designed piping system. For systems requiring relatively short time lags and having large capacities, fairly simple control systems are adequate, normally permitting the use of high-sensitivity controllers with quick-opening valves. For this type of service, flow characteristics become secondary considerations, and valve characteristics such as size and material become primary.

As the time lag increases and the capacity decreases, required control schemes become more complex. In general, the larger the time lag, the smaller and more precise must be the variations in the flow rate of the control medium for given changes in process conditions.

Common control valves are illustrated in Figures 7-28 through 7-32. Table 7-14 provides brief descriptions of various control valves.

Knowledge of where the various parts are located in different valve types ensures proper installation and helps to maintain them. Preventive maintenance, such as regular lubrication, tightening or replacing packing, or replacing composition discs, or resurfacing seats and discs on a regular basis will greatly extend the life of a valve. Figures 7-33 and 7-34 show the basic parts of common valves.

A brief description of the major valve types and guidelines for their installation and maintenance follows.

Valves and Their Selection

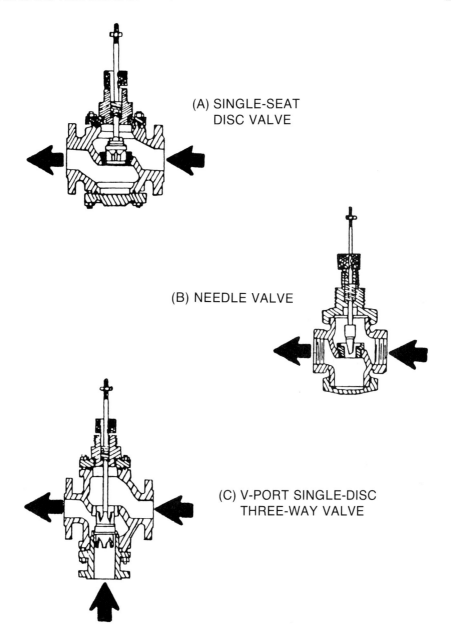

Figure 7-28 Various valve bodies: (A) single-seat disc valve, used mainly on simple, high-sensitivity, small time-lag applications and for open-and-shut service; or stable load conditions and constant line pressures; (B) needle valve, used for control of very small flows; employed generally where requirements are for less than ½-in. valve size; (C) V-port, single-disc three-way valve, suitable for mixing service.

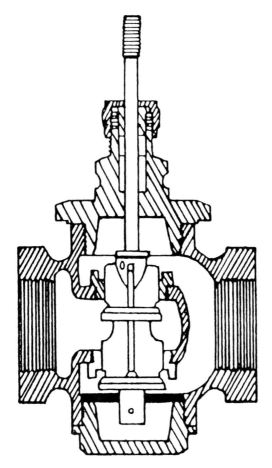

Figure 7-29 Double-seated, quick-opening inner valve with guided discs. These are used where close throttling flow is needed. Often used on low-sensitivity applications.

Gate Valves

As implied by the name, when the wheel is turned to lift the wedge of a gate valve it is literally like opening a flood gate. Fluid flows freely through a full pilot with little pressure drop. As with all standard valves, turning the handwheel counterclockwise opens it, while turning it clockwise closes it. Gate valves are neither meant to be throttling valves, nor are they designed for frequent operating cycles. Operation of a gate valve with the wedge partially open can cause vibration and rapid wear of the wedge and seat surfaces, leading to possible valve failure. Nonrising stem gate valves are particularly useful when there are space limitations. The gate itself is threaded and rides up on the stem when the stem is rotated. Other variations include inside screw, resisting stem and outside screw and yoke (see Figure 7-33). U-bolt gate valves are designed for quick disassembly and are recommended for use with viscous fluids where the fluid tends to collect in the bonnet structure.

Unless specially constructed, gate valves can be installed with either face to the inlet side. Normally, discoloration or residue on the bonnet indicates a packing leak. The packing nut should be tightened or the packing replaced. Packing material must be compatible with the fluid in service. The packing nut must not be overtightened, and gland nuts must be tightened evenly to prevent a cocked gland flange.

If valves are to be installed in a dirty or dusty environment, leather or metal stem jackets are recommended. A clean handwheel provides a firm grip when the valve is to be opened or closed.

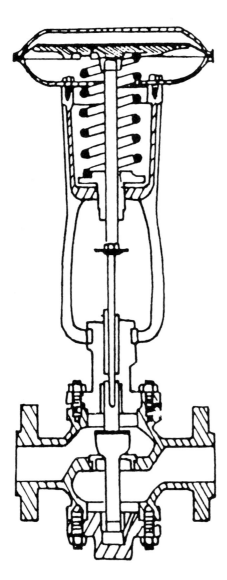

Figure 7-30 Single-seated, globe-type body with solid-plug inner valve. The throttle plug controls flow by means of variable port openings, as determined by curvature of the plug proper.

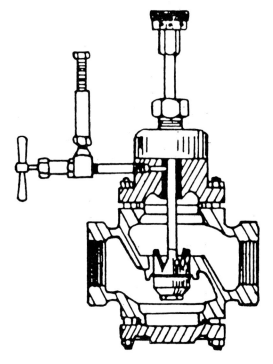

Figure 7-31 Single-seat, V-port inner valve

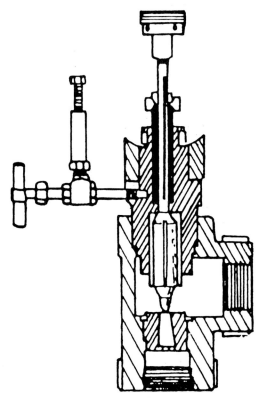

Figure 7-32 Angle valve

Valves that require frequent use should be lubricated regularly. Most outside screw (OS) and yoke (Y) valves have a special grease fitting for this purpose.

Gate valve wedges and seats usually wear together; therefore, they should be replaced together. Most wear occurs on the outside of the wedge. A temporary repair can be made by reversing the wedge.

The interior of the valve body should be inspected periodically for corrosion or scale, which may impede the flow. Scale deposits on smooth interior walls can promote eddies, which accelerate corrosion.

Check the stem for signs of dezincification (a form of corrosion peculiar to bronze gate valves, which leaves the metal soft and spongy). Replace the stem immediately if dezincification is detected.

Globe Valves

Globe valves are used most often for throttling or modulation. The choice of the disc and seat material must be based on its compatibility with the fluid in service. Regrind/renewable-type globes have removable bronze or corrosion-resistant stainless steel seats, which are threaded into the valve body and may be reground or replaced.

Regrindable globes have permanent or integral seats, which can be reground by using a regrinding tool or by using the metal valve disc pinned to the stem.

Renewable-composition disc globe valves are used when normal wear results from frequent throttling. The disc is changed by removing the old disc from the disc holder and slipping on a new disc, which is held in place by the disc nut (see Figure 7-33).

About 90 percent of all globe valves are installed with the inlet pressure beneath the seat. This allows the valve to be opened somewhat more easily. Under severe operating conditions (for example, in high-pressure steam service), some mechanical advantage may be obtained in closing the valve by installing it with pressure flow above the seat.

The following procedure should be followed in replacing a composition disc in a globe valve.

1. Isolate the valve to be repaired by closing the main valve and draining the line.
2. Open the valve slightly to separate the disc from the seat and remove the bonnet using a flatjawed wrench.
3. Slip the disc holder off the stem and replace the disc by removing the disc nut.
4. Reassemble the valve by reversing the operation.

TABLE 7-14 DESCRIPTIONS OF COMMON CONTROL VALVE TYPES

Valve type	Description	Applications	Limitations/disadvantages
Single-Seated	Quick-opening disc type.	Used on simple, high-sensitivity, small time-lag applications/open-and-shut service.	Impractical for throttling control if line pressures fluctuate widely, unless controller sensitivity is very high. Not recommended for throttling control above 2-in. size.
Double-Seated	Upstream pressure enters body between the seats, tending to create equal upward and downward forces. Balanced valve eliminates line pressure fluctuations.	Used for close throttling control, low-sensitivity applications.	Seating surfaces subject to wear. Cannot supply tight shutoff.
Throttle-Plug	Double-seated quick-opening inner valve with guided discs.	For large load changes. Handles large pressure drops well.	Limited to larger sizes.
V-Port	—	Used where widely varying flow rates are encountered and where full throttling control over entire flow range is needed. Has greatest controllable range.	

Single-Seat Throttling	In sizes smaller than 1 in., they are generally preferred to double-seat type. Provide tight shutoff of controlling medium.	Same as single-seated type.
Needle-Type	Practically any required flow characteristic is possible.	For control of very small flows. Limited to small sizes.
Three-Way Type	Provide highly stable flows.	Service requirements include: • mixing fluids (e.g., mixing hot and cold water) • feeding two fluids to a vessel or pipeline, used as a throttling controller • to divert fluid from one vessel or pipeline to two-load demands
Angle Valves	Basically single-port valves.	Used to facilitate piping or where a self-draining piping system is needed. Used for handling fluids containing solids, slurries, flashing fluids.
Butterfly Valves	Can be fitted with either manual or automatic valve positions. Main body may be diaphragm, cylinder, float, electric motor, or solenoid operated.	For control of low-pressure, low-velocity fluids. Subject to shamming.

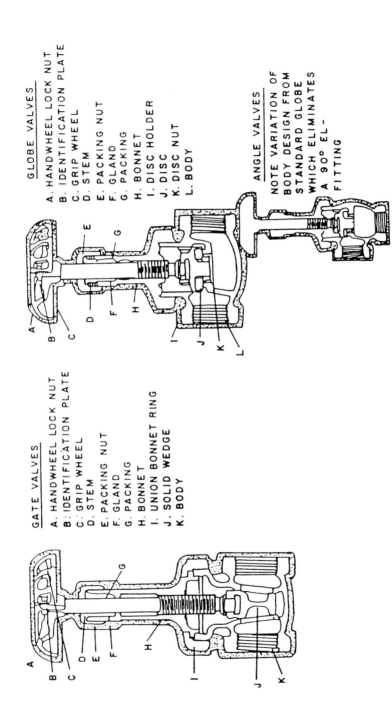

Figure 7-33 Basic parts of common valves

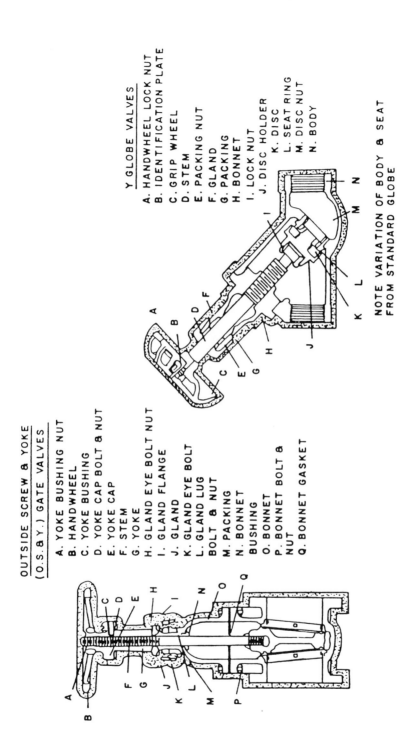

Figure 7-33 Basic parts of common valves (*continued*)

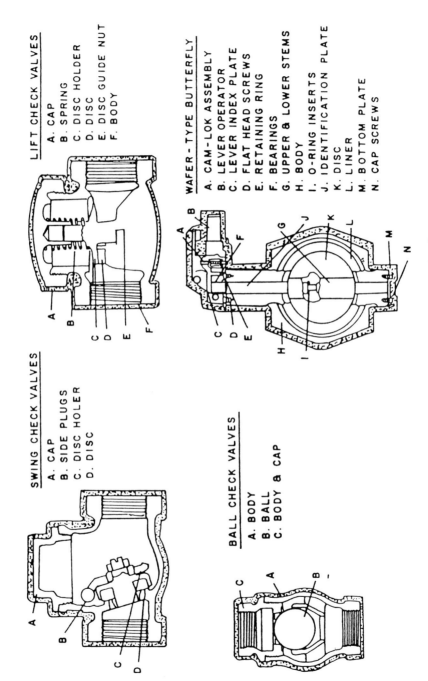

Figure 7-33 Basic parts of common valves *(continued)*

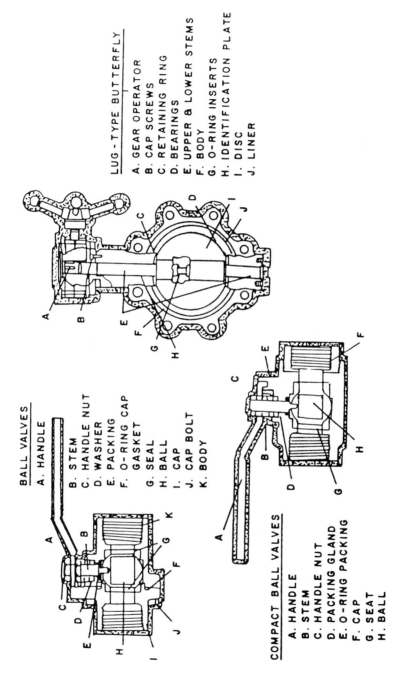

Figure 7-34 Basic parts of common valves

The following procedure should be used in repairing regrind/renewable discs and seats:

1. After the valve has been isolated and the line drained, remove the bonnet and place a small amount of grinding compound on the metal disc.
2. Position a pin in the disc nut notch and stem hole; reinsert the trim and screw down the union ring. Then back off one full turn.
3. Employ the stem as your grinding tool, exercising care not to overgrind. When finished, clean away the grinding compound, remove the pin, lubricate the threads, and reassemble the valve.

Check Valves

There are several types of check valves. Swing check valves are similar to gate valves in that they ensure almost free flow. Fluid velocities are generally low and nonpulsating; otherwise, the swing disc will flap and pound on the seat causing excessive wear and possible failure.

Horizontal lift checks have a diaphragm setting that resembles that of globe valves. Generally these are used to prevent backflow in steam, air, gas, or water lines having higher flow velocities than swing check valves.

Ball checks are similar to lift checks; however, a metal ball replaces the metal or composition disc. These valves are used in heavy, viscous fluid service.

For all installations, flow through a check valve must be from under the seat; otherwise, normal line pressure will force the disc against the seat causing a complete stoppage of flow. Swing checks can be installed in horizontal lines or in vertical lines where the flow is in the upward direction only. Lift checks should be installed only in horizontal lines.

Check valves usually trap fluid downsteam of the valve which must be drained when maintaining the line. A drain valve should be installed downstream or the valve body should be drilled, tapped, and boss plugged.

Check valves should be installed as far from pumps as possible. Pulses will cause the disc to flap or pound on the seat, causing possible damage to the valve.

Many lift checks have replaceable seat rings. Discs are composition or metal and may be replaced or reground. The regrinding technique used is similar to that of globe valves, except that a screwdriver is used in a milled slot on top of the disc holder.

Most swing checks are provided with an access opening in line with the seat. This allows regrinding of the seat and disc with a screwdriver.

Ball Valves

Ball valves are used widely because of their small pressure drop due to the straight-through hole. A quarter turn will fully open or close such a valve.

Elastomeric seats are used to provide a tight closure against a polished metal ball. The elastomeric material also serves as a lubricant, promoting easy operation.

Note that elastomeric seats, as well as the body and ball material, must be compatible with the fluid. In selecting a ball valve, features such as a blowout-proof stem and ample packing box should be included.

When installing ball valves, pipe alignment is critical since strain on the valve itself may pinch or bind the ball. When soldering bronze ball valves, some manufacturers recommend complete disassembly to avoid heat damage to the seats. A constant check on valve body temperature should be maintained during soldering operations.

Check for possible leaks at two locations: the packing or seats. If there is a packing nut under the handle, tighten or replace the packing. Care should be taken not to overtighten the packing or the seam may be ruined.

When installing threaded ball valves in a line, a pipe wrench should be placed on the pipe and the body of the valve screwed onto the pipe and tightened with a flatjawed wrench on the body and hex flats. Then the other pipe should be screwed into the cap while holding the cap hex flats stationary with a flatjawed wrench. The pipe cap then should be tightened with a pipe wrench. The cap should not be allowed to turn and cause excessive pressure on the ball and seats.

When reassembling a ball valve, one should torque up bolts or cap screws evenly. The ball should be rotated during the final tightening to avoid binding. One should then check that the stem is squarely in the milled slot in the ball for proper alignment.

Butterfly Valves

Butterfly valves are best suited to relatively low-pressure service. Generally, the fluid friction coefficient is low. Because of the narrow design of the body, sediment buildup is usually minimal. There are two basic types of butterfly valves:

1. The wafer design, which fits between pipe flanges and inside the circle of bolts.
2. The lug type, which can be connected directly to a flange or flanges.

Butterfly valves operate with a quarter turn. Manual operators may be lever type or geared. The position of the disc is noted by an indicator on the gearbox or from the position of the handle.

During installation, care should be taken not to damage the elastomeric liner. To install the wafer type, fit the bottom lug bolts loosely in the pipe flanges. Then drop the valve in place on top of the bolts, insert the remaining bolts and tighten (using the alternate, cross-over method to ensure even distribution of torque).

The valve always should be closed during installation or maintenance to maintain the disc within the body and to avoid possible damage. Gaskets generally are not necessary since the valve liner maintains a tight seal between the valve and pipe flanges.

Valve Selection

Each valve type has its specific uses. Even the highest-quality valve installed in the wrong service will require frequent attention or may result in failure, causing unnecessary downtime and lost production. Therefore, selection must be based not only on materials, size, and pressure rating, but on intended application as well.

The simplest fluid control application is on/off service. This type of control is required on the main water supply to a building or feed-water control to a boiler system. In both cases, it is necessary to provide the fullest possible flow with a minimal pressure drop.

Gate valves are used most often for this type of service. Gate valves provide almost full flow, minimal pressure drop, and minimum turbulence and fluid trapped in the line. These valves often are used where operation is infrequent and should never be used for throttling.

For modulation or throttling, the amount of fluid or gas in the service must be controlled or varied. Whenever the flow is reduced, there is a corresponding drop in pressure across the valve. Globe, Y, angle, and needle valves are best suited for this type of service. The globe family of valves is designed specifically for throttling; hence, particular attention must be given to the several types of seating materials available to avoid unnecessary wear. Whether they consist of composition or metal, they must be compatible with the fluid in service.

There are many services in which the reversal of flow could have disastrous results. One example is the failure in a boiler system, which could force live, superheated steam back into a city water system. Check valves are designed to prevent backflow by automatically seating when the direction of flow is reversed. The four basic types are swing, check, lift check, and ball check.

Quarter-turn valves display features such as versatility and speed of operation, economy, and lightweight service. Such valves are fully opened or closed by turning the hand 90° (hence, the name *quarter-turn valves*). Ball valves probably are the most versatile of all quarter-turn valves because of the variety of metals and different services in which they are used. In these designs, a highly polished ball seats against elastomeric seals. They are easier to operate than their older counterpart, the plug valve. Butterfly valves often are used to save weight and, thus, cost. They are used in larger low-pressure services. The body is usually iron, with a choice of disc metals to meet required service conditions. The tight closure is provided by a one-piece elastomeric liner seal.

Valves are manufactured primarily from bronze, iron, and steel. Proper body material must be combined with the right seating and packing materials to resist wear and corrosion throughout the life of the valve.

Bronze is the oldest material used to make valves. It is used most often in valves for hot and cold water, and low-pressure steam systems, and for other noncorrosive services. Bronze is durable, noncorrosive, and easily fabricated. It is often used as seating services in larger iron body valves to ensure tight closure by eliminating rust in this critical area.

Standard iron valves generally are used in medium-to-large pipelines to control noncorrosive fluids and gases where pressures do not exceed 250 psi at 450°F, or 500 psi in cold water, oil, or gas. Variations in trim combinations and seating elements make them suitable for long service in many mildly corrosive services.

Carbon steel is a strong, ductile material. Valves made from this metal are used in higher-pressure services, such as steam lines up to 600 psi at 850°F. Many steel valves are available with butt-weld ends for economy. They are used generally in high-pressure steam service, as well as other higher-pressure, non-corrosive services.

There are hundreds of stainless steel alloys developed for specific uses. Generally, 316SS can be used for more than 90 percent of corrosive services.

Valves forged from carbon steel are used in services up to 2,000 psi and temperatures up to 1,000°F in gate, globe, and check valves. Forged steel ball valves may be used in high-pressure services, but are limited in terms of temperature because of their elastomeric seals.

Valve Ratings

Valves generally have several identifying marks that provide information on the valve class, its type of service, pressure, and temperature ratings. Information identifying a valve and its class may be found in two places on a valve: cast into the body or stamped onto an identification plate, which is found under the handwheel nut on bronze, iron, and forged steel valves or fastened to the yoke or yoke bonnet on steel valves.

Information cast on the body usually includes the valve size and pressure class (primary and secondary ratings).

Identification plates include the name of the manufacturer, figure number, and other important data, such as size, body, stem, disc and seat material, pressure class, and secondary pressure rating and temperature rating, if it is different from the standard. The primary rating indicates the safe operating pressure for a valve at some elevated temperature. This temperature is dependent on the materials used and the fabrication of the valve. When specific data are not available, a rule of thumb to follow is the temperature of saturated steam or the primary rating indicated on the valve body. For example, if a valve has a primary rating of 150 psi, as indicated by 150S, then it has a maximum operating temperature of 367°F (temperature of saturated steam at 150 psi). The safe operating pressure of a valve is always relative to the working temperature. Basically, as the pressure increases there must be a proportional decrease in temperature to remain within the safe operating limits of the valve.

For example, a cast carbon steel valve rated at 150 psi may be operated safely up to 100°F at 275 psi. This same valve may sustain a working temperature of 800°F if the working pressure does not rise above 92 psi.

Manufacturers publish charts for a broad range of working pressures and temperatures in their comprehensive valve catalogs. Note that these are the maximum allowable nonshock pressures at the tabulated temperatures.

Valve Installation

Threaded and flanged joints

Long pipe threads permit the pipe to enter the valve too deeply and press against the seats, preventing proper seating of the disc or wedge and damaging the valve. Pipe-sealing compounds should be applied to the pipe-end only and not inside the valve body, where it can cause failure of the valve to the seat. The following steps should be followed when installing valves with threaded and flanged joints.

1. Ream out all burrs that can impede the flow. Then blow out welding scale and other foreign matter before making up the joints. Wipe away all chips and dirt from the pipe threads.
2. Line up the flanged joints carefully. Misaligned flanges can place unnecessary strain on a valve. Use a thread lubricant on the flange bolts and pull up the joint snugly. Pipe hangers are necessary to support valves and piping and to minimize pipeline stress, which can adversely affect the proper operation of valves.
3. Use a pipe wrench on the pipe near the valve and a flatjawed wrench on the valve itself on the end nearest the pipe. Keep the valve closed tightly during installation to prevent twisting and distorting. Place the pipe in a pipe vise (never the valve).
4. On systems with flanged valves and piping, tighten bolts with the alternate-side cross-over method, which gives a much better bearing over the flange faces and reduces stress on the valve parts.

5. Check for proper flow direction. In many cases a valve can be mounted either way. However, particularly for globe and check valves, the fluid flow direction must be correct.
6. Whenever possible, install valves with the bonnet up to prevent dirt, condensate, and sediment from collecting in the bonnet structure and causing unnecessary wear on the stem thread.
7. Locate valves where they can be reached conveniently. If this is not possible, install chain-wheel operators extension stems or powered actuators. Be sure to allow clearance for rising-stem valves to open completely.
8. Insulate piping systems containing hot fluids to prevent injury from contact and to prevent freezing of lines exposed to subfreezing temperatures.

Soldered joints

The following steps should be followed when installing valves with soldered joints:

1. Carefully measure the tubing run between fittings or valves and cut the tubing with a sharp hacksaw or round-wheel tubing cutter. The cut should be at a perfect right angle to the tubing.
2. Place the end of the tubing into the valve or fitting to double-check your measurement. If the joint is to be made in an existing line, make sure it is drained completely and dry. Heat from the torch will turn any water present into steam, preventing a perfect solder joint.
3. Remove all burrs from the outside and inside edges of the cut and clean it with a piece of fine sandpaper, emery cloth, or steel wool. Then wipe clean both the tubing and the inside of the valve or fitting.
4. Coat the outside end of the tubing and the inside end of the valve or fitting lightly with flux and twist the tubing or fitting at least 180° once or twice to distribute the flux evenly.
5. Apply the flame to the fitting end and allow it to reach the proper temperature. The flame should not touch the solder or tubing directly. Fill the joint completely and wipe off the excess with a damp cloth to make a tight, good-looking, professional joint.

NOMENCLATURE

c	speed of sound (fps)
C_1, C_2, C_3	allowance coefficients (m)
C_p	amount of compression (in.)
D	diameter (m)

E^t	modulus of elongation at temperature t (N/m²)
F	cross-sectional area (m²)
l	length (m)
n	number threads per inch
P	pressure (N/m²)
q	specified load (MN/m²)
R_{av}	average bending radius (m)
R_m	expansion joint rated moment (in.)
S'	thickness (m)
S_c	design thickness of soft packing (mm)
T	temperature (°F)
t	temperature (°C)
W	shock wave velocity (fps)
Z	number of tubes
α	coefficient of thermal expansion (in./in./°F)
Δ	displacement (m)
Δ_x	expansion (in.)
δ	laminate wall thickness (in.)
ε	modulus of elasticity (psi)
η	safety factor
$\dot{\theta}$	elapsed time (sec)
$\dot{\mu}$	Poisson's ratio
σ_{allow}	allowable stress (kg/m²)
X	reinforced wall thickness (ft)

8

Seals

INTRODUCTION

In the design of process equipment, we are concerned with numerous connections that are very important to the chemical and process technology. Apertures in equipment may have varied diameters and purposes, placing different requirements on the sealing closures used. The sealing connections—flanged, threaded, bayonet, and bow—must be tough, tight, detachable, and immobile. These connections must be strongest for equipment requiring frequent assembly, for those working with great fluctuations of temperature and pressure, and especially in corrosive environments.

Obturation is defined as the tightening of immovable detachable connections and may be classified in the following manner.

1. By the tightening force:
   1. Forced sealing at the expense of stress developed by bolts.
   2. By *unsupported-area* seal (a self-energized seal), which develops a force reaction on an area smaller than the area exposed to the pressure medium.
2. By the kind of tightening material:
   1. Sealing with soft metal gaskets working for crumpling.

2. Sealing with rigid metal rings or lens working in the region of elastic deformation.
3. By the design:
   1. Sealing with soft metal gasket.
   2. Cone sealing.
   3. Double-cone sealing.
   4. Sealing with elastic metal ring, lens sealing.

Obturation is achieved by pressing down with a certain force, providing the hermetic nature of surfaces to be tightened directly with each other or through gaskets from softer material located between them. This pressing is accomplished by the use of bolts or studs (in flanged connections), with threads (in threaded connections), with bows (in bow connections), and so on. Depending on the action of pressure in the apparatus, there are two types of obturation:

1. *Supported-area* obturation (not self-energizing), in which the pressing force of tightened surfaces decreases under operating conditions.
2. *Unsupported-area* obturation (self-energizing), in which the pressing force of the tightened surfaces is increased or, in some cases, does not change under the conditions of operating pressure.

Therefore, the initial compression at the assembly of the connection for a supported-area obturation should differ from that of unsupported-area obturation. In the first case, the preliminary compression has to be maximum; in the second case it can be minimum.

The unsupported-area obturation is more reliable for providing tightness. Besides, it is more effective because it demands less force at initial compression. However, the obturation used in process equipment is mostly of the unsupported-area type because it is often impossible to accomplish such sealing from design considerations.

The unsupported-area type is used mainly at high pressures (for connecting covers, different inlets, and so on) for some connections in equipment of medium and low pressures, such as connections at the top tube sheet with shells in coil heat exchangers, as well as in vacuum connections.

The principal and the most commonly used types of obturations in process equipment are presented in Table 8-1, where three groups are noted according to the method of sealing: with gaskets, without gaskets, and special type.

The gasket-type obturation is most common. It is used to make connections of low, medium, and high pressures, as well as at vacuum.

Obturation of seals without gaskets is applied basically for small diameters of apertures at high pressures, and in cases in which it is impossible to use gaskets because of temperature limitations.

Special seals have limited applications mostly to high-pressure operations. Selection of the proper type of seals requires information on equipment

Introduction

TABLE 8-1 PRINCIPAL TYPES OF SEALS USED IN PROCESS EQUIPMENT FOR DETACHABLE, IMMOVABLE, TOUGH-TIGHT CONNECTIONS

Type	Sketch	Items	Fluid	P MN/m^2	t, °C
			Operating conditions		
		With gaskets			
I	A, B	With a gasket between flat surfaces	Inert, low aggressive	≤2.5	≤500
II		With a gasket in male-and-female flange		≥2.5 up to 20	
III	A, B	With a gasket in tongue-and-groove flange	aggressive, toxic flame and explosion	≤100 vacuum	≤800 ≤250
IV	A, B	With a gasket in the lock	any	>2.5 up to 100 vacuum	≤800 ≤250
V		With a gasket in the groove	Inert, low aggressive		≤100
VI		With a gasket of dovetail shape		<0.1 and vacuum	
VII		With a gasket between two conical surfaces	any	≥20 up to 150	≥−50 up to 800
VIII	WITH SELF-SEALING / WITHOUT SELF-SEALING	With a gasket in the lock	dangerously explosive and fire hazard	≥10 up to 100	≤500

(*continued*)

294   Seals   Chap. 8

TABLE 8-1 *Continued*

				Operating conditions	
Type	Sketch	Items	Fluid	P MN/m²	t, °C
		Without Gaskets			
IX		With flat sealing surfaces		≤150	in material resistance limits
X		With flat and spherical or two spherical sealing surfaces	any	>10 up to 100	
XI		With conical and spherical sealing surfaces		≤20	
XII		With two conical sealing surfaces	≥10		
		Special			
XIII		With obturator having double cone		≥30 up to 100	≥−50 up to 500
XIV		With obturator having outside and inside cone	any	≥30	in material resistance limits

operation and maintenance (O&M) needs. Gasket-type obturation is preferred when frequent equipment disassembling is needed without changing gaskets. This requirement demands the use of highly elastic materials. Disassembly can be done many times without changing gaskets with such materials as paronite, fiber, fluoroplastic, corrugated metal, asbestos insert, solid flat metal, rings of iron or soft steel. Obturation without gaskets and obturation with lens gaskets demand additional polishing after each disassembling. The shape of sealing in

Sealing with Gaskets

all types of obturation almost always is circular. However, obturation of types I, II, III, and IV, along with circular shapes, also may have rectangular and profile configurations.

SEALING WITH GASKETS

Wherever permitted by temperature, corrosion, and other conditions, sealing with gaskets is used to provide better leak-proof connections and decrease the compression force, which is necessary to apply for sealing surfaces. Gaskets should be fabricated of material that is milder than the sealed surfaces. Gaskets may be fabricated of metal, organic, and nonorganic materials. The principal types used in process equipment are given in Table 8-2.

The sizes of gaskets of rectangular cross section (type I, except the rubber gaskets for obturation type V), serrated (type III), and corrugated (type IV) are given in Table 8-3.

Obturation types I-A are less by 1–2 mm than the width of sealed surfaces (the outside sizes of a gasket and surface are equal). In obturation types I-B, III-A, and IV-A, the width of the gasket is equal to that of the sealed surface or a groove. The sizes of gaskets with round cross sections are given in Table 8-4.

After selecting the obturation type and constructive elements, it is necessary to determine the design seating force to provide a tight seal. The design seating force for gaskets types 1-S and 7 is determined as follows:

$$P'_n \geq 3D_n b_{ef} q, \quad N$$

where D_n = central diameter of the sealing, m

b_{ef} = effective width of the sealing, m

q = specific load on the sealing area, N/m²

The force P_n for rectangular or profile shape of the sealing is calculated from the formula above if the average length of the sealing surface is assumed in place of $3D_n$.

The values of design-effective width of the sealing, b_{ef}, for different types of specific loads, q, depending on gasket material, are given in Tables 8-5, 8-6, and 8-7.

The design width of the sealing, b', in obturation types III and IV with metal gaskets types 1 and 2, and for obturation type VIII, is based on the condition that there are no residual bearing deformations on the sealing surfaces and is determined from the following formulas: In connections without self-sealing,

$$b' \geq \frac{0.33 D'_p}{\sigma_y - 0.33p' - 1.1q'}, \quad m$$

TABLE 8-2 PRINCIPAL TYPES OF GASKETS USED IN PROCESS EQUIPMENT

Type	Sketches	Item	Material	Operating conditions	
				Type of obturation	P, MN/m²
1		Rectangular section	Rubber, rubber with cotton fabric insertion fiber, asbestos, fluoroplastic paronite, aluminum, copper, lead, nickel, steel, Monel.	I; II; III-A; IV-A; V	≤ 10
2		Round section	Aluminum, brass, lead, steel, monel, vacuum rubber.	III-A and IV-A	>10 ≥ 10 Vacuum ≥ 2.5
3		Serrated section	Steel.	III-B and IV-B	≥ 10
4		Corrugated section	Asbestos, jacketed with steel, aluminum, nickel, copper, brass.	II and III-A in connection of pipes and fittings.	at $t \geq 200°C$
5		Profile section	Rubber, plastic.	I and II	>0.3 to 10 at $t \geq 400°C$ up to $500°C$
6		Lens section	Carbon and alloy steels.	VI	≤ 0.1 and vacuum
7		Trapezoid section	Aluminum, copper, nickel, steel, Monel.	VII	≥ 20 up to 150
				VIII	>10 up to 100

TABLE 8-3 SIZES OF GASKETS OF RECTANGULAR CROSS SECTION, SERRATED AND CORRUGATED SHAPES

Type of gasket[a]	1			2	3	4
Material of Gasket	Rubber[b], Fiber, Asbestos, Vegetable Fiber, Plastic, Fluoroplastic Rubber with Cotton Fiber	Paronite	Cardboard	Metals		Asbestos Board Jacketed with Metal
at D, thickness (mm)						
<100	2	2	1.5	2.5	2.5	3
100–300	2.5	2	2	3	3	3
300–1,000	3	3	2	4.5	4	3
>1,000	4	4				

[a] From Table 8-2.
[b] Except obturation type V.

TABLE 8-4 SIZES OF GASKETS WITH ROUND CROSS SECTIONS

	Diameter of gasket cross section[a]	
D, mm	Metals	Rubber
≤6	2	
>6 up to 50	3	
>50 up to 300	4–5	
>300 up to 1,000	5–8	6–8
>1,000		10–12

[a] The gasket mean diameter is equal to that of the sealed surface.

TABLE 8-5 DESIGN-EFFECTIVE WIDTH OF SEALING, b_{ef}, FOR DIFFERENT TYPES OF GASKETS

Design-effective width of sealing	Type of gasket (from Table 8-2)						
	1; 2; 4; 7		3 (see Figure 8-1)		See Table 8-7		
	Width of gasket						
	$b \leq 10$	$b > 0$	$b \leq 10$	$b > 10$	A	B	C
b_{ef}	b	$3.16\sqrt{b}$	$b_0 Z$	$\dfrac{3.16 b_0 Z}{\sqrt{b}}$	3.5	5.5	8

TABLE 8-6 VALUES OF SPECIFIC LOADS FOR GASKETS OF DIFFERENT MATERIALS

	Rubber, at p MN/m²	Fluoroplastic	Rubber with cotton fiber	Asbestos, plastics	Paronite	Fiber polyethylene	Metals
q, MN/m²	≤ 3 $1.7p + 1.5$ >3 $0.7p + 4.5$	$p + 4$	10	20	30	40	$1.2\sigma_y$ MN/m²

In connections with self-sealing,

$$b' \geq 0.5D \left(1 - \sqrt{1 - 1.1 \frac{p'}{\sigma_y}}\right), \quad m$$

where D = the inside diameter for supported-area connection seals and outside diameter for unsupported-area seals

σ_y = yield point of sealed surfaces (N/m²)

The unsupported-area seal conditions with gaskets type I–V and VII are provided if the following inequality is fulfilled:

$$b_{ef} < 0.5D \left(1 - \sqrt{1 - \frac{p}{q}}\right)$$

The design seating force for gasket type VI is determined from the following formula:

$$P'_n \geq 3D_k q_{line} \frac{\sin(\alpha + \rho_{f2})}{\cos \rho_{f2}}$$

where D_k = diameter of sealing (contact) of the lens (m)

q_{line} = specific load along the line of sealing (from Figure 8-2)

TABLE 8-7 DESIGN ELEMENTS OF PROFILE GASKETS (TYPE V)

		D_n	b	h	R
			mm		
	A	≤ 200	7.2	6.5	2.7
	B	>200 to 500	11.0	10.0	4.0
	C	>500	16.5	15.0	6.5

Sealing with Gaskets

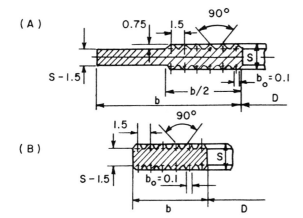

Figure 8-1 Constructive elements of serrated gaskets (type III): (A) obturation type II; (B) obturation type III. The number of teeth depends on value b.

α = slope angle of conical surface for a lens, in grades

ρ_{f2} = 8–10° friction angle of the lens along the conical surface, in grades

The relationship between D_k and q_{line} is given in Figure 8-2. The lens gasket should be checked for tensile strength with the following formulas:

At operating conditions,

$$\frac{Dph}{F} \leq \frac{\sigma_y^t}{1.2}, \quad N/m^2$$

At hydrostatic test,

$$\frac{Dp'h}{F} \leq \frac{\sigma_y}{1.2}, \quad N/m^2$$

where σ_y = yield point for sealed surfaces (N/m²)

F = diametral cross-sectional area of the lens (m²)

Values for D and h are given in Tables 8-8 and 8-9.
 The design seating force from Table 8-1 is

$$p'_n = L_{av} \times b_{ef} \times q = 2.08 \times 0.0141 \times 1.67 \times 10^6 = 4.9 \times 10^4 \text{ N}$$

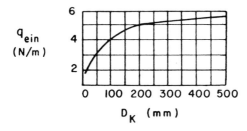

Figure 8-2 Diagram for determination of specific load along the sealing line

TABLE 8-8 CONSTRUCTION ELEMENTS FOR LENS GASKETS (mm)

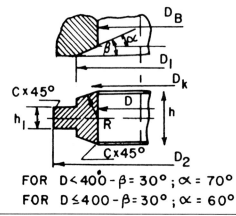

FOR D<400 - β = 30°; α = 70°
FOR D≤400 - β = 30°; α = 60°

D_n	<25	≥25 to 50	>50 to 100	>100 to 200	>200 to 400	>400
D	D_6	$D_6 + 1$	$D_6 + 2$		$D_6 + 3$	
D_k	$D + 4$	$D + 5$	$D + 7$	$D + 10$	$D + 14$	$D + 20$
D_1						
D_2			$\geq D_1 + 2(r + c + 1)$			
R			$\dfrac{D_k}{2 \sin \beta}$			
h			$\geq 2 \left(\sqrt{R^2 - \dfrac{D^2}{4}} - \sqrt{R^2 - \dfrac{D_1^2}{4}} \right) + h_1 + 2r + 2$			
h_1	4	$\dfrac{D_n}{5}$	$\dfrac{D_n}{6}$	$\dfrac{D_n}{7}$	30	10
r	1	2	3	4	5	6
c	0.25	0.5		1.0	1.5	2.0

TABLE 8-9 CONSTRUCTION ELEMENTS FOR SEAL TYPE I (mm)[a]

D_y	h	c
≤300	5	0.5
>300 to 1,000	6 ÷ 8	1.0
>1,000	10	1.5

[a] The selection of $D_b \approx D_n$ is made from design considerations.

SEALING WITHOUT GASKETS

Gasketless sealing is used for high-pressure connections and in cases in which it is impossible to apply gaskets from temperature, corrosion, and other considerations. Tables 8-10, 8-11, 8-12 show construction elements of various seals. The recommended construction elements of basic types of gasketless seals are presented in Tables 8-13, 8-14, 8-15, and 8-16.

Seal type IX with flat polished sealing surfaces may be used for all flanged connections at $D_n > 50$ mm and $p < 150$ MN/m².

Seal type X is used for threaded connections at $D_n < 50$ mm at $p > 10$ up to 100 MN/m².

Seal type XI unsupported-area and supported-area seals are used in flanged and threaded connections at $D_n < 500$ mm and $p > 20$ MN/m². The recommended taper of sealed surfaces is $1/0.866(2\alpha = 60°)$.

Seal type XII unsupported-area and supported-area seals are used for flanged and threaded connections at $D_n < 100$ mm and $p > 10$ MN/m². The taper of sealed surfaces is taken within the limits $2\alpha = 30 - 60°$ for the female part. This provides better sealing in the supported-area seal connections, which should have taper of 2° in excess of the end. The unsupported-area seal connections should be 2° or less.

The hardness of the material of the connecting parts in all gasketless obturations should be different: for the ends, they should be less hard than the nests (females). In obturation type X for pipe connections, both sealed surfaces made according to sketch A (see Table 8-14) are chosen from the same material.

The design seating force of sealed surfaces in obturation type IX is determined from a preceding equation where $q = 4p$ (for $p > 10$ MN/m²) and b_{ef} is taken from Table 8-15 for gasket type I.

The design width of sealing in this obturation is based on the assumption of absence of residual deformations on the sealing surfaces and is determined from the following formula:

$$b' \geqslant \sqrt{\frac{0.33(D + b_0)^2 p'}{\sigma_y - 4.4p} + 0.25D^2} - 0.5D$$

The design seating force of sealing surfaces in seal types X–XII is determined, where q_{line} is taken from Figure 8-2.

In obturation type XII (Table 8-16), the diameter of sealing (contact) D_k is assumed: in connections without self-sealing, $D_k = D_1$.

The design diameter of the end of the next D_1 for supported-area seal obturation is determined from the following formula:

$$D_1' \geqslant D \sqrt{\frac{1.32p' + 4.4 \frac{q_{line}}{D}}{\sigma_y} + 1}, \quad m$$

and for unsupported-area seal connection, from the following formula:

$$D_1' \geqslant D + \frac{0.66p'D + 2.2q_{line}}{\sigma_y - 0.66p'}$$

TABLE 8-10 CONSTRUCTION ELEMENTS OF SEALING SURFACES FOR SEAL TYPE III[a]

P, MN/m²		D	a	b	b_1	h
			mm			
	For nonmetal gaskets of rectangular cross section					
>1.6 to 10		≥400 to 500	8	13	14	5
		>500 to 800	9			
		>800 to 1,000	10	15	16	6
		>1,000 to 1,200	12	16	18	
		>1,200	14	18	20	7
	For metal gaskets of rectangular cross section					
≥10		<50	5	3 ÷ 8		5
		≥50 to 100	6			
		>100 to 200	7	4, 5 ÷ 12		6
		>200 to 400	8	6 ÷ 18		
		>400 to 600	9	8 ÷ 24		8
		>600	10	10 ÷ 30		10

P, MN/m²		D_n	a	b	b_1	h
			mm			
	For gaskets of round cross section					
Vacuum	Nonmetal	≤1,500	Depending on D_n, see above	d see Table 8-4	at $d ≤ 4$ $d + 0.5$	at $d > 1$ $d + 1$
≥2.5 to 10	Metal	≤1,000				$d + 2$

[a] The selection $D_b ≈ D_n$ is made from design considerations. Value $b = b_1$ is determined from formula.

Sealing Without Gaskets

TABLE 8-11 CONSTRUCTION ELEMENTS OF SEALING SURFACES FOR SEAL TYPE IX[a,b]

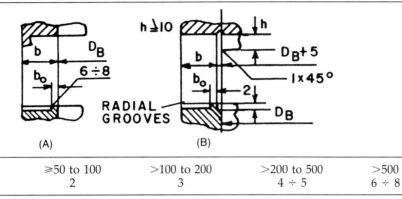

D_n, mm	≥50 to 100	>100 to 200	>200 to 500	>500
b_0, mm	2	3	4 ÷ 5	6 ÷ 8

[a] The selection $D_b \simeq D_n$ is made on the basis of design considerations.
[b] b' is determined by calculation.

TABLE 8-12 CONSTRUCTION ELEMENTS OF SEALING SURFACES FOR SEAL TYPE X (mm)[a,b,c]

	D_n	≤10	>10 to 20	>20 to 50
	D_k	$D_b + 4$	$D_b + 6$	$D_b + 8$
	R	4	5	6
	b	4	6	8

[a] The sealing surfaces of pipe connections are fabricated according to (A).
[b] For connecting pipe to apparatus, the surface from the pipe side is made according to (A) and, from apparatus side, according to (B).
[c] The selection $D_b \simeq D_n$ is made on the basis of design considerations.

TABLE 8-13 CONSTRUCTION ELEMENTS OF SEALING SURFACES FOR SEAL TYPE XI (MM)[a]

D_k	$D_b + 1.5(0.05 D_b + 5)$
R	$0.578 D_k$
D_1	$D_b + 2(0.05 D_b + 5)$
h	$0.05 D_b + 8$
h_1	$0.5 R + 0.4(0.05 D_b + 5)$
h_2	$0.167(0.05 D_b + 5) + 3$

[a] The selection $D_b \simeq D_n$ is made from design considerations.

TABLE 8-14 CONSTRUCTION ELEMENTS OF SEALING SURFACES FOR SEAL TYPE XII[a]

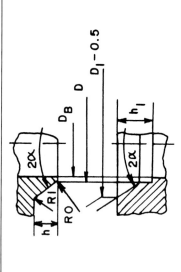

Values		For connection cups, cover, and so on	For connection pipes, nipples, and so on
α	Grades	$15 \div 30$	
α_1		For connection without self-sealing—$\alpha - 1$.	
		For connection with self-sealing—$\alpha + 1$.	
D		D_b	$D_b + (0.05\, D_b + 4)$
D_1		For connection without self-sealing—	
		For connection with self-sealing—no less $D + 5$.	
h	mm	$\geq \dfrac{D_1 - D}{2\,\text{tang}\,\alpha} + 5$	
h_1			$\geq \dfrac{(D_1 - D) + (0.05\, D_b + 4)}{2\,\text{tang}\,\alpha}$

[a] The selection of $D_b \simeq D_n$ is made on the basis of design considerations.

Special Obturation

TABLE 8-15 DESIGN ELEMENTS OF SEAL TYPE XIII[a]

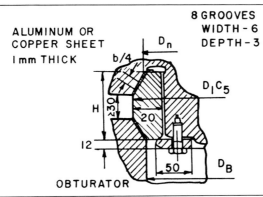

D_n, mm	500 to 800		>800 to 1,000 D		>1,000	
P, MN/m²	<70	≥70	<70	≥70	<70	≥70
s	30	35	40	50	50	60
b			From formula but no less than 40			
	mm					
D_1			$D_b + b$, but no less than 40			
D_n			$\dfrac{D_1 + D_b}{2}$			
H			$b \tang 60° + 30$			

[a] The selection of $D_b > D_n$ is made on the basis of design considerations.

The unsupported-area seal obturation types XI and XII provide a reliable performance when the following condition is fulfilled:

$$q_{line} < 0.25 D_k p \frac{\cos \rho_{fr}}{\sin(\alpha + \rho_{fr})}, \quad N/m$$

SPECIAL OBTURATION

Special obturation is used mostly for high-pressure connections. Performance under high pressure requires careful assembly of the apparatus. The quality of this assembly depends completely on the quality of the manufactured parts of the installation. Especially strict requirements are placed on threads and sealing surfaces. Threads should be rounded, that is, without sharp crests. To reduce the danger of so-called *seizing*, threads should not be dimensioned the same as the sealing surface. Usually, the separation distance left should be 3–5 mm.

Sealing surfaces require equally careful manufacture. The slightest scratch, groove, or crack will break the hermetic sealing of the obturator. Hence, sealing surfaces usually are polished.

It is extremely important that the plane of the sealing element be strictly perpendicular to the axis of the thread; that is, the tightening device causes the sealing surface to contact the part simultaneously at all points on the circumference. This requirement must be met more strictly for the harder parts being sealed. The better this requirement is fulfilled, the less force that is required for sealing.

The internal surfaces of high-pressure apparatus sealing rings also should be made with extreme care. Some applications require polishing to within accuracies of thousandths of a millimeter. The higher the pressure, the greater the possibility for penetration of the pressure-transmitting medium into microfissures, gaps, and irregularities on surfaces. Finally, stresses are concentrated at fissures, cracks, sharp angles, and so on, which may lead to the rupture of parts.

Special attention should be given to concentricity or eccentricity of mating apertures. Strict concentricity will provide proper operation of the obturator. The cylindricity or conicity of holes or parts is no less important. If the tolerances of conicity are not observed, the operation of sealing lenses is impeded.

Inasmuch as the sealing will be compressed when it moves toward the smaller diameter, it will be ineffective in the opposite direction of movement.

SPECIAL SEALS

Special seals are used mostly for high-pressure connections. Performance under high pressure requires careful assembly of the apparatus. The quality of this assembly depends completely on the quality of the manufactured parts of the installation. Especially strict requirements are placed on threads and sealing surfaces. Threads should be rounded, that is, without sharp crests. To reduce the danger of seizing, threads should not be dimensioned the same as the sealing surface. Usually, the separation distance left should be 3–5 mm.

Sealing surfaces require equally careful manufacture. The slightest scratch, groove or crack will break the hermetic sealing of the obturator. Hence, sealing surfaces usually are polished.

It is extremely important that the plane of the sealing element be strictly perpendicular to the axis of the thread; that is, the tightening device causes the sealing surface to contact the part simultaneously at all points on the circumference. This requirement must be met more strictly for the harder parts being sealed. The better this requirement is fulfilled, the less force that is required for sealing.

The internal surfaces of high-pressure apparatus sealing rings also should be made with extreme care. Some applications require polishing to within accuracies of thousandths of a millimeter. The higher the pressure, the greater the possibility for penetration of the pressure-transmitting medium into microfissures, gaps, and irregularities on surfaces. Finally, stresses are concentrated at fissures, cracks, sharp angles, and so on, which may lead to the rupture of parts.

Special attention should be given to concentricity or eccentricity of mating apertures. Strict concentricity will provide proper operation of the obturator. The cyclindricity or conicity of holes or parts is no less important. If the tolerances of conicity are not observed, the operation of sealing lenses is impeded.

Inasmuch as the sealing will be compressed when it moves toward the smaller diameter, it will be ineffective in the opposite direction of movement.

First we will consider the sealing with gaskets of soft metal. For example, in the closure presented in Figure 8-3 the gasket fabricated of annealed copper, aluminum, or Armco iron is crushed with main studs. This connection is simple in fabrication and operation but may be used for cold apparatuses with diameters up to 600 mm, which are opened approximately once a year. The closure with a machined copper gasket of triangular cross section, where a screw socket is used instead of studs, is shown in Figure 8-4. The upper part of the screw is milled (Figure 8-5), forming a bayonet lock. This permits rapid opening and closing of the head by turning it at the angle equal to that of the milled sector. The location of socket and head after assembly is fixed with a slide block.

Crushing the gasket is accomplished by special studs that are screwed in the head and the gasket pressed into the ring. This design, operating up to 850

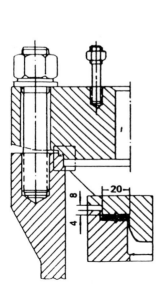

Figure 8-3 Closure with a flat metal gasket sealed with main studs

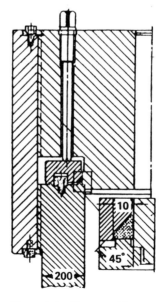

Figure 8-4 Closure with a metal gasket of triangular cross section sealed with special studs to the main fixture with a screw socket

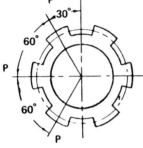

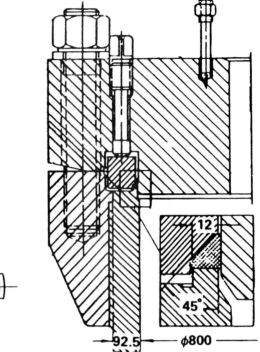

Figure 8-5 Milling of thread in the upper part of the socket

Figure 8-6 Closure with a metal gasket of triangular cross section sealed with studs of main fixing with short studs for columns up to 1,200 mm in diameter and pressures up to 350 atm

atm, permits application of narrow gaskets 8–12 mm wide and, as a result, it is possible to decrease the cross section of studs that crushes the gasket.

The closure operating up to 350 atm with a machined copper gasket for a column 800 mm in diameter is given in Figure 8-6, which is somewhat simpler in design than the closure in Figure 8-4.

Figure 8-7 shows a closure in which the stress from the gas pressure acting on the top head is absorbed with socket consisting of two fixed halves. The sealing is achieved owing to the crushing of a copper gasket with a pressing ring, which, in turn, is under action of a pressing bolt. The height of the socket in this design is lower than that of Figure 8-4 because in the bayonet lock half the thread is cut away.

The unsupported-area seals (self-energizing seals have found a wide application in high-pressure closure designs) develop a force reaction on an area smaller than that exposed to the pressure medium (see Figure 8-7). Therefore, the pressure received by the projected area of the head is transmitted through

Special Seals

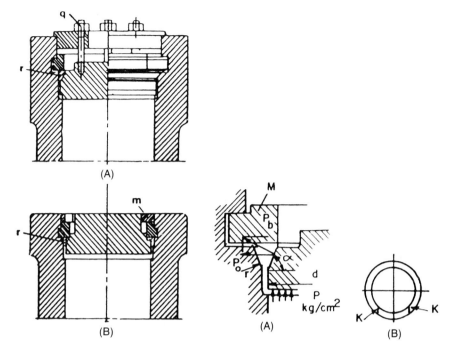

Figure 8-7 Closure of unsupported-area type

Figure 8-8 Diagram of forces acting on a trapezoidal ring

the head to the area of sealing that is less than the area of the head; consequently, the stresses generated in the obturator are greater than the working pressure. In this case, the greater the pressure the greater the sealing of the connection. The force of internal pressure presses down the head of the apparatus to the sealing ring, r, of the trapezoidal square section. The shape of the sealing ring determines the appearance of radial force, p_o, acting on the shell of the apparatus through the sealing ring and generating in the last tensile and crumpling stresses (shown in Figure 8-8).

The force $p_o = P_1$ tan may reach large values at high values of angle and, in the final analysis, must be accepted by the shell of the apparatus due to the ring deformation.

The force p_1 is equal to

$$p_1 = p \frac{\pi d^2}{4} \times \frac{1}{\pi d} = p \frac{d}{4}, \quad \text{kg/cm}$$

where p is the internal pressure (kg/cm²).

The force, P_1, induces thrust collar M to work by bending and shear (Figure 8-8). The thrust collar is detachable and consists of three to four parts, depending on the size of an apparatus. This collar enters into the groove of the shell and disassembles easily. Besides the collar, for example, consisting of three parts, two planes of parting are made not in the radial direction but as is shown by *K* in Figure 8-8B. Before starting the unit for generating internal pressure, it is necessary to provide a preliminary tightening, which is made with studs, *q* (Figure 8-7A) or with a nut, *m* (Figure 8-7B) at small diameters of the apparatus.

In the generation of the internal pressure, all the load from the nuts is taken away. The unsupported-area seals can withstand especially high pressures because the seating force increases with pressure. Such an obturation provides reliability and hermetic nature at high temperatures and lowers the price of the design.

The deterioration of tightness is possible along two cone surfaces of ring 2 (Figure 8-8A). However, the force, p_o, is usually so great that it provides the safety sealing.

Sometimes the contact of the sealing surfaces is made along the obturating band close to the contact line, or gasket spherical side surface and cone nest, or at two cone surfaces with different taper, or at two spherical surfaces with

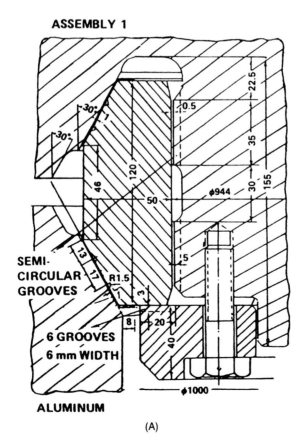

Figure 8-9 (A) Seal with a double-cone lens for the cover of a high-pressure vessel

Special Seals

different radii of spheres. In all these cases, there arises a considerable specific pressure in the place of sealing.

The closures presented in Figures 8-9 and 8-10 are used widely in high-pressure equipment. These closures have obturators, K with two cone surfaces, 1, and two ring grooves, p, on each surface. Each seal ring is furnished with aluminum gaskets 1 mm thick to provide tightness.

The stress in aluminum gaskets is higher than the yield point, and the aluminum-filling microgap provides the tightness of connection. The obturators are fixed to the covers of the apparatus with the rings, m, and bolts, r, in Figure 8-10. There is a gap, c, for providing the desired screwing up between the cover and end of an apparatus. A small increase in an obturator diameter induced by the internal pressure provides a better tightening of all connections.

Connection with double-cone seals permits a reduction to a minimum in the bending moment acting on the flange because of a decrease in the lever arm

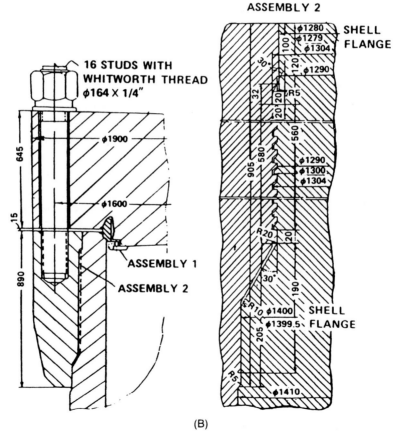

Figure 8-9 (B) Seals with a double-cone lens for the cover of a high-pressure vessel

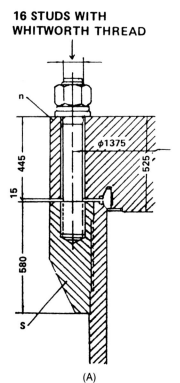

Figure 8-10 (A) Seal with a double-cone lens for the cover of a high-pressure vessel

between the lens and bolt circle and a decrease in the vertical component of the reactive lens force.

The application of threaded flanges eliminates the influence of flange bending moments on the cylindrical shell of the vessel and makes possible the fabrication of cylindrical parts of the shell without reinforcement (that is, it is then possible to make the shell out of one wall thickness along the vessel).

The recommended construction elements typical for unsupported-area obturations with an obturator located between the sealed surfaces (which are mostly used for connection covers of apparatus) are given in Tables 8-15 and 8-16.

Obturation type X11 is used at $D_b \geq 500$ mm up to 1,500 mm for $p \geq 30$ up to 100 MN/m². The material of the sealing surfaces (shell and cover) and obturator may be the same for $\sigma_y \geq 6p$.

The design seating force for a given obturation is determined from the following formula:

$$P'_n \geq 3D_n b_{eq} q \frac{\sin(\alpha + \rho_{f_2})}{\cos \rho_{f_2}}, \quad N$$

Special Seals

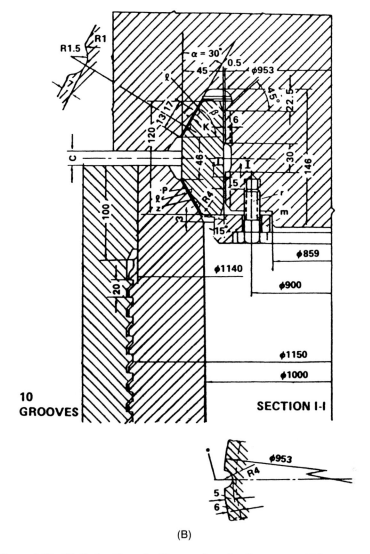

Figure 8-10 (B) Seal with a double-cone lens for the cover a high-pressure vessel

where D_n = mean diameter of sealing (m)

b_{eq} = 0.75b (m) (see Table 8-17)

$q = p$ (N/m²)

pf_2 = 8–10° friction angle of lens along the cone surface, grades

α = slope angle of conical surface for the lens

The design width of sealing, b', is determined from the condition or absence of residual deformations of crumpling on the sealed surfaces from the following formula:

$$P'_0 \geq 3D_1 b_0 \sigma_y \frac{\sin(\alpha + \rho_{f_2})}{\cos \rho_{f_2}}$$

Obturation type XIV (see Table 8-18) is used for $D_b > 50$ mm and $p \geq 30$ MN/m². The material of the obturator with $\sigma_y \geq 2p$ should be softer than the sealed surfaces (shell and cover). It can be used in several assemblies, but a preliminary polishing of sealed surfaces is required before each assembly. The sequence of assembly is as follows: (1) a slight pressing down of the obturator from outside, then (2) a pressing of the obturator with the thrust on it at the same time. As a result, the obturator is expanded in the radial direction providing the preliminary sealing on both cone surfaces.

The pressure of the medium in the apparatus with the thrust on the obturator from outside will evenly wedge up the obturator and thereby improve the tightness of the connection.

The design force of the preliminary pressing down of the cover to the obturator is determined from the following formulas:

TABLE 8-16 CONSTRUCTION ELEMENTS OF SEAL TYPE XIV[a]

α			$18 \div 24$	D_2		$D_k + (2 \div 3)$
α_1		Grade	$\alpha - 1$	D_3		From formula
β			$12 \div 15$	a	mm	$1 \div 2$
β_1			$\beta - 1$	h		$\dfrac{D_3 - D_2}{2 \tang \alpha} + h_1$
D			From formula	h_1		$3 \div 5$
D_k		mm	$D_b + (2 \div 5)$	h_2		$5 \div 8$
D_1			$D_2 - 2a$			
				F	m²	$\approx \dfrac{(D_1 - D)^2*}{4 \tang \alpha}$

[a] The selection $D_b \simeq D_n$ is made on the basis of design considerations.

$$P_1 = \frac{0.78 D_1^2 p' \cos \rho_{f_2}}{\sin(\alpha + \rho_{f_2})}, \quad N$$

$$P_2 = \frac{\cos(\alpha + \rho_{f_2})}{\cos(\beta + \rho_{f_2})} - 6.3 F \sigma_y \frac{\cos \rho_{f_2}}{\cos(\beta - \rho_{f_2})}, \quad N$$

$$P_4 = P_1 \frac{\sin(\alpha + \beta) \cos \rho_{f_2}}{\cos(\beta - \rho_{f_2})} + P_3 - 6.3 \sigma_y \tan(\beta - \rho_{f_2}), \quad N$$

$$P_3 = 0.78 (D_2^2 - D_1^2) p', \quad N$$

$$D \leq \sqrt{D_1^2 - \frac{1.4 P_1 \sin \alpha}{\sigma_y}}, \quad m$$

$$D_3 \geq \sqrt{D_2^2 + \frac{1.4 P_2 \sin \beta}{\sigma_y}}, \quad m$$

NOMENCLATURE

b_{ef}	effective design width of sealing (m)
D_k	diameter of sealing (m)
D_n	central diameter of sealing (m)
F	area (m²)
h	height (m)
p	design pressure (N/m²)
q	specific load (N/m²)
Z	number of bolts
σ	stress (N/m²)
σ_y	yield point of sealed surfaces (N/m²)
α	slope angle of conical surface for a lens, grade
ρ_{f_2}	friction angle of lens, grades

9

Flanges and Threaded Connections

REQUIREMENTS FOR FLANGE CONNECTIONS

Among the variety of attachments and accessories essential to vessels and piping, flanges are perhaps the most widely used items. They are employed on the shell of a vessel to provide for disassembly and removal or cleaning of internal parts. There may be usually several flanged openings in any vessel: inlets and outlets for products and heat carriers, manholes for installing working devices and inspection, and so on. Flanges also are typically used for making connections for piping and for nozzle attachments of openings larger than 1 1/2 in. nominal pipe size.

The need for detachable connections results from operating considerations, but also arises for reasons of equipment fabrication (such as the need in casting or forge-and-pressing shops) machine sizes, crane lifting capacity, and others. Even when production possibilities permit fabrication of a vessel of a desired size in one piece, it is often made of detachable parts for convenience of assembly and maintenance.

Flanges used for vessels and piping must conform to the following requirements:

1. They must provide tightness under given operating conditions.
2. They must provide sufficient strength.

Flange Selection

TABLE 9-1 CARBON-STEEL FLANGE PRESSURE RATINGS IN WATER, STEAM, AND OIL SERVICE (VALUES ARE FOR FLANGES WITH STANDARD FACINGS; EXCLUDED ARE RING JOINTS)

		\multicolumn{7}{c}{Primary service pressure ratings (psi)}
		150	300	400	600	900	1,500	2,500
		\multicolumn{7}{c}{Maximum hydrostatic-shell-test pressures (temperature of test water should not exceed 125°F) (psi)}
		350	900	1,200	1,800	2,700	4,500	7,500
\multicolumn{2}{l}{Service temperatures}							
(°F)	(°C)	\multicolumn{7}{c}{Maximum nonshock service-pressure ratings (psi)}
100	37.8	230	600	800	1,200	1,800	3,000	5,000
200	93.3	210	580	770	1,160	1,740	2,900	1,830
300	148.9	190	560	740	1,120	1,680	2,800	4,660
400	204.4	170	540	710	1,075	1,615	2,690	4,475
500	260.0	150[a]	500	665	1,000	1,500	2,500	4,465
600	315.6	130	445	590	890	1,330	2,220	3,700
700	371.1	110	380	500	760	1,140	1,900	3,160
800	426.7	92	300[a]	400[a]	600[a]	900[a]	1,500[a]	2,500[a]

[a] Primary service pressure ratings.

3. They must permit quick and frequent assembly and disassembly.
4. They must provide mass production, with standard specifications at reasonable cost.

To avoid fabricating flanges for each service pressure, pipe, and shell diameter, a rating system was developed. This rating is as follows: 150-, 300-, 400-, 600-, 900-, 1,500-, and 2,500-lb flanges.

The ratings corresponding to service pressures and temperatures are given in Table 9-1. The table shows a decrease in the rating of flanges with temperature. For example, a 150-lb flange has a rating of 230 psi at 100°F and a rating of 150 psi service pressure at 500°F. A 2,500-lb flange has a rating of 5,000 psi at 100°F and 2,500 psi at 800°F.

FLANGE SELECTION

A variety of types and sizes of flanges is available for various temperature and pressure services. Flanges differ in design, method of attachment to the pipe or shell, and form (round, square, irregularly shaped), as well as in their facing surfaces.

The flanges from forged steel or cast iron (for low-pressure service) are fabricated in accordance with American Standard Association. Forged-steel flanges are manufactured in the following standard types for all pressure ratings:

- Welding-neck flanges
- Slip-on flanges
- Screwed flanges
- Lap-joint flanges
- Blind flanges

Welding-Neck Flanges

Figure 9-1 and its accompanying table (Table 9-2) show a flange sectional view and list the dimensions of several standard 150-lb welding-neck flanges from 1/2-in. to 24-in. nominal pipe size.

Welding-neck flanges have a long tapered hub, which provides a gradual transition from the flange ring to the pipe wall. This design decreases the discontinuity stresses and increases the strength of the flange. This type of flange works together with a shell (pipe). Because of the integral connection with a shell, it is more unloaded and may be fabricated thinner than slip-on flanges.

Welding-neck flanges are used for extreme service conditions, such as large mechanical loads, wide fluctuations in process parameters, and high and low temperatures. These flanges are preferred for handling flammable or explosive fluids, where failure or leakage of a flange joint might bring disastrous consequences.

Slip-On Flanges

This type of a flange has no rigid connection with a pipe; therefore, it is more loaded and, consequently, somewhat thicker than a welding-neck flange. The slip-on type of flange, which is presented in Figure 9-2 and Table 9-3 (the sketch

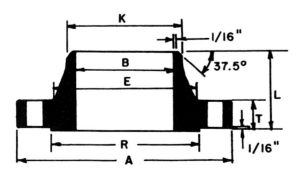

Figure 9-1 Dimensions of welding-neck flanges (refer to Table 9-2)

TABLE 9-2 DIMENSIONS OF STANDARD 150-LB STEEL WELDING-NECK FLANGES, AMERICAN STANDARD-ASA B17e-1939—FORGED AND ROLLED STEEL—ASTM A 181

Nominal pipe size (in.)	Outside diameter of flange A	Thickness of flange (minimum) T	Outside diameter of raised face R	Diameter of hub at base E	Diameter of hub at point of welding K	Length through hub L	Inside diameter of standard wall pipe B	Number of holes	Drilling template Diameter of holes	Diameter of bolts	Bolt circle	Approximate weight each (lb)
1/2	3 1/2	7/16	1 3/8	1 3/16	0.81	1 7/8	0.62	4	5/8	1/2	2 5/8	2
1	4 1/4	9/16	2	1 15/16	1.32	2 3/16	1.05	4	5/8	1/2	3 1/8	2
1 1/2	5	11/16	2 7/8	2 9/16	1.90	2 7/16	1.61	4	5/8	1/2	3 7/8	4
2	6	3/4	3 5/8	3 1/16	2.38	2 1/2	2.07	4	3/4	5/8	4 3/4	6
2 1/2	7	7/8	4 1/8	3 9/16	2.88	2 3/4	2.47	4	3/4	5/8	5 1/2	8
3	7 1/2	15/16	5	4 1/4	3.50	2 3/4	3.07	4	3/4	5/8	6	10
4	9	15/16	6 3/16	5 5/16	4.50	3	4.03	8	3/4	5/8	7 1/2	15
6	11	1	8 1/2	7 9/16	6.63	3 1/2	6.07	8	7/8	3/4	9 1/2	24
8	13 1/2	1 1/8	10 5/8	9 11/16	8.63	4	7.98	8	7/8	3/4	11 3/4	39
10	16	1 3/16	12 3/4	12	10.75	4	10.02	12	1	7/8	14 1/4	52
12	19	1 1/4	15	14 3/8	12.75	4 1/2	12.00	12	1	7/8	17	80
14	21	1 3/8	16 1/4	15 3/4	14.00	5	13.25	12	1 1/8	1	18 3/4	102
16	23 1/2	1 7/16	18 1/2	18	16.00	5	15.25	16	1 1/8	1	21 1/4	127
20	27 1/2	1 11/16	23	22	20.00	5 11/16	19.25	20	1 1/4	1 1/8	25	170
24	32	1 7/8	27 1/4	26 1/8	24.00	6	23.25	20	1 3/8	1 1/4	29 1/2	260

TABLE 9-3 DIMENSIONS OF STANDARD 150-LB FORGED SLIP-ON FLANGES (REFER TO FIGURE 9-2), AMERICAN STANDARD-ASA B16.3e-1939—FORGED AND ROLLED STEEL—ASTM A 181

Nominal pipe size (in.)	Outside diameter of flange A	Thickness of flange (minimum) T	Outside diameter of raised face R	Diameter of hub at base E	Length through hub L	Diameter of bore B	Number of holes	Drilling template		Bolt circle	Approximate weight each (lb)
								Diameter of holes	Diameter of bolts		
1/2	3 1/2	7/16	1 3/8	1 3/16	5/8	0.88	4	5/8	1/2	2 3/8	2
3/4	3 7/8	1/2	1 11/16	1 1/2	5/8	1.09	4	5/8	1/2	2 3/4	2
1	4 1/4	9/16	2	1 15/16	11/16	1.38	4	5/8	1/2	3 1/8	2
1 1/2	5	11/16	2 7/8	2 9/16	7/8	1.97	4	5/8	1/2	3 7/8	3
2	6	3/4	3 5/8	3 1/16	1	2.44	4	3/4	5/8	4 3/4	5
2 1/2	7	7/8	4 1/8	3 9/16	1 1/8	2.94	4	3/4	5/8	5 1/2	7
3	7 1/2	15/16	5	4 1/4	1 3/16	3.56	4	3/4	5/8	6	8
4	9	15/16	6 3/16	5 5/16	15/16	4.56	8	3/4	5/8	7 1/2	13
6	11	1	8 1/2	7 9/16	1 9/16	6.72	8	7/8	3/4	9 1/2	19
8	13 1/2	1 1/8	10 5/8	9 11/16	1 3/4	8.72	8	7/8	3/4	11 3/4	30
10	16	1 3/16	12 3/4	12	1 15/16	10.88	12	1	7/8	14 1/4	43
12	19	1 1/4	15	14 3/8	2 3/16	12.88	12	1	7/8	17	64
14	21	1 3/8	16 1/4	15 3/4	2 1/4	14.19	12	1 1/8	1	18 3/4	85
16	23 1/2	1 7/16	18 1/2	18	2 1/2	16.19	16	1 1/8	1	21 1/4	93
20	27 1/2	1 11/16	23	22	2 7/8	20.19	20	1 1/4	1 1/8	25	155
24	32	1 7/8	27 1/4	26 1/8	3 1/4	24.19	20	1 3/8	1 1/4	29 1/2	210

Flange Selection

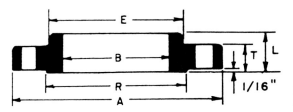

Figure 9-2 Dimensions of forged slip-on flanges

of a standard 150-lb slip-on flange and dimensions for such flanges of 1/2-in. to 24-in. nominal pipe sizes), is widely used because of its greater ease of alignment in welding assembly and because of its low initial cost.

The strength of a slip-on flange as calculated from internal pressure considerations is approximately two thirds that of a corresponding welding-neck type of flange. The use of this type of flange should be limited to moderate services in which pressure and temperature fluctuations, vibrations, and shock are not expected to be severe. It also is used for equipment and piping from mild nonferrous metals, such as aluminum and copper, or for brittle material (ferrosilicon, ceramics), as well as in those cases in which it is necessary to save expensive construction materials as, for example, high-alloy steels and nonferrous alloys.

Lap-Joint Flanges

The lap-joint flange (Figure 9-3) is used for extreme service conditions, to high pressures of up to several hundred atmospheres and temperatures up to 530°C. The principal advantage of this flange is that the bolt holes are easily aligned. This simplifies the erection of large-diameter vessels and unusually stiff piping.

However, the lap-joint flange costs about 30 percent more than a welding-neck flange of the same size and rating, but has only about 10 percent the fatigue life.

Screwed Flanges

Screwed flanges (Figure 9-4) are used in extremely high-pressure service with alloy steel, which is not easily welded. They are employed in applications involving no bending or thermal cycles. For these conditions, the flanges are susceptible to leakage.

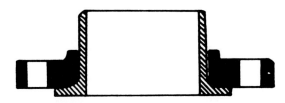

Figure 9-3 Sectional view of steel lap-joint flange and lap-joint stub

Figure 9-4 Sectional view of a screwed flange

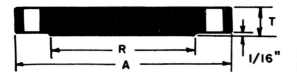

Figure 9-5 Dimensions of a blind flange (refer to Table 9-4)

Blind Flanges

This flange (Figure 9-5 and Table 9-4 give dimensions for such flanges of 1/2-in. to 24-in. nominal pipe size) is used for blanking off pressure vessel openings. The blind flange also is used to block off the ends of piping and valves. A valve followed by a blind flange permits addition to the line which it is "on stream."

NONSTANDARD FLANGES

Nonstandard large-diameter welding-neck or slip-on flanges are fabricated by rolling a hot annular blank in sizes ranging from 26 in. to 96 in. in a variety of ratings. As an example, Figure 9-6 and Table 9-5 give a sectional view of a welding-neck flange and the dimensions for welding-neck and slip-on flanges, respectively.

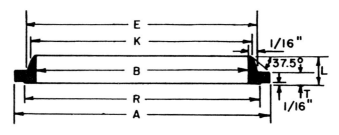

Figure 9-6 Nonstandard flange (refer to Table 9-5)

TABLE 9-4 DIMENSIONS OF STANDARD 150-LB FORGED SLIP-ON FLANGES (REFER TO FIGURE 9-2), AMERICAN STANDARD-ASA B16.5e-1939—FORGED AND ROLLED STEEL—AST A 181

Nominal pipe size (in.)	Outside diameter of flange A	Thickness of flange (minimum) T	Outside diameter of raised face R	Drilling Template					Approximate weight each (lb)
				Number of holes	Diameter of holes	Diameter of bolts	Bolt circle	
1/2	3 1/2	7/16	1 3/8	4	5/8	1/2	2 3/8	2
3/4	3 7/8	1/2	1 11/16	4	5/8	1/2	2 3/8	2
1	4 1/4	9/16	2	4	5/8	1/2	3 1/8	2
1 1/2	5	11/16	2 7/8	4	5/8	1/2	3 7/8	3
2	6	3/4	3 5/8	4	3/4	5/8	4 3/4	4
2 1/2	7	7/8	4 1/8	4	3/4	5/8	5 1/2	7
3	7 1/2	15/16	5	4	3/4	5/8	6	9
4	9	15/16	6 3/16	8	3/4	5/8	7 1/2	17
6	11	1	8 1/2	8	7/8	3/4	9 1/2	26
8	13 1/2	1 1/8	10 5/8	8	7/8	3/4	11 3/4	45
10	16	1 3/16	12 3/4	12	1	7/8	14 1/4	70
12	19	1 1/4	15	12	1	7/8	17	110
14	21	1 3/8	16 1/4	12	1 1/8	1	18 3/4	311
16	23 1/2	1 7/16	18 1/2	16	1 1/8	1	21 1/4	170
20	27 1/2	1 11/16	23	20	1 1/4	1 1/8	25	272
24	32	1 7/8	27 1/4	20	1 3/8	1 1/4	29 1/2	411

TABLE 9-5 DIMENSIONS OF NONSTANDARD LARGE-DIAMETER FLANGES, FORGED AND ROLLED STEEL

Size (in.)	Outside diameter of flange A	Thickness of flange T	Outside diameter of raised face R	Compressed asbestos gasket size	Taper-hub diameters E	K	Length through hub L	Diameter of bore B	Drilling Template: Number of holes	Diameter of holes	Bolt circle	Approximate weight each (lb)
50-lb Pressure at 100°F (welding-neck type)												
26	31½	1¼	28⅝	27¾ × 28⅝	27⅛	26½	3	26	32	1	29⅝	98
30	35½	1¼	32⅝	31¾ × 32⅝	31⅛	30½	3	30	36	1	33⅝	112
34	40¼	1¼	37	35⅞ × 37	35⅜	34⅝	3¼	34	40	1⅛	38⅛	149
36	42¼	1¼	39	37⅞ × 39	37⅜	36⅝	3¼	36	40	1⅛	40⅛	157
42	49	1¼	45½	44⅛ × 45½	43¾	42¾	3½	42	48	1¼	46¾	209
48	55	1¼	51½	50⅛ × 51½	49¾	48¾	3¾	48	52	1¼	52¾	214
54	61¼	1⅜	57¾	56¼ × 57¾	56	54⅞	4	54	61	1¼	59	312
60	67¼	1⅝	63¾	62¼ × 63¾	62	60⅞	4⅜	60	72	1¼	65	398
66	74	1⅞	70⅛	68¾ × 70⅛	68	67	4⅞	66	72	1⅜	71½	556
72	80	2¼	76⅛	74⅜ × 76⅛	74	73	5¼	72	80	1⅜	77½	705
50-lb Pressure at 100°F (slip-on type)												
26	33	1¼	30	29⅛ × 30	28½	28	2¼	26$\frac{11}{16}$	32	1	31	122
30	37	1¼	34	33⅛ × 34	32½	32	2¼	30$\frac{11}{16}$	36	1	35	148
31	41½	1¼	38¼	37⅛ × 38¼	36⅝	36⅛	2½	34$\frac{13}{16}$	40	1⅛	39⅜	181
36	43½	1¼	40¼	39⅛ × 40¼	38⅝	38⅛	2½	36$\frac{13}{16}$	44	1⅛	44⅜	191
42	50	1¼	46½	45⅛ × 46½	44¾	44¼	2¾	42$\frac{15}{16}$	48	1¼	47¾	234
48	56	1¼	52½	51⅛ × 52½	50¾	50¼	2⅞	48$\frac{15}{16}$	56	1¼	53¾	269
54	62½	1⅜	59	57½ × 59	57¼	56½	3⅛	55$\frac{1}{16}$	68	1¼	60¼	335
60	68½	1⅝	65	63½ × 65	63¼	62½	3⅝	61$\frac{1}{16}$	72	1¼	66¼	451
66	75½	1¾	71⅝	69⅞ × 71⅝	69½	68¾	4	67$\frac{3}{16}$	72	1⅜	73	591
72	81½	2	77⅝	75⅞ × 77⅝	75½	74¾	4½	73$\frac{3}{16}$	80	1⅜	79	728

Flanges of Special Types 325

FLANGES OF SPECIAL TYPES

Square, oval, and irregularly shaped flanges sometimes are used for pipe connections and nozzle attachments of openings no larger than 1-in. nominal pipe size. This type of flange is common in high-pressure and cryogenic equipment.

Welded Around Flange Connections

In vessels used for processing highly toxic substances any leakage is unacceptable. Therefore, it is advisable to design all welded equipment without flanges and attach the piping to the equipment by welding. However, if in such equipment the flange connection is inevitable, such as for a manhole, for example, it is carried out without gaskets but with the use of a thickening seam (Figure 9-7).

For disassembly, the seam is broken through and then welded again for assembly. The thickening seam holds out six to ten assemblies. It is then necessary to change the lap-joint stubs. In designing such connections, the strength of a seam is not taken into consideration as it serves only for sealing. The trapezoidal form of the flange section (Figure 9-7) is made to decrease the flange weight. In this case, the flange strength is only slightly decreased because the outer flange fibers are less loaded.

Detachable Flanges

To disassemble a vessel, it is sometimes necessary to detach the flange from a pipe. Screwed flanges often are used in this case. However, it is generally not the best solution to this problem because flanges become so rusted that they cannot be unscrewed. It is often better to use detachable flanges with a split ring set in a pipe groove, as shown in Figure 9-8. The detachable flanges for lap-joint pipes are fabricated sectionally, either in two- or in four-ring parts, as shown in Figure 9-9.

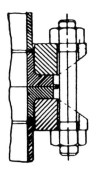
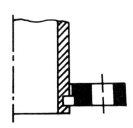

Figure 9-7 Flange with a thickening seam

Figure 9-8 Detachable flange with a split ring

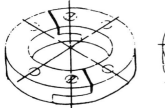 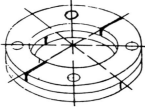

Figure 9-9 Detachable flange in two and four sections

This flange design is simpler in fabrication and less susceptible to corrosion than are the screwed flanges.

Flanges for Pipes and Shells from Fragile Materials

It is not advisable that flanges for pipes and vessels from ferrosilicon, ceramics, glass, porcelain, polyvinyl chloride (PVC) plastics, and similar materials be fabricated integrally with equipment because of their inadequate flexing resistance. The ends of pipes, shells, and heads from such materials are made with thick conical parts on which special flanges are placed. They, as well as the detachable flanges, are made in two fashions: (1) detachable flanges with two-ring parts, and (2) flanges with split rings.

Figure 9-10 shows the detachable flange from malleable cast iron, where its two halves are fixed by bolts.

Figure 9-11 shows a sectional view of a steel flange with a bore somewhat larger than a thicker part of a pipe (shell), which leans against the two cast-iron

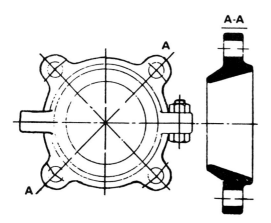

Figure 9-10 Detachable flange for ceramic pipe

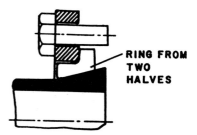

Figure 9-11 Flange with a split ring for pipes for siliceous cast iron

Flange Facings

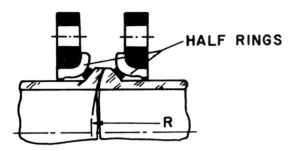

Figure 9-12 Connection for glass pipe

half rings. Flanges with split rings have higher strength and are cheaper than the detachable flanges. They are, however, less compact.

The detachable flanges from malleable cast iron are used in the case in which smaller sizes permit design of a more compact vessel, for example, a sprinkling cooler from siliceous cast-iron pipes, and so on. A flange design with split rings for glass pipe connections is presented in Figure 9-12. The half rings are made from polyamide plastics; the gasket between the polished pipe sections and thick pipe ends is made from rubber. The spherical edge joints permit pipe connections to be made at an angle extending to 20°.

FLANGE FACINGS

Standard flanges are used with a variety of machined faces, as shown in Figure 9-13 (corresponding dimensions are given in Table 9-6). The raised face flanges are used extensively because of the simplicity of their design and adequacy for average service conditions. Flanges with ratings of 750 lb and 300 lb have 1/16-in.-high raised faces, and those for higher ratings have 1/4-in.-high raised faces. For better yield and holding, a gasket is placed over the raised fall of a flange, which is machined with spiral or concentric grooves (approximately 1/64 in. deep).

The male and female facings (3/16-in.-deep recess on the female face and 1/4-in.-high raised face on the male part) confine the gaskets and minimize the possibility of blowout. However, this type of facing has no protection against forcing the gasket into the vessel.

The tongue-and-groove face gives protection against deforming soft gaskets into the interior of the vessel. This facing is less subject to errosive or corrosive contact with the working fluid.

The ring-joint type of facing offers the greatest protection under severe service conditions or with the use of hazardous fluids. While using this type of facing, the internal pressure in a vessel acts on the ring, increasing the sealing force of the joint.

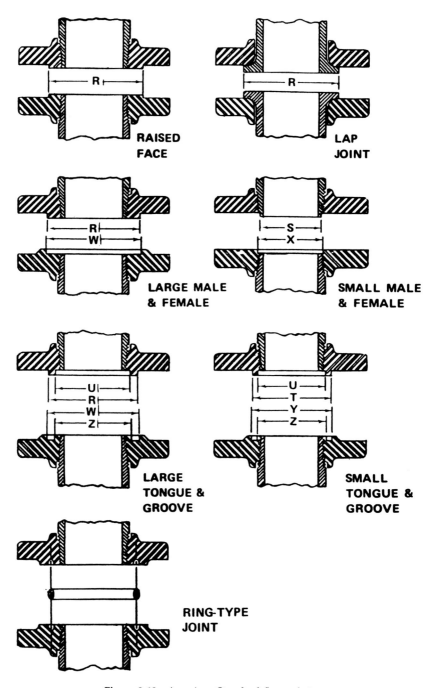

Figure 9-13 American Standard flange facings

TABLE 9-6 AMERICAN STANDARD FLANGE FACINGS FOR 150-, 300-, 400-, 600-, 900-, 1,500-, and 2,500-LB FLANGES

Nominal pipe size (in.)	Outside diameter[a]: Raised face, lap-joint, large male, and large tongue[b] R	Outside diameter[a]: Small male[b,c] S	Outside diameter[a]: Small tongue[b] T	Inside diameter of large and small tongue[a,b] U	Outside diameter[a]: Large female and large groove[b] W	Outside diameter[a]: Small female[b,c] X	Outside diameter[a]: Small groove[b] Y	Inside diameter of large and small groove[a,b] Z	Raised face, 150- and 300-lb standards[d]	Height raised face, large and small male and tongue, 400-, 600-, 900-, 1,500- and 2,500-lb standards[e]	Depth of groove or female
1/2	1 3/8	23/32	1 3/8	1	1 7/16	25/32	1 7/16	15/16	1/16	1/4	3/16
3/4	1 11/16	15/16	1 11/16	1 5/16	1 3/4	1	1 3/4	1 1/4	1/16	1/4	3/16
1	2	1 3/16	1 7/8	1 1/2	2 1/16	1 1/4	1 15/16	1 7/16	1/16	1/4	3/16
1 1/2	2 7/8	1 3/4	2 1/2	2 1/8	2 15/16	1 13/16	2 9/16	2 1/16	1/16	1/4	3/16
2	3 5/8	2 1/4	3 1/4	2 7/8	3 11/16	2 5/16	3 5/16	2 13/16	1/16	1/4	3/16
2 1/2	4 1/8	2 11/16	3 3/4	3 3/8	4 3/16	2 3/4	3 13/16	3 5/16	1/16	1/4	3/16
3	5	3 5/16	4 5/8	4 1/4	5 1/16	3 3/8	4 11/16	4 3/16	1/16	1/4	3/16
4	6 3/16	4 5/16	5 11/16	5 3/16	6 1/4	4 3/8	5 3/4	5 1/8	1/16	1/4	3/16
6	8 1/2	6 3/8	8	7 1/2	8 9/16	6 7/16	8 1/16	7 7/16	1/16	1/4	3/16
8	10 5/8	8 3/8	10	9 3/8	10 11/16	8 7/16	10 1/16	9 5/16	1/16	1/4	3/16
10	12 3/4	10 1/2	12	11 1/4	12 13/16	10 9/16	12 1/16	11 3/16	1/16	1/4	3/16
12	15	12 1/2	14 1/4	13 1/2	15 1/16	12 9/16	14 5/16	13 7/16	1/16	1/4	3/16
14	16 1/4	13 3/4	15 1/2	14 3/4	16 5/16	13 13/16	15 9/16	14 11/16	1/16	1/4	3/16
16	18 1/2	15 3/4	17 5/8	16 3/4	18 9/16	15 13/16	17 11/16	16 11/16	1/16	1/4	3/16
20	23	19 3/4	22	21	23 1/16	19 13/16	22 1/16	20 15/16	1/16	1/4	3/16
24	27 1/4	23 3/4	26 1/4	25 1/4	27 5/16	23 13/16	26 5/16	25 3/16	1/16	1/4	3/16

[a] A tolerance of plus or minus 0.016 in. (1/64 in.) is allowed on the inside and outside diameters of all facings.

[b] Gaskets for male-female and tongue-groove joints shall cover the bottom of the recess with minimum clearances taking into account the tolerances prescribed in footnote *a*.

[c] Care should be taken in the use of joints of these dimensions (they apply particularly on lines where the joint is made on the end of pipe) to ensure that the pipe used is thick enough to permit sufficient bearing surface to prevent crushing the gasket. Threaded companion flanges are furnished with plain face and are threaded with American Standard Locknut Thread.

[d] Regular facing for 150- and 300-lb steel flanged fittings and flange standards is a 1/16-in. raised face included in the minimum flange thickness. A 1/16-in. raised face also is permitted on the 400-, 600-, 900-, 1,500-, and 2,500-lb flange standards, but it must be added to the minimum flange thickness.

[e] Regular facing for 400-, 600-, 900-, 1,500-, and 2,500-lb flange standards is a 1/4-in. raised face not included in minimum flange thickness dimensions.

GASKET SELECTION

A gasket is interposed between the flange facings to provide better tightness of a flange connection and to decrease the necessary compressive force (*seating* force). The gasket must conform to the following requirements:

1. The gasket must be sufficiently plastic to interpose its material through deformation under load and seal the minute surface irregularities to prevent leakage of the fluid.
2. The gasket must not change its elasticity during operation.
3. The gasket must not change the flange facing.

The decision as to which gasket material is to be selected depends on temperature, pressure, and the aggressiveness of the sealed medium.

Gaskets usually made in thicknesses from 1/64 in. to 1/8 in. are commonly fabricated from such materials as paper, rubber, cloth, asbestos, plastics, copper, lead, aluminum, nickel, Monel, and soft iron. Paper, cloth, and rubber gaskets are used below 250°F. Asbestos-composition gaskets may be used up to 650°F, and ferrous and nickel-alloy metal gaskets may be used up to the maximum temperature rating of the flanges.

Laminated gaskets (with a metal jacket and soft filler, usually asbestos) require less bolt load to seat and keep tight than solid metal, flat-ring gaskets. They can be used for temperatures of up to about 850°F.

Serrated solid metal gaskets having concentric grooves machined into the faces greatly reduce the contact area on initial tightening, thereby reducing the bolt load. Serrated gaskets are used extensively in power stations in which soft or laminated gaskets are unsatisfactory and the bolt load is excessive with a flat-ring metal gasket.

Corrugated gaskets with asbestos filling have a rigid surface with concentric rings. These gaskets require less seating force than laminated or serrated gaskets and are widely used in low-pressure fluid operations. The same gaskets, but without asbestos, are used extensively in sealing water, steam, gas, oil, acid, and other chemicals at higher temperatures.

Ring-joint gaskets of oval and octagonal cross sections are used for high-pressure service. These gaskets are fabricated of solid metal—soft iron, soft steel, Monel, and stainless steels. Before using alloy-steel, they should be heat treated to soften them. The flange connection plastic rings may be used for corrosion resistance and as a means of electrically insulating in the case of low-temperature services.

The amount of force (*seating* or *yield* force) that must be applied to the gasket to cause its material to flow and seal the surface irregularities depends on the gasket's width and thickness, roughness of the facings, and on the type of the gasket itself. As a result of applying force to the gasket there arises the yield stress that represents the minimum load that ensures the tightness of connection for whatever pressures are used in the vessel.

Gasket Selection

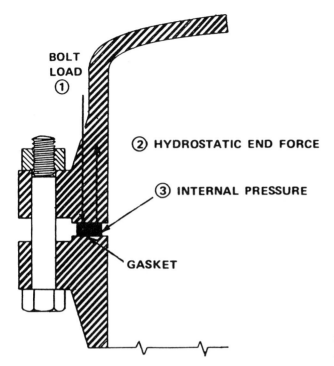

Figure 9-14 Major forces acting on a gasket

For pressure vessels (Figure 9-14), an end force tends to separate the flanges and decrease the stress on the gasket. Leakage will occur in this case if the hydrostatic end force is great enough that the difference between it and the bolt-load force decreases the gasket load below a critical value.

The *gasket factor* is defined as the ratio of the gasket stress to the internal vessel pressure. It is a property of the gasket material and construction and is independent of the internal pressure. In selecting a gasket, it is necessary to determine the total amount of force required to make the gasket yield and to keep a tight seal under operating conditions. Some common types of gaskets, gasket factor, minimum design seating stress, and recommended facings are given in Table 9-7. The effective width of the gasket, b, for various types of facings is presented in the Figure 9-15.

Because the yield force at a given value of yield stress is proportional to the gasket surface area, it is irrational to place the gasket over the total flange facing (Figure 9-16). Although such a layout makes it easier for the assembly of a flange connection and unloads the flanges from the bending moments, the bolt load will be increased unnecessarily; further, the diameter and number of bolts become so great that the load fails to provide the proper distribution over the flange. The higher the pressure in the vessel, the narrower the gasket required. For high-pressure service, the sealing area of metal gaskets or rings is determined by the strip width of elastic deformation of compressed parts.

TABLE 9-7 GASKET FACTORS (m) FOR OPERATING CONDITIONS AND MINIMUM DESIGN SEATING STRESS (y)[a]

Gasket material		Gasket factor (m)	Minimum design seating stress (y)	Sketches and holes	Facing limitations (Refer to figure 9-5)
Rubber without fabric or a high percentage of asbestos fiber:					
below 75, Shore durometer		0.50	0		
75 or higher, Shore durometer		1.00	200		Use 1, 4, 6 only
Asbestos with a suitable binder for the operating conditions	1/8 in. thick	2.00	1,600		
	1/16 in. thick	2.75	3,700		
	1/32 in. thick	3.50	6,500		
Rubber with cotton-fabric insertion		1.25	400		
Rubber with asbestos-fabric insertion, with or without wire reinforcement	3-ply	2.25	2,200		None
	2-ply	2.50	2,900		
	1-ply	2.75	3,700		
Vegetable fiber		1.75	4,400		1, 4, 6
Spiral-wound metal, asbestos filled	Carbon	2.50	2,900		
	Stainless	3.00	4,500		
Serrated steel	Asbestos filled	2.75	3,700		
Currugated metal, asbestos inserted or Corrugated metal, jacketed asbestos filled	Soft aluminum	2.50	2,900		Use 1a only
	Soft copper or brass	2.75	3,700		
	Iron or soft steel	3.00	4,500		
	Monel or 4–6% chrome	3.25	5,500		
	Stainless steels	3.50	6,500		
	Soft aluminum	2.75	3,700		
	Soft copper or brass	3.00	4,500		

Gasket material		m	y	Sketch	Notes
Corrugated metal	Iron or soft steel	3.25	5,500		
	Monel or 4–6% chrome	3.50	6,500		
	Stainless steels	3.75	7,600		
Flat metal, jacketed, asbestos filled	Soft aluminum	3.25	5,500		
	Soft copper or brass	3.50	6,500		
	Iron or soft steel	3.75	7,600		Use 1a, 2[b] only
	Monel	3.50	8,000		
	4–6% chrome	3.75	9,000		
	Stainless steels	3.75	9,000		
Grooved iron or soft steel with or without metal jacket	Soft aluminum	3.25	5,500		
	Soft copper or brass	3.50	6,500		
	Iron or soft steel	3.75	7,600		Use 1, 2, 3, only
	Monel or 4–6% chrome	4.00	8,800		
	Stainless steels	4.25	10,100		
Solid flat metal	Soft aluminum	4.00	8,800		
	Soft copper or brass	4.75	13,000		
	Iron or soft steel	5.50	18,000		None
	Monel or 4–6% chrome	6.00	21,800		
	Stainless steels	6.50	26,000		
Ring joint	Iron or soft steel	5.50	18,000		
	Monel or 4–6% chrome	6.00	21,800		Use 8 only
	Stainless steels	6.50	26,000		

[a] This table gives a list of many commonly used gasket materials and contact facings with suggested design values of m and y that have generally proved satisfactory in actual service when using effective gasket seating width, b, given in Figure 9-16.
[b] The surface of a gasket having a lap should be against the smooth surface of the facing and not against the nubbin.

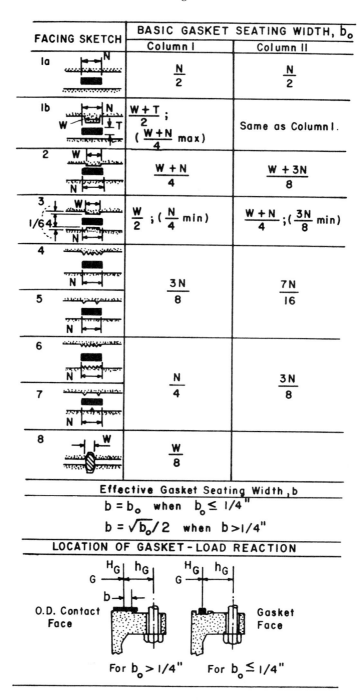

Figure 9-15 Effective gasket width and location of gasket load reaction. (Note that gasket factors listed apply only to flanged joints where the gasket is contained entirely within the inner edges of the bolt holes.)

Gasket Selection

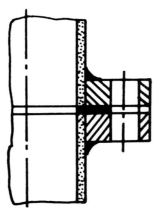

Figure 9-16 Gasket resting on total flange facing

The type of facings greatly influences the flange performance and the amount of bolt load. Due to the turning of the flange section under the bolt load, the outer gasket part is more compressed than the inside gasket part. With this in mind, the effective gasket width (which is less than its actual width) is used in flange designing.

A preliminary estimate of gasket proportions may be obtained if we neglect the elastic deformations of bolts, gasket and flanges:

(Gasket seating force) − (Hydrostatic pressure force) = (Residual gasket force)

The residual gasket force cannot be less than that required to prevent leakage of the fluid from the vessel under operating pressure:

$$\frac{\pi}{4}(d_o^2 - d_i^2)y - \frac{\pi d_o^2}{4}p = \frac{\pi}{4}(d_o^2 - d_i^2)pm$$

where y = yield stress (lb/in.^{2}) (see Table 9-7)

p = internal pressure (lb/in.^{2})

m = gasket factor (see Table 9-7)

d_o = outside diameter of gasket (in.)

d_i = inside diameter of gasket (in.)

Rewriting the preceding equation, we obtain

$$\frac{d_o}{d_i} = \sqrt{\frac{y - pm}{y - p(m + 1)}}$$

When it is desirable to reduce the gasket width, a gasket seating stress slightly in excess of y should be used (very large excesses of y can crush the gasket or cause it to be squeezed out between the flange faces).

TABLE 9-8 BOLT DATA

Bolt size (d) (in.)	Standard thread		8-Thread series		Bolt spacing[a]		Minimum radial distance (R) (in.)	Edge distance (E) (in.)	Nut dimension (across flats) (in.)	Maximum fillet radius (r) (in.)
	Number of Threads	Root Area	Number of Threads	Root Area	Minimum B_s (in.)	Preferred (in.)				
1/2	13	0.126	No. 8		1 1/4	3	13/16	5/8	7/8	1/4
5/8	11	0.202	thread		1 1/2	3	15/16	3/4	1 1/16	5/16
3/4	10	0.302	series		1 3/4	3	1 1/8	13/16	1 1/4	3/8
7/8	9	0.419	below 1"		2 1/16	3	1 1/4	15/16	1 7/16	3/8
1	8	0.551	8	0.551	2 1/4	3	1 3/8	1 1/16	1 5/8	7/16
1 1/8	7	0.693	8	0.728	2 1/2	3	1 1/2	1 1/8	1 13/16	7/16
1 1/4	7	0.890	8	0.929	2 13/16	3	1 3/4	1 1/4	2	9/16
1 3/8	6	1.054	8	1.155	3 1/16	3	1 7/8	1 3/8	2 3/16	9/16
1 1/2	6	1.294	8	1.405	3 1/4		2	1 1/2	2 3/8	5/8
1 5/8	5 1/2	1.515	8	1.680	3 1/2		2 1/8	1 5/8	2 9/16	5/8
1 3/4	5	1.744	8	1.980	3 3/4		2 1/4	1 3/4	2 3/4	5/8
1 7/8	5	2.049	8	2.304	4		2 3/8	1 7/8	2 15/16	5/8
2	4 1/2	2.300	8	2.652	4 1/4		2 1/2	2	3 1/8	11/16
2 1/4	4 1/2	3.020	8	3.423	4 3/4		2 3/4	2 1/4	3 1/2	11/16
2 1/2	4	3.715	8	4.292	5 1/4		3 1/16	2 3/8	3 7/8	13/16
2 3/4	4	4.618	8	5.259	5 3/4		3 3/8	2 5/8	4 1/4	7/8
3	4	5.621	8	6.324	6 1/4		3 5/8	2 7/8	4 5/8	15/16

[a] B_s = Center-to-center distance between bolts, inches.

OPTIMUM BOLT SELECTION

The following forces load the bolts: the force required to seat the gasket and the force required to withstand the internal pressure and to maintain the gasket-factor pressure (mp) at the same time. The greater of these two forces is taken as the designing force. The required bolting area is determined by dividing the designing force by the allowable bolting stress.

The minimum bolt spacing based on wrench clearances limits the number of bolts that can be placed in a given bolt circle. The maximum bolt spacing is limited by the flange deflection, providing a no-leak joint. The following empirical formula for maximum bolt spacing has been recommended:

$$B_{s(max)} = 2d + \frac{6t}{m + 0.5}$$

where $B_{s(max)}$ = maximum bolt spacing for a tight joint (in.)

d = bolt diameter (in.)

t = flange thickness (in.)

m = gasket factor (see Table 9-7)

It is desirable to use a minimum diameter bolt circle, which may be determined from Table 9-8. This diameter either will be the diameter necessary to satisfy the radial clearances, $d = B + 2(g_1 + R)$, or the diameter necessary to satisfy the bolt-spacing requirement, $d = (NB_S/\pi)$, whichever is greater. The optimum design is achieved when these two controlling diameters are approximately equal.

TABLE 9-9 SELECTION OF OPTIMUM BOLT SIZE

Bolt size	Root area	Minimum number of bolts	Actual number (N)	B_s	R	$\frac{NB_s}{\pi}$	$B + 2(g_1 + R)$
¾	0.302	73.7	76	3	1⅛	72.5	36¾
⅞	0.419	53.3	56	3	1¼	53.4	37
1	0.551	40.4	44	3	1⅜	42.0	37¼
1⅛	0.728	30.6	32	3	1½	30.5	37½
1¼	0.929	24.0	24	3	1¾	22.9	38

PRINCIPLES OF FLANGE DESIGN

Despite the simplicity of a flange connection, there is no universally adopted method of design. Approximate methods that had been acceptable cannot be used for high-temperature, high-pressure designs involving large equipment.

Flanges designed according to the old formulas cannot provide sufficient tightness even for large vessels working under standard conditions. At the same time, the sizes of most units demand economy in design without any compromise for safety. The result has been a steadily increasing need for deriving a number of different methods for flange design that are correct over a wide range of operating conditions.

Only the analysis of Timoshenko is described here. This analysis applies to both ring flanges and hubbed flanges. Let us first analyze a loose flange. A loose flange is a ring of the rectangular section loaded by torque moments evenly distributed on the circumference (Figure 9-17). Suppose that the deformation consists of a rotation of the square section on the angle, ν, without any distortion of the rectangular flange cross section. R_i and R_o are inside and outside ring radii, and r is the fiber radius. For the fiber under investigation with radius r, its relative elongation as a result of rotation on the angle, ν, will be

$$\varepsilon = \nu y / r$$

and the stress corresponding to this elongation is

$$\sigma = \frac{E\nu y}{r}$$

According to equilibrium conditions for a half ring, the sum of all normal forces acting on its cross section must equal zero, and the moment of these forces with

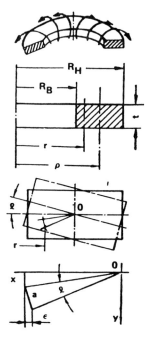

Figure 9-17 Sketch of flange design

Principles of Flange Design

respect to the x-axis must be

$$M = M_t p$$

where M_t = the torque moment per unit length of the base line

p = radius of the base line

The equilibrium conditions for a ring of any cross section may be written as follows:

$$\int_F \frac{E\nu y}{\rho} dF = 0$$

$$\int_F \frac{E\nu y^2}{\rho} dF = 0$$

The integration is performed on all cross sections of the ring, F. The equilibrium conditions for the ring with a rectangular cross section are expressed by the following equations:

$$\int_{-t/2}^{+t/2} \int_c^d \frac{E\nu y}{r} dr dy = 0$$

$$\int_{-t/2}^{t/2} \int_{R_i}^{R_o} \frac{E\nu y}{r} dr dy = M$$

Integrating the last equation, we obtain

$$\frac{E\nu t^3}{12} \ln \frac{R_o}{R_i} \ M$$

Hence, we obtain the value of the rotation angle:

$$\nu = \frac{12M}{Et^3 \ln R_o/R_i} = \frac{12M_t \rho}{Et^3 \ln R_o/R_i}$$

Substituting ν in the equation for stress, we obtain

$$\sigma = \frac{E\nu y}{r} = \frac{12My}{rt^3 \ln(R_o/R_i)}$$

The maximum stresses will arise at the inside angles of the ring cross sections. In this case,

$$r = R_i$$

and

$$y = \pm \frac{t}{2}$$

and maximum stress is

$$\sigma_{max} = \frac{6M}{t^2 R_i \ln(R_o/R_i)} = \frac{6M_t \rho}{t^2 R_i \ln(R_o/R_i)}$$

It is possible to directly determine the stress and the thickness of the loose flange:

$$\sigma = \frac{6Wl}{\pi t^2 D_i 2.3 \log_{10}(D_o/D_i)} = 1.66 \frac{Wl}{t^2 D_i \log_{10}(D_o/D_i)}, \quad kg/cm^2$$

and

$$t = 1.29 \sqrt{\frac{Wl}{\sigma_{allow} D_i \log(D_o/D_i)}}$$

where W = bolting load (kg)

l = lever arm of moment

The design of integral flanges connected homogeneously with a pipe is more complicated. In this case, the flange under action of forces, P, not only rotates on the angle, ν, but bends the pipe itself (Figure 9-18). The force, P, is related to the length of the inside circle of the pipe. The force related to the outside pipe circle is PR_i/R_o. Because the pipe and the flange each have different rigidity at the point of the junction of the flange with the pipe, there will arise cutting forces, P_o, and bending moments, M_o.

We will find the values P_o and M_o related to the unity length of the pipe inside the circle from the condition of continuity at the point of junction between the flange and pipe. Considering that the flange is extremely rigid in the radial direction, the radial displacement resulting from forces P_o can be neglected. The rotation angle of the pipe rim is equal to the rotation angle of the flange itself, ν. Determining the values of the boundary factors for a flange connected to a pipe, we obtain the following equations for M_o and P_o:

$$\frac{1}{2K^3 D}(P_o - KM_o) = 0$$

$$-\frac{1}{2K^2 D}(P_o - 2KM_o) = \nu$$

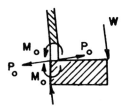

Figure 9-18 Analysis of forces and moments in an internal flange

Principles of Flange Design

The design of integral flanges connected homogeneously with a pipe is more complicated. In this case, for the flange under action of forces, P, the coefficient is

$$K = \sqrt{\frac{3(1 - \bar{\mu}^2)}{R_i^2 S^2}}$$

where S = the pipe thickness at the connection point with the flange

R_i = inside pipe radius

$$M_o = 2KDv$$

$$P_o = 2K^2Dv$$

The torque moment is related to the unity of the flange center line and arises under action of forces, W:

$$M_1 = \frac{R_i}{\rho}[W(R_o - R_i) - M_o - P_o t/2]$$

$$= \frac{R_i}{\rho}[W(R_o - R_i) - M_o - KM_o t/2]$$

Substituting this value in our equation for determining the rotation angle, v, we will find its value, and by substitution of 1 we obtain

$$M_o = 2KDv$$

and, hence,

$$M_o = 2KD \frac{12 R_i}{Et^2 \ln(R_o/R_i)} \left[W(R_o - R_i) - M_o - KM_o \frac{t}{2} \right]$$

Introducing the value of cylinder rigidity for a pipe,

$$D = \frac{ES^3}{12(1 - \bar{\mu}^2)}$$

Finally, we obtain

$$M_o = W(R_o - R_i) \frac{1_t}{1 + \dfrac{Kt/2}{2} + \dfrac{1 - \bar{\mu}^2}{2KR_i}(t/S)^3 \ln(R_o/R_i)}$$

The maximum stress from bending the pipe is

$$\sigma = \frac{6M_o}{S^2}$$

ADDITIONAL COMMENTS AND NOTES ON FLANGE DESIGN

The following factors should be considered when designing flanges:

1. All piping and vessels' nozzle flange templates must comply with the ASME code. However, in designing main joint flanges and heads there is no reason to follow the code in the flange templates. It is better to design the connection so that the distance between the gasket and bolt is at a minimum and, consequently, the bending moments, flange sizes, and weight of the connection are held at a minimum.
2. The number of bolts must be a multiple of four.
3. In general it is desirable not to space the bolts on the principal axes.
4. The gasket should be entirely within the bolt circle.
5. The higher the internal pressure, the narrower the gasket should be. The more narrow the gasket, the less the bolt load for tightening the flange connection is required.
6. The flange bolts and studs must work only for tension. Nuts and bolt heads should not sit on fillets or sit crookedly on the stud or bolt. The nut must sit on the flange over its entire surface. In critical cases the flange surface should be faced. A less costly approach is to have the face under each nut reamed, but not deeper than 1–2 mm.

THREADED CONNECTIONS

Threaded connections are commonly used in a variety of process equipment for joining smaller-sized, high-pressure pipes (diameters, d, typically less than 20 mm), pipes in service of moderate and low pressures and temperatures ($d < 32$ mm), and other preferentially metal parts (flanges, caps, nipples, and so on), or plastics.

Such joints do not pose any special fire or explosion hazards, cause damage to valve seats, or allow plugging of small orifices by permitting particles to enter into the piping system. They are not appreciably affected by changes in temperature. Major disadvantages are that they result in a weakening of the joint and increase the potential for corrosion.

Components consist of threaded male and female ends screwed together to provide satisfactory service without excessive cost. The proper manner in which the connection should be threaded is illustrated in Figure 9-19.

Normally, 3/4 in./ft of thread length is recommended. The length of effective thread, F, can be determined from the following formula:

$$F = (0.80D + 6.8)\tilde{P}$$

Symbols F and P are shown in Figure 9-19.

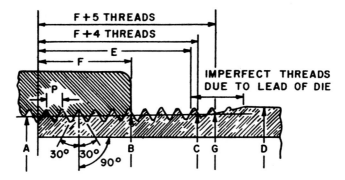

Figure 9-19 American standard taper pipe thread: *A*. pitch diameter at end of male thread; *B*. pitch diameter at end of female thread; *C*. maximum pitch diameter of straight male locknut thread; *D*. outside diameter of pipe; *E*. length of effective thread; *F*. normal engagement by hand between male and female thread; *G*. minimum pitch diameter of straight female locknut thread; *P*. pitch of thread.

The unified thread form shown in Figure 9-20 is now the standard in the United States, Canada, and Great Britain. This form is used for practically all threaded fasteners in the United States; therefore, no other thread types will be discussed.

Pitch diameters for the thread given in Figure 9-19 may be determined as follows:

$$A = D - (0.050D + 1.1)P$$

$$B = A + 0.0625F$$

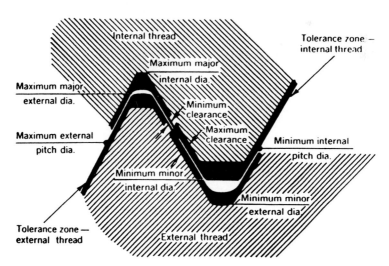

Figure 9-20 Unified thread form

where A = pitch diameter of thread at end of connection

B = pitch diameter of thread at gauging notch

D = outside diameter of pipe

F = normal engagement by hand between external and internal threads

$\tilde{P}$ = pitch of thread

The pitch of the thread is the distance the axis will advance in one revolution, that is, $1/n$, where n is the number of threads per inch.

The taper of the thread is 1 in 16 as measured on the diameter; that is, there is a change in diameter. Threaded connections may be divided into three groups:

1. **Pipe to pipe.** This type of joint is required where it is necessary to install a section of pipe between pieces of equipment, valves, or fittings.
2. **Pipe to fittings.** This connection is required to permit change in the direction or the size of the piping, or to provide a means for branch connections to the system.
3. **Pipe to equipment or valves.** This type is required for the proper operation of the system.

Pipe-to-pipe threaded connections are made with either couplings or screwed flange unions. Typical designs of detachable threaded connections used in manufacturing chemical equipment are illustrated in Figure 9-21.

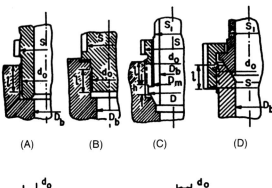

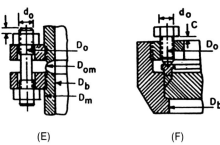

Figure 9-21 Main typical designs of threaded connections used for chemical equipment

Screwing parts I, II, IV, and V (Figure 9-21) is accomplished with tensile strength indicating the axial load; designs III and VI are screwed without pulling. Devices should be provided for that will prevent rotation (for example, hexagonal, flattened spots for a wrench, and so on) in all connections where there is an axial load.

SCREWED FITTINGS

Screwed fittings include elbows, tees, crosses, 45° ells, Y-branches or laterals, return bends, and reducers in various patterns and weights. Equipment and valves generally require joints similar to those used with fittings that are tapped for American Standard pipe threads. For normal requirements, couplings are threaded with right-hand threads for the full length and have plain ends. When the connection is made up, the ends of the pipe should not touch (Figure 9-22).

When it is necessary to install a piece of pipe close to the space between two pipe sections already in place, without using a union or flanges, one end of the filler piece must be threaded with a left-hand end. The coupling is threaded with one right-handed end and the other with a left-handed thread. This makes it possible to screw the coupling on the ends of the two adjacent pieces at the same time, making the connection tight.

Couplings operating under high pressure for services in which shocks are frequent and for use on API line pipe have the ends of the couplings recessed to cover the imperfect threads on the pipe. This is illustrated in Figure 9-23. While this design does not strengthen the joint, it does protect the thread against dirt and moisture and, therefore, reduces the danger of corrosion.

Unions consist of three pieces and, where used for connecting two adjacent pieces of pipe, have female ends threaded for American Standard pipe threads. The collar piece normally has a straight thread. The joint between two adjacent sections may be designed for use with a gasket or for metal-to-metal contact.

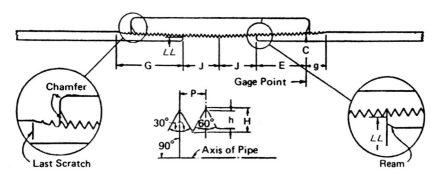

Figure 9-22 Thread specifications for coupling

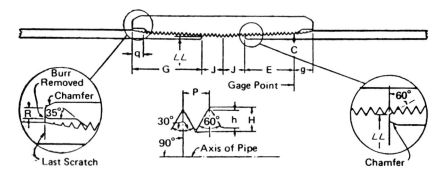

Figure 9-23 Couplings for high-pressure service

The application of unions to a typical layout is given in Figure 9-24. Unions without gaskets may have beveled male and female adjoining surfaces with seats made of all iron, all steel, iron or steel to bronze, or bronze to bronze. Unions with one male end and one female end provide a simple means of making a joint to equipment that must be disconnected frequently and is not equipped with flanged connections.

When selecting material for threaded parts, one should choose a higher hardness and strength for the male part. This can be achieved by choosing different materials or by corresponding heat treatment of the same material. Usually, the diameter of a thread is selected based on design considerations. Threads are selected from the standards with a definite pitch, t, and designed for the proper length of screwing.

Sources for stresses that may be imposed on male or female parts of threaded connections are the external load encountered in service, the load developed to make the connection tight, and the load involved in the tightening process. These loads produce the following stresses:

1. A *bearing stress* is caused by the male threads bearing on the threads of the female part of the connection.
2. A *shear stress in the threads* is produced, that is, a tendency to strip the threads. A full-depth thread has its minor diameter associated with the male, while the female part has a nearly equal diameter; the standard unified thread is based on a 75 percent thread depth. When tightened, a regular female part drilled out to only 50 percent of a full-depth thread will break the male part rather than strip the threads of the female part. A standard thread will develop the full tensile strength of the part when engaged at a distance of about one half the diameter of the male part of the threaded connection. The thickness of the female part is almost equal to the diameter of the male part. Thus, the shear stress in the threads generally will not cause problems.

Screwed Fittings

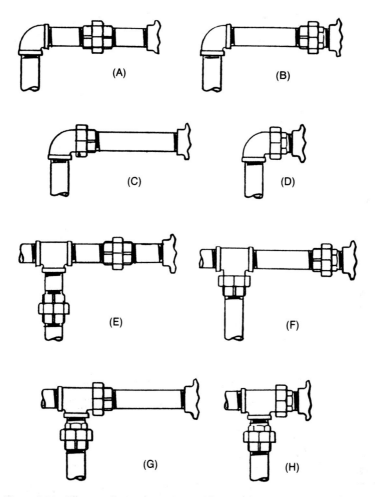

Figure 9-24 Elbow and tee connections with union joints: (A) 1 piece of pipe, 2 fittings, 5 pipe threads, 6 joints assembled; (B) 2 pieces of pipe, 2 fittings, 1 pipe thread, 5 joints assembled; (C) 2 pieces of pipe, 1 fitting, 3 pipe threads, 4 joints assembled; (D) 1 piece of pipe, 1 fitting, 1 pipe thread, 3 joints assembled; (E) 5 pieces of pipe, 3 fittings, 8 pipe threads, and 10 joints assembled; (F) 3 pieces of pipe, 2 fittings, 4 pipe threads, 7 joints assembled; (G) 3 pieces of pipe, 2 fittings, 4 pipe threads, 7 joints assembled; (H) 2 pieces of pipe, 2 fittings, 2 pipe threads, 6 joints assembled.

3. A *shear stress in the body* of the male part is produced when it is used in shear. This becomes a simple problem of selectng the male part that has sufficient area in shear.
4. A *tensile stress* is produced in the male part of the threaded connection. If the male part is used in an application such that there is no load due to

tightening and only a static applied load, the problem reduces to proper selection of a male part, with adequate area at the root diameter of the thread. If the male part is tightened to the extent that a tensile stress is developed that is almost equal to the yield strength, any load that induces a stress less than this will have no effect on the male part.

5. A *torsion stress* in the male part is produced due to tightening. As a male part is tightened, the tensile load that is induced forces the threads of the male and female parts together, thus increasing the friction. Consequently, to overcome this friction and to develop the tension load, it is necessary to apply a torque, which will disappear when tightening is completed.

The required torque to produce a given tension load is

$$T = CDP$$

where T = torque

C = coefficient = 0.2

D = diameter of a male part

P = tension load in male part

10

Equipment Supports

Installation of process equipment on foundations or on special supporting structures is done with supports. Equipment with flat bottoms, operating during filling conditions, are installed directly onto the foundation. In all cases, the recommended average specific loads, q, on supporting surfaces are given in Table 10-1.

The support structures located between the vessel and the foundation may be divided into two groups: supports for vertical vessels and supports for horizontal vessels. Figure 10-1 gives typical designs of welded supports for vertical vessels.

Support types I–V are located underneath an apparatus, rigidly fixed with it and designated for a cylindrical apparatus. Support types VI–IX are located on the sides of an apparatus, and also rigidly fixed with it and designated for cylindrical and box-type apparatuses.

Support types I–III, VI, and VII are unitized for supporting constructions, and support types IV, V, VIII, and IX are separate lugs, the number of which should be no less than three for one piece of equipment. For small-sized equipment, in some cases it is possible to install only two lugs. In cast equipment the supports are fabricated integrally with a shell or head. The support design in this case may be similar to types IV–IX.

The size and shape of supports depend basically on the amount and character of load material of the equipment, its weight, as well as its location in

TABLE 10-1 AVERAGE VALUES OF SPECIFIC LOADS ON SUPPORTING SURFACES

	$q(\text{MN/m}^2)$
Wooden Decking	≤ 2.0
Stonework	≤ 0.8
Concrete	≤ 2.0
Steel and Cast-Iron Surface	≤ 100.0

space. If the vessel is to be installed on the floor inside a building, then support types IV or V are recommended. If the apparatus is to be suspended on the bearing structures between the intermediate floor, then support types VI–IX are recommended. Skirt supports (types I–III) are used mainly for installation on foundations outside buildings, especially if the ratio of equipment height to diameter is greater than 5.

If the equipment is subjected to vibrations and dynamic forces such as moving machinery, then its supports are fabricated as a massive rigid frame. Its function is not only to transfer the weight of the equipment on the bearing foundation surface, but with the use of anchor bolts to involve all the foundation mass in absorbing dynamic forces and vibrations that take place during operation.

An alternative design may involve the use of lugs or brackets (Figure 10-1, types IV–IX) attached to the vessel and resting on columns or beams.

Certain horizontal cylindrical equipment such as storage vessels, heat exchangers, and so on are commonly supported by saddle supports or cradles. The typical designs of welded supports of such equipment are shown in Figure 10-2.

Support types I and II are located underneath equipment and may be detachable (left-hand side) and welded to the apparatus (right-hand side). Support types III and IV are located on the sides of an apparatus and are welded to it. The number of support types I and II and doubled support types III and IV for an equipment may be two or more, depending on the length of the apparatus.

Support types I and II (saddles) decrease the bending stresses in the wall of the equipment as compared to types III and IV (lugs). In addition, they insulate the foundation from the equipment. To decrease the bending stresses in the walls of the apparatus, the angle of the bearing or of the shell on saddle should be no less than 120°.

Equipment used in hot service generally undergoes elongation. If thermal elongation or expansion is not provided for, the walls may be subjected to thermal stresses:

$$\sigma = E\alpha(t_1 - t_2)$$

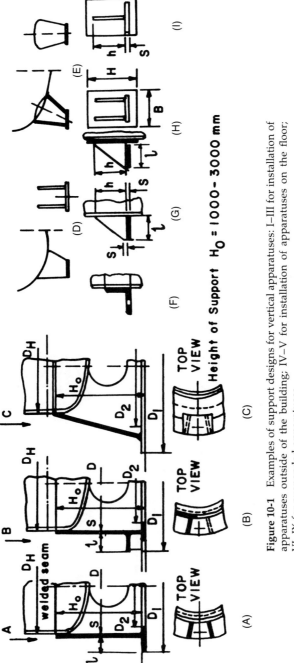

Figure 10-1 Examples of support designs for vertical apparatuses: I–III for installation of apparatuses outside of the building; IV–V for installation of apparatuses on the floor; VI–IX for suspended apparatuses

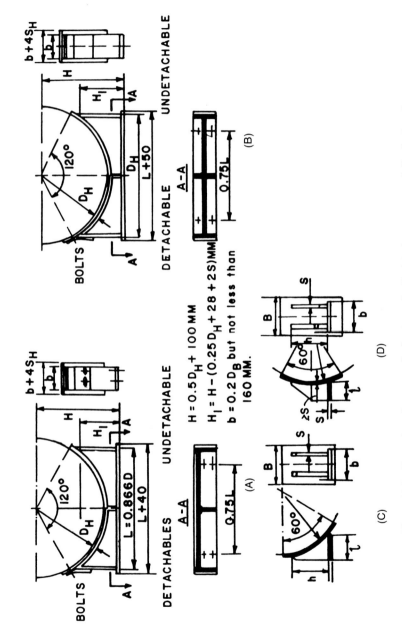

Figure 10-2 Examples of supports for horizontal cylindrical apparatuses: I, II—installation on saddles; III, IV—for installation on lugs

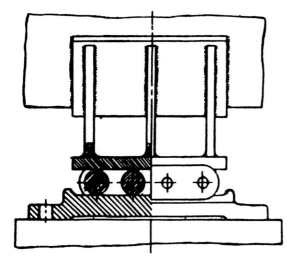

Figure 10-3 Saddle support with rollers

where E = modulus of elasticity

σ = coefficient of thermal expansion

t_1 = temperature of cold apparatus walls

t_2 = temperature of walls at service conditions

In application of detachable supports, one of the end supports of the equipment is securely fixed while the others are provided with relative displacement of the equipment along its axis due to thermal expansion. For this reason, oval holes are provided in the lugs on shells with which the apparatus is fastened with bolts to the support (Figure 10-2, element V). In these cases, all the supports are rigidly connected to the foundation. If there are inseparable supports, one of them is rigidly tightened to the foundation, whereas the others must have a free displacement relative to the foundation. This is achieved by inserting the support steel plates underneath, thus permitting the sliding of supports over them. To decrease friction, cylindrical rollers are inserted between the plate and the support (Figure 10-3); the number of rollers depends on the load accepted by the support.

DESIGN OF SUPPORTS FOR VERTICAL VESSELS

Skirt Supports

Tall vertical vessels usually are supported by skirts. Because a cylindrical skirt has an all-metal area located at a maximum distance (for a given diameter) from the neutral axis, the section modulus, W, is at a maximum and the induced stress for the metal involved is at a minimum. Thus, the cylindrical skirt is an economical design for the support of a tall vertical vessel.

The skirt may be welded directly to a bottom dished head, flush with the shell, or to the outside of the shell. If the skirt is welded flush with the shell, the weight of the vessel in the absence of wind and seismic loads places the weld in compression. On the other hand, if the skirt is welded to the outside of the vessel, the weld seam is in shear.

The wall thickness of cylindrical skirt types I and II (Figure 10-1) and conical skirt type III (Figure 10-1) is selected from design considerations in the range $S = 6$ to 12 mm and then checked by the bending stress in the wall of the skirt as follows:

$$\sigma = \frac{GD + 4M_w}{\pi D_i^2 S} \leq \sigma_{b,allow}, \ N/m^2$$

where S = assumed wall thickness of the skirt (m)

G = the maximum load on the support from dead weight under conditions of operation and hydrostatic test

D_i = internal diameter of the support, which is (for support type III assume the lesser diameter) equal to the outside diameter of the shell, D(m)

M_w = wind-load moment (N − m)

In addition, the assumed wall thickness of the skirt should be checked for stability.

The bottom of the skirt should be anchored securely to the concrete foundation by anchor bolts to prevent overturning from the bending moments induced by wind or seismic loads. The bending moment and the weight of the vertical vessel result in loading conditions on the concrete foundation, as shown in Figure 10-4. These induce a tensile load on the upwind anchor bolts and a compressive load on the downwind anchor bolts.

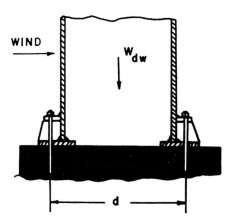

Figure 10-4 Sketch of loading of anchor bolts

Design of Supports for Vertical Vessels

The maximum compression stress on the surface of a bearing ring is

$$\sigma_{max} = \frac{G_{max}}{0.785(D_1^2 - D_2^2)} + \frac{M_w D_1}{0.1(D_1^4 - D_2^4)}$$

This stress should not exceed the allowable value for concrete, which is given in Table 10-1.

The outside and inside diameters of the skirt-bearing ring, D_1 and D_2, are taken from Figure 10-1 and should be checked for maximum bearing stress of the foundation under ring surface, σ_{max}, from the preceding formula. The diagrams of stresses from weight and wind loads, as well as a diagram of total stresses, are given in Figure 10-5.

If the condition in the preceding equation is not fulfilled, the inside diameter, D_2, of the supporting ring should be decreased correspondingly. It is also recommended to add gusset plates inside the support.

The thickness of the skirt-bearing ring may be calculated from the following approximate procedure: The thickness of the bearing ring is determined by the compression load on the downwind side of the vertical vessel. The minimum required width of the bearing ring was determined previously. The maximum compressive stress between the bearing ring and the concrete occurs at the outer periphery of the bearing ring. As shown in Figure 10-5, the compressive stress varies from a maximum to a lesser value at the junction of the skirt and the bearing ring. The value at the bolt circle may be used for simplicity in evaluating the required thickness of the bearing ring (Figure 10-6).

A bearing ring is assumed to be a uniformly loaded cantilever beam, with δ_c (max. induced) being the uniform load. The maximum bending moment for this

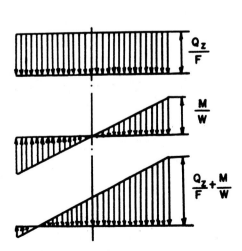

Figure 10-5 Stress diagrams from weight and wind loads

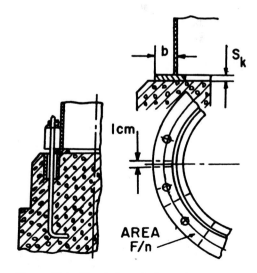

Figure 10-6 Sketch for calculating bearing ring thickness

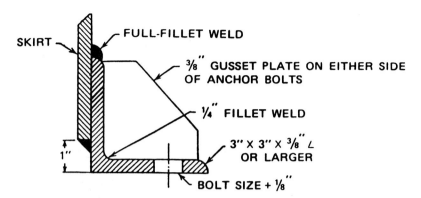

Figure 10-7 Rolled angle bearing plate

beam takes place at the junction of the skirt and the bearing ring for a unit circumferential length. Figure 10-6 shows a sketch for calculating bearing-ring thickness.

For short vessels in which the calculated thickness of a bearing plate is 1/2 in. or less, a steel angle rolled to fit the outside of the skirt may be lap welded, as shown in Figure 10-7.

If the design thickness of the bearing plate is 1/2–3/4 in., a single-ring bearing may be applied, as shown in Figure 10-8. If the designed bearing-plate thickness is 3 in. or greater for the design in Figure 10-8, a bolting *chair* can be applied, as shown in Figure 10-9. Figure 10-10 shows a typical external bolting chair.

Figure 10-11 shows a vessel skirt with external blotting chairs.

Figure 10-12 illustrates a skirt with a continuous compression ring with straps.

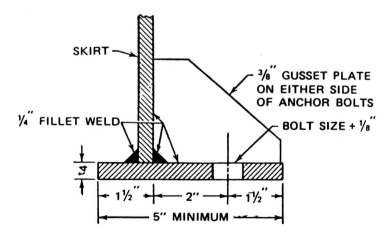

Figure 10-8 Single-ring beam plate with gussets

Design of Supports for Vertical Vessels

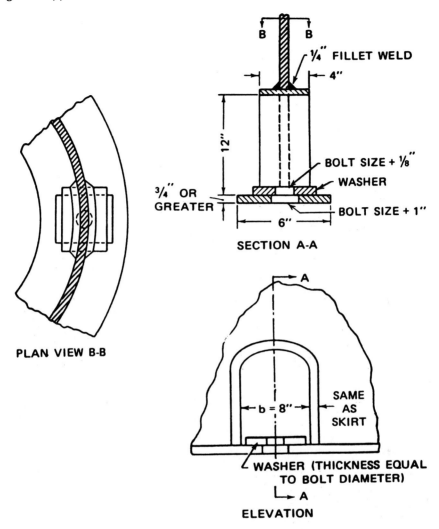

Figure 10-9 Centered anchor-bolt chair

Figure 10-13 shows the insertion of the reinforcing pad between the lug and thin-walled vessel.

Figure 10-14 is a sketch of a vessel on four-lug supports with horizontal-plate stiffness.

Lug Supports

Brackets, or lugs (Figure 10-1, types VI–IX), offer many advantages over other types of supports. They can absorb diametrical expansions, are easily attached to the apparatus, and easily leveled or shimmed in the field.

358 Equipment Supports Chap. 10

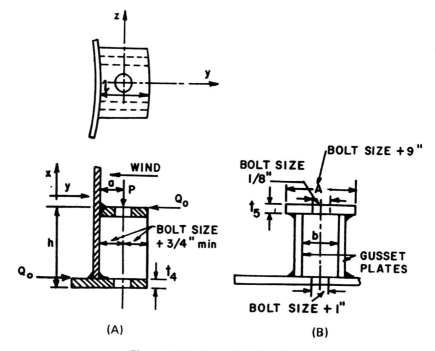

Figure 10-10 External bolting chair

Due to the eccentricity of a lug support, compressive, tensile, and shear stresses are induced in the wall of the apparatus. The tensile and compressive stresses are combined with pressure stresses circumferentially and longitudinally. The forces induced by these shear stresses are relatively small and can be ignored.

Lug supports are very favorable for thick-walled vessels because the thick wall has a considerable moment of inertia and, therefore, is capable of absorbing flexural stresses due to the eccentricity of the loads.

The eccentricity of the loads applied to thin-walled vessels is not convenient unless the proper reinforcements are used and sufficiently tightened to the vessel (Figure 10-12).

If a vessel with lugs is located outside, the wind load as well as the deadweight load should be taken into account. Because lug-supported vessels are usually smaller in height than their skirt-supported vessels, wind loads may be a minor consideration.

The highest compressive stresses in the supports occur on the leeward side when the vessel is full (because dead load and wind load are additive). The highest tensile stresses are induced on the windward side when the vessel is empty. The dead load in this case is subtracted from the wind load. As a result,

Design of Supports for Vertical Vessels

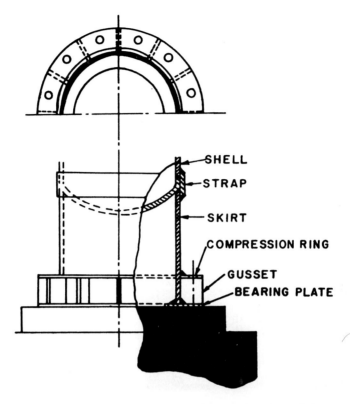

Figure 10-11 Skirt with continuous compression ring and strap

Figure 10-12 Insertion of reinforcing pad between the lug and thin-walled vessel

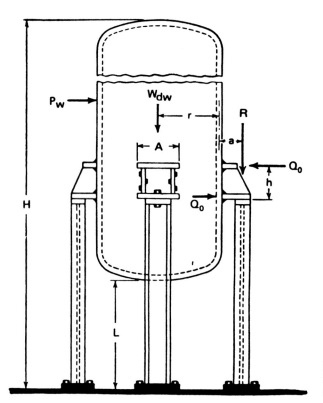

Figure 10-13 Sketch of a vessel on four-lug supports with horizontal-plate stiffeners

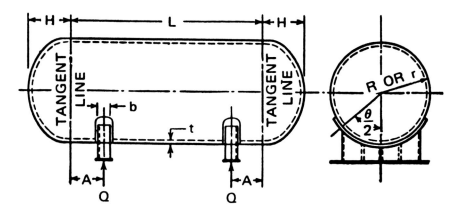

Figure 10-14 Sketch of a horizontal vessel with two saddle supports

stresses on the leeward side are the determining factors in the design of supports.

The lugs that are of essentially the same design as that shown in Figure 10-10 for external chairs are shown in the figure. These lugs use to their advantage the axial stiffness and strength of the cylindrical shell to absorb the bending stresses from the supports. The bottom and top plates have continuous welded seams because the maximum compressive and tensile stress takes place in these two plates, respectively. These seams and the intermittent welds of the vertical gussets to the shell carry the vertical shear load.

SUPPORTS FOR HORIZONTAL VESSELS

In all cylindrical equipment resting on saddle supports horizontally, bending stresses arise in the wall of the shell resulting from the weight of an apparatus and its contents.

When resting on saddle supports such as shown in Figure 10-14, horizontal vessels behave as beams. In designing for this situation, it is necessary to determine the critical compression stress in the shell cross sections induced by the supports, which results in the loss of local stability (that is, formation of wrinkles and hollows in the compression zone).

It is also necessary to check the stability of the cylindrical shell under the action of concentrated loads of the supports. It is important, therefore, to determine the most favorable sizing between supports, the rigidity of the cross section of the vessel, and its diameter. Determining and checking these parameters are important to the structural design of horizontal cylindrical equipment.

Index

A

abrasion resistance, 70
acetal resin, 118
acetylene welding, 165
acrylics, 116, 219
acrylonitrile, 115
adhesion, 125
adhesion mechanical, 125
adhesive, 1, 125
adsorbing, 14
aging, 125
agitating, 13
air chambers, 255
air separating, 15
alkyd plastics, 125
allowable leakage, 266
aluminizing, 60
aluminum, 84, 90, 213
aluminum alloys, 90
aluminum castings, 54
aluminum gaskets, 311

American Society for Testing Materials, 32
amino plastics, 125
anchor arrangements, 239
anchor bolts, 354
anchor pattern, 60
anchor-bolt chair, 356
angle valves, 279
angle view, 276
anion, 60
anneal, 125
annealed lead, 89
anode, 60
anode polarization, 60
anodic inhibitor, 60
anodic metallic coating, 60
anodic protection, 60
anodizing, 60
arc welding, 165
arc-flush steel welds, 179
argon-arc welding, 165
assembly time, 125

Index

asymmetrical shapes, 57
austenite steels, 78
austenitic cast iron, 70
austenitic steels, 77, 197
automatic welding, 169
available space, 10

B

B-stage, 125
backpressure-relief port, 125
ball checks, 284
ball valves, 285
base potential, 60
beam plate, 356
bearing stress, 346
bend radius for pipes, 189
binder, 125
biological corrosion, 30
biological treatment processes, 15
blade welding, 198
blanket, 125
blast peening, 61
blending, 13
blind flanges, 322
blind holes, 182
blister, 126
blocking, 126
bloom, 126
bolster, 126
bolt insertion, 192
bolt withdrawal, 193
bond, 126
box sections, 151
brackets, 356
brass, 84, 208
brazing, 38, 54
breakaway corrosion, 61
bricks, 104
bricks and tiles, 104
bubble cap, 212
bubble-cap tray tower, 147
bulk factor, 126
buried installations, 262
buried pipe, 258
buried pipeline, 56
butadiene-styrene, 115
butterfly valves, 279, 285

C

C-stage, 127
C-veil, 127
capillary tubing, 8
capital, 26
capital investment, 1
carbon, 102
carbon ferritic, 76
carbon martensitic, 76
carbon packing, 55
carbon steel pipes, 227
case harden, 126
cast equipment, 149, 155
cast film, 126
cast iron, 67, 148, 215
cast ribs, 154
cast steel, 150
casting alloys, 92
catalyst, 126
cathode, 61
cathodic inhibitor, 61
cathodic protection, 61
cation, 61
cavitation corrosion, 28
cavity, 126
cellular plastic, 126
cellular striation, 126
cellulosic plastics, 126
cements, 104
centrifugal casting, 126
centrifugal filtering, 14
centrifugal settling, 15
ceramic pipe, 326
ceramic-lined equipment, 215
chalking, 126
chamber settling, 14
check valves, 284
chemical addition, 16
chemical processing, 1
chemical resistance, 111, 242
chemically formed plastic, 126
chipping, 13
chlorinated polyether, 116
chlorinated polyvinyl chloride, 145
chlorinated PVC, 115
chlorinated rubber paints, 122
chlorosulfated polyethylene, 121
choker cable, 262

chrome-nickel steel, 201
chromium, 76, 78, 80
chromium diffusion, 101
chromium steels, 76
clad metals, 54
clamping plate, 126
classifying, 14
clean metal samples, 32
closed-cell foam, 126
closures, 263
cohesion, 127
cold bending, 189
cold flow, 127
cold molding, 127
cold pressing, 127
cold slug, 127
cold slugwell, 127
collection, 15
compaction, 264
compensating capacity, 269
compensators, 267
complex shape components, 165
complexity of castings, 163
composite plate, 61
compression ring and strap, 359
concentration cell corrosion, 28
concentric holes, 223
condensate filters, 58
condensation, 127
condensers, 58
conduits, 264
conical nozzle, 218
connections, 291
consistency, 127
construction, 5
contact molding, 127
control devices, 10
control valve types, 278
controller, 8
converter, 6
conveying, 13
copolymer, 127
copper, 82, 206
copper alloy pipes, 55
copper alloys, 82
copper equipment, 212
copper rivets, 211
copper shell, 206
copper tubes, 208

core and separator, 127
core pin, 127
core-pine plate, 127
corrosion, 27
corrosion fatigue, 30
corrosion fatigue limit, 61
corrosion potential, 61
corrosion rate, 61
corrosion rates of steel, 50
corrosion resistance, 70, 73, 87, 89
corrosion resistance of metals, 42
corrosion resistance of nonmetals, 46
corrosion terms, 60
corrosion-resistant service, 247
corrosive medium, 32
corrosiveness, 10
corrugated gaskets, 303
cost evaluation, 25
cost of equipment, 1
cost-reduction, 26
couple, 61
crane boom, 262
creep, 127
creep strength, 75
crevices, 57
critical humidity, 61
cross-linkage, 127
crushing, 13
cryogenic applications, 92
cryogenic liquid, 92
crystalizing, 14
cull, 127
cupro-nickels, 87
cure, 127
current density, 61
curve radius, 260

D

deactivation, 61
decarburization, 30
decision tree, 21
deformation, 149
degradation, 127
degree of crystallinity, 114
delamination, 127
deposit attack, 61
design decision, 23

Index

design factors, 9
design guidelines, 52
design limitations, 41
design of pipes, 229
design of supports, 353
design principles, 146
design thickness, 229
detachable flanges, 325
deterioration, 127
dialyzing, 15
diaphragm gate, 128
diaphragm motor, 8
die adaptor, 128
die block, 128
die body, 128
dielectric separation, 54
dielectric strength, 61
differential aeration, 61
diffusing, 13
diffusion coating, 62, 100
diffusion of carbon, 203
dip coating, 100
direct heating, 13
disc globe valves, 277
dished, 128
dispersant, 128
dispersing, 13
displacement-type flow meter, 8
disposal technologies, 21
dissimilar metals, 52
dissimilar structures, 173
dissolving, 13
distilling, 14
doping, 128
double-cone lens, 313
double-cone seals, 311
dowel, 128
draft, 128
drain nozzles, 192
draining, 14
drilling schemes, 182
drying, 14
durometer hardness, 128

E

economic factors, 41
effective gasket width, 334

ejector pin, 128
elastic modulus, 172
elasticity, 128
electrical leads, 8
electro-slag welding, 165
electrodeposition, 99
electrogalvanizing, 62
electrolysis, 62
electrolyte, 62
electrolytic cleaning, 62
electrophoresis, 14
electrophoretic plating, 62
electroplating, 62
electrostatic separating, 14
emergency shutdown, 4
emulsifying, 13
enameled equipment, 215
enameling, 215
encapsulating, 54
epoxy plastics, 128
epoxy resin coatings, 123
epoxy resin paints, 122
epoxy resins, 121
equipment design, 2
equipment standards, 7
erosion corrosion, 28
ethylene, 111
evaporating, 14
envenomation, 128
excessive agitation, 58
exfoliation, 62
exfoliation corrosion, 29
exotherm, 128
expandable plastics, 128
expansion joint tie-in, 235
extender, 128
external bolting, 358
extracting, 14
extraction, 128
extrusion, 129

F

fabricating assemblies, 227
fabrication, 34, 146
fabrication parameters, 38
fail closed, 6
fasteners, 54, 192

fatigue, 172
fiberglass resins, 119
filament winding, 129
filament-wound pipe, 232, 234
filiform corrosion, 62
filler, 129
filler materials, 120
filler rod, 71
fillet welds, 169
final control element, 6
finishing, 129
finishing paints, 122
fisheye, 129
flame plating, 62
flange bending, 312
flange connections, 316, 325
flange design, 203, 337
flange facings, 327, 328
flange pressure ratings, 317
flange rating, 9
flange selection, 317
flanged joints, 288
flanges, 316
flanges for pipes, 326
flash, 129
flash corrosion, 62
flexible compensators, 267
flexible connections, 201
flotation, 14
flow indicator, 8
fluid control, 286
fluids handling, 13
fluorinated plastics, 115
fluorinated rubbers, 121
flush welds, 179
foamed plastic, 129
force plate, 129
formability, 38
fouling, 62
fragile materials, 326
freezing, 14
fretting corrosion, 28
furane plastics, 129
furane resins, 121
fusion, 129
fusion-welding cast iron, 71

G

galvanic corrosion, 28, 52
galvanic potentials, 52
galvanizing, 62
gas absorbing, 14
gas cleaning equipment, 22
gas cleaning system, 25
gas diffusing, 15
gaseous corrosion, 31
gasket factor, 331
gasket insertion, 55
gasket selection, 330
gasket-type obturation, 292
gasketless sealing, 301
gaskets, 330
gate valves, 274, 286
gel, 129
gel point, 129
gelation, 129
geometric tolerances, 186
glass, 103, 129
glass pipe, 327
glass sight gauges, 67
glass transition, 129
globe valves, 277
gold, 99
grades of copper, 83
graphite, 54, 55, 102
gravity filtering, 14
gray cast iron, 68, 150
grinding, 13
grooving, 30
gum, 129
gunmetals, 86

H

half-finished parts, 180
halocarbon plastics, 130
hand layup piping, 247
hanger support, 237
hard solders, 208
hardener, 130
hardening stainless steel, 77
hazardous waste treatment, 21
head designs, 223

Index

heat exchangers, 57, 208
heat resistance, 65, 73
heat transfer, 13
heat treat, 130
heat treatment, 34
heat-exchanger inlets, 59
heat-exchanger surfaces, 65
heat-resistant nickel alloys, 81
heat-treatable alloys, 91
heavy castings, 163
hermetic seal, 62
high-alloy steels, 78, 197
high-carbon, 75
high-carbon steels, 74
high-impact PVC, 115
high-pressure vessel, 313
high-temperature corrosion, 31
holes for bolts, 223
hollow shaft, 197
horizontal lift checks, 284
horizontal-plate stiffeners, 360
hydrocarbon plastics, 130
hydrogen blistering, 30
hydrogen embrittlement, 30
hydrostatic design pressures, 113
hypalon, 121

I

impellers, 57
impingement of fluids, 58
impinging, 15
implementing process, 3
indicator, 6
indirect heating, 13
inflation, 23
injection, 40
inlet pipes, 59
inserted bolts, 192
installed cost, 10
installing pipe, 264
instrument air lines, 8
instrument codes, 8
instrument loop, 6
instrumentation, 10
instrumentation codes, 5
instrumentation flow diagrams, 5

interest rates, 23
intergranular corrosion, 29, 78
interior of piping, 58
internal threads, 182
ion erosion, 62
iron rot, 62
isotactic, 130
isotropic, 130

J

jigging, 15
joint deflection, 260

K

kinetic pumping, 13
kneading, 13

L

labor, 10
laboratory immersion, 144
laminar scale, 62
laminate, 130
laminated cross, 130
laminated gaskets, 330
laminated parallel, 130
lamination, 40
lap joints, 199
lap-joint flanges, 321
laps, 57
leaching, 14
lead, 87
lead alloys, 87, 89
lens compensators, 268
life expectancies, 34
light alloys, 161
lignin plastics, 130
line pipe, 130, 226
line testing, 265
liquid metal corrosion, 31
liquids settling, 15
localized attack, 62

location of weld joints, 173
longitudinal weld joints, 174
low temperature, 74
low-alloy steel, 74, 75
low-carbon, 74
low-carbon ferritic, 76
low-carbon steels, 72
low-temperature ductility, 75
lug supports, 356
lugs, 356
lyophilic, 130

M

machinability, 38
machining, 180
machining allowance, 161
machining surfaces, 181
magnetic separating, 14
male and female facings, 327
malleable cast iron, 69, 150
manganese bronzes, 84
manual welding, 168
maragining steels, 78
masticating, 13
material selection, 65
materials evaluation, 32
materials handling, 13
materials selection, 35
mechanical characteristics, 37
mechanical tubing, 130, 226
mechanically foamed, 130
melamine plastics, 130
melting, 14
mercury, 54
metal accumulation, 150
metal cladding, 62
metal gasket, 308
metal lens compensators, 269
metallic coatings, 62, 99
metallizing, 62
metals, 35, 40
metastable, 131
mild steel, 72
mill scale, 63
mixing, 13
modulation, 277
mold base, 131

molding, 40
molding bag, 131
molding blow, 131
molding compression, 131
molding contact pressure, 131
molding high pressure, 131
molding injection, 131
molding low pressure, 131
molding materials, 219
molding transfer, 131
molybdenum iron, 80
Monel, 56
monomer, 131

N

neoprene, 121
nickel, 78, 79
nickel alloys, 79
nickel austenitic steels, 76
nickel chromium, 81
nickel silicon, 81
nitrile rubbers, 121
noble, 63
noble metals, 52
noble potential, 63
nodular cast iron, 70
nondestructive evaluation, 33
nondetachable pipe, 227
nonmetallic materials, 34
nonmetals, 35, 40
nonrigid plastic, 131
nonstandard flanges, 322
novolak, 131
nozzles, 175
nucleation, 58
nylon, 116, 117
nylon plastics, 131

O

obturation, 291
on-site technology, 17
operating characteristics, 1
operating costs, 24, 26
operating guidelines planning, 4

optimum bolt selection, 337
organic acids, 93
organic coatings, 122
organosol, 131
oxidation, 36
oxidation resistance, 75, 79
oxygen concentration, 32

P

paints, 122
parting, 63
parts fabrication, 186
passivation, 56, 63
pattern shape, 159
peen plating, 63
percolating, 14
performance, 24
permanent distortion, 173
personnel protection, 4
petroleum, 1
phenolic plastics, 131
phenolic resins, 119
physical characteristics, 35, 37
physical processes, 16
physical-chemical processes, 16
pickling, 56, 63
piling pipe, 131
pipe movements, 239
pipe threads, 288
pipes, 226
piping, 9, 55, 226
piping systems, 58
pitting corrosion, 92
planning, 3
planning projects, 2
plant age, 10
plant design projects, 3
plasma plating, 63
plastic, 105
plastic deformation, 188
plastic materials, 219
plastic pipe, 144, 230, 249
plastic PVC, 115
plastic terms, 123
plastic welding, 132
plasticate, 132
plasticize, 132

plastics, 105
plastisol, 132
plastisol coatings, 124
platinum, 97, 99
pollution control, 1, 13
polyamide, 116
polyamide plastics, 132
polycarbonate, 118
polychloroprene, 121
polyester glass-flake linings, 122
polyester plastics, 132
polyester resin systems, 242
polyester resins, 119
polyesters, 231
polyethylene, 111, 132
polyethylene polymers, 113
polymer, 132
polymerization, 132
polyolefins, 111
polyphenylene oxide, 118
polypropylene, 111, 132, 145
polystyrene, 132
polysulfone, 118
polyvinyl acetate, 132
polyvinyl chloride, 114, 145
polyvinyl chloride acetate, 132
polyvinyl fluoride, 116
polyvinylidene fluoride, 145
pot life, 132
precious metals, 94, 98
prefabricated sheets, 219
preform, 132
prepolymer, 132
pressing, 14
pressing of studs, 224
pressure filtering, 14
pressure grouting, 264
pressure pumping, 13
pressure recorder, 8
pressure tubing, 132, 227
pressure vessels, 56
primary element, 6
primary treatment, 18
prime, 133
process and control summary, 19
process engineering, 13
process industries, 11
process variable, 6
promoted resin, 133

properties of steels, 72
properties of thermoplastics, 112
propylene, 111
protective coating, 39
protective paints, 122
pulley hub, 181
pulverizing, 13
punch-press frame, 161

Q

quality of finish, 39
quarter-turn valves, 286

R

radiation transmission, 33
rapid surging, 58
rash rusting, 63
reduction, 63
regrindable globes, 277
reinforced plastic, 133
reinforced plastic pipe, 230, 247
remote, 6
residual stresses, 173, 174
resin, 133
rib welding, 200
rib-wall construction, 233
rigid PV, 114
ring-joint gaskets, 330
ring-joint type of facing, 327
riveted connections, 210
road crossings, 263
rolled copper, 206
roller, 133
rope or cable, 259
roving, 133
rubber, 121
rubber linings, 121
rust, 36
rust creep, 63

S

saddle supports, 360
sampling, 4

sand castings, 150
saran plastics, 133
scaling, 63
screening, 14
screwed fittings, 345
screwed flanges, 321
sealing, 54
sealing surfaces, 302
sealing with gaskets, 295
sealing without gaskets, 301
seals, 291
seals without gaskets, 292
season cracking, 63
secondary treatment, 18
sedimentation, 15
selective leaching, 29
separation, 14
service environment, 34
shear stress, 346, 347
sharing strength, 172
sheet bending, 188
sheet lead, 88
sheet materials, 188
sheet rolling, 189
shelf life, 133
shell heat exchangers, 102
shock waves, 256
short vessels, 356
shredding, 13
shutdown procedure, 4
sieving and bolting, 14
silicon alloy, 81
silicon bronzes, 87
silicone plastics, 133
silver, 98
silver solder, 209
size reduction, 13
sizes of gaskets, 295
skirt supports, 353
slip-on flanges, 318
sludge treatment and disposal, 16
soft metals, 58
soft solder, 206
soft soldering, 207
softening range, 133
soldered joints, 206, 289
soldering, 38
solvation, 133
special obturation, 305
special seals, 292, 306

specifications, 24
split ring, 326
sprayed coatings, 100
spraying, 13
stainless steel, 56
standard flanges, 327
standard pipe, 133, 227
startup, 4
steel bearing cap, 161
stop-knots, 259
storing, 13
stoved phenolics, 122
stray current corrosion, 31
stress, 63
stress concentration, 174
stress crack, 78, 133
stress intensities, 174
stress-corrosion cracking, 29, 78
structural material, 146
structural pipe, 133, 227
structural shapes, 133
structural strength, 65
stuffing-box, 267
stuffing-box compensators, 271
styrene plastics, 134
styrene-rubber plastics, 134
subliming, 14
support arrangements, 237
support spacing, 238
support structures, 349
support types, 349
supports, 349
supports for horizontal vessels, 361
surface preparation, 63
surface treatment, 63
suspending, 13
swing check valves, 284
syneresis, 134

T

tabling, 15
tanks and vessels, 58
tantalum, 94, 96, 97
taper pipe thread, 343
tee connections, 347
temperature effects, 92
temperature resistance, 71
temperature well, 8

tensile stress, 347
tension load, 182
thermal expansion, 200
thermal stresses, 149
thermoelasticity, 134
thermoforming, 134
thermogalvanic corrosion, 28
thermoplastic materials, 105, 106
thermoplastic pipe, 113
thermoplastics, 135, 219
thermoset, 134
thermosetting plastics, 118
thixotropy, 134
thread engagement, 182
thread relief, 182
threaded connections, 316, 342
threaded flanges, 224, 312
throttling, 277
thrust blocks, 263
tightness, 310
tiles, 104
tin bronzes, 85
tinning of copper, 57
titanium, 94, 95
tongue-and-groove face, 327
torsion stress, 348
tracer yarn, 134
transducer, 6
transient distortion, 173
transient stresses, 173
transmitter, 7, 8
transporting, 13
trench bedding, 258
trench configuration, 265
trench floor, 258
tube joints, 148
tube plate, 199
tube soldering, 209
tuberculation, 31, 64
tubing, 226
turbulence, 58
types of corrosion, 27
types of gaskets, 296

U

unalloyed titanium, 96
undercutting, 224
undrainable, 57

uniform corrosion, 27
union joints, 347
unit cleaning, 4
unit operations, 11, 13
urea plastics, 134
urethane plastics, 134
utilities, 10

V

vacuum deposition, 64
vacuum forming, 134
vacuum service, 253
valve bodies, 273
valve installation, 288
valve ratings, 287
valve selection, 286
valves, 272
variations in flow, 9
vertical vessels, 353
vessel codes, 9
vessel drainage, 59
vinyl acetate plastics, 134
vinyl alcohol plastics, 134
vinyl esters, 231, 232
vinyl plastics, 134
vinylidene plastics, 134

W

wall thickness, 150
washing, 14
waste disposal, 4
waste streams, 9

wastewater processes, 18
water cooling, 13
water jetting, 264
weathering, 134
weld decay, 64
weld joint designs, 176
weld metal, 175
weldability, 39
welded equipment, 164
welded head, 148
welded joints, 168, 197
welded units, 58
welding a nozzle, 200
welding a shell, 198
welding cast iron, 72
welding method, 71, 165, 173
welding strains, 175
welding stresses, 172
welding vessels, 169
welding-neck flanges, 318
white cast iron, 69
white iron, 69
wooden patterns, 164
wooden skids, 259

Y

yield value, 134

Z

zinc, 50
zinc diffusion, 101
zirconium, 94, 98